THE LIFE OF NORTH AMERICAN SUBURBS

Imagined Utopias and Transitional Spaces

Edited by Jan Nijman

This book chronicles and explains the role of suburbs in North American cities since the mid-20th century. Examining 15 far-flung case studies – from New York to Vancouver, Atlanta to Chicago, Montreal to Phoenix – *The Life of North American Suburbs* traces the insightful connection between the evolution of suburbs and the cultural dynamics of modern society. Suburbs are uniquely significant spaces: their creation and evolution reflect the shifting demographics, race relations, modes of production, cultural fabric, and class structures of society at large. The case studies presented here investigate the place of suburbs within their wider metropolitan constellations, examining the crucial role they play in the cultural, economic, political, and spatial organization of the city. Together, the chapters paint a compelling portrait of North American cities and their dynamic suburban landscapes, which have evolved to defy traditional conceptions of the postwar suburban neighborhood.

(Global Suburbanisms)

JAN NIJMAN is director of and distinguished professor at the Urban Studies Institute at Georgia State University and a professor of geography at the University of Amsterdam.

GLOBAL SUBURBANISMS

Series Editor: Roger Keil, York University

Urbanization is at the core of the global economy today. Yet, crucially, suburbanization now dominates 21st-century urban development. This book series is the first to systematically take stock of worldwide developments in suburbanization and suburbanisms today. Drawing on methodological and analytical approaches from political economy, urban political ecology, and social and cultural geography, the series seeks to situate the complex processes of suburbanization as they pose challenges to policymakers, planners, and academics alike.

For a list of the books published in this series, see page 387.

EDITED BY JAN NIJMAN

The Life of North American Suburbs

Imagined Utopias and Transitional Spaces

UNIVERSITY OF TORONTO PRESS
Toronto Buffalo London

© University of Toronto Press 2020
Toronto Buffalo London
utorontopress.com

ISBN 978-1-4875-0109-9 (cloth)   ISBN 978-1-4875-1247-7 (EPUB)
ISBN 978-1-4875-2077-9 (paper)   ISBN 978-1-4875-1246-0 (PDF)

Library and Archives Canada Cataloguing in Publication

Title: The life of North American suburbs : imagined utopias and transitional spaces / edited by Jan Nijman.
Names: Nijman, Jan, editor.
Series: Global suburbanisms.
Description: Series statement: Global suburbanisms | Includes bibliographical references and index.
Identifiers: Canadiana (print) 20190211067 | Canadiana (ebook) 20190211075 | ISBN 9781487501099 (cloth) | ISBN 9781487520779 (paper) | ISBN 9781487512460 (updf) | ISBN 9781487512477 (epub)
Subjects: LCSH: Suburbs – North America. | LCSH: Cities and towns – North America – Growth. | LCSH: Urbanization – North America.
Classification: LCC HT352.N7 L54 2019 | DDC 307.76097 – dc23

University of Toronto Press acknowledges the financial assistance to its publishing program of the Canada Council for the Arts and the Ontario Arts Council, an agency of the Government of Ontario.

 Canada Council   Conseil des Arts
for the Arts   du Canada

Contents

List of Figures, Maps, and Tables   vii

Preface   xi

1 Introduction: Elusive Suburbia   3
JAN NIJMAN

Part 1: Questioning North American Suburbia

2 Using Toronto to Explore Three Suburban Stereotypes, and Vice Versa   23
RICHARD HARRIS

3 Mexico City: Elusive Suburbs, Ubiquitous Peripheries   45
LIETTE GILBERT

4 Searching for Suburbia in Metropolitan Miami   65
JAN NIJMAN AND TOM CLERY

5 Spatial Transformations in the Suburbs of the North Carolina Piedmont Region   88
FANG WEI AND PAUL KNOX

Part 2: Changing Political Economies of Suburbanization

6 The Strange Case of the Bay Area   109
RICHARD WALKER AND ALEX SCHAFRAN

7 Vancouverism as Suburbanism   129
ELLIOT SIEMIATYCKI, JAMIE PECK, AND ELVIN WYLY

8 Montreal: An Ordinary North American Metropolis? 149
CLAIRE POITRAS AND PIERRE HAMEL

9 New York's Suburbs in a Globalized Metropolitan Region 170
JAMES DEFILIPPIS AND CHRISTOPHER NIEDT

Part 3: Race, Ethnicity, and the Remaking of Suburbia

10 Diverging Racial Geographies in Phoenix's Postwar and Post–Civil Rights Suburbs 201
DEIRDRE PFEIFFER

11 Suburbanization and the Making of Atlanta as the "Black Mecca" 223
KATHERINE HANKINS AND STEVEN R. HOLLOWAY

12 Edmonton, Mill Woods, *Amiskwaciy Waskahikan* 245
ROB SHIELDS, DIANNE GILLESPIE, AND KIERAN MORAN

13 Economic Development and the New Immigrant Segregationist Politics in Suburban Chicago 269
DAVID WILSON

Part 4: Contested Suburbs

14 Governance, Politics, and Suburbanization in Los Angeles 289
ROGER KEIL AND DEREK BRUNELLE

15 Reaching Suburbia: Towards a Socially Just Transit System for Ottawa 307
CAROLINE ANDREW AND ANGELA FRANOVIC

16 Contested Spaces: Suburban Development in Halifax and Other Midsized Canadian Cities 328
JILL L. GRANT

Epilogue: Suburbs as Transitional Spaces 349
JAN NIJMAN

Contributors 363

Index 369

Figures, Maps, and Tables

Figures

3.1 Ciudad Satélite, Naucalpan, State of Mexico 51
3.2 Ciudad Santa Fé, Federal District 52
3.3 San Buenaventura, Ixtapaluca, State of Mexico 55
3.4 Xico, Chalco Solidaridad, State of Mexico 58
4.1 The combined population and municipal incorporations of Miami–Dade County and Broward County 67
4.2 Central cities versus suburbs in Miami–Dade and Broward counties, 1990, 2000, and 2010 77
4.3 Street view of Florida City, 2013 81
4.4 Street view of Pinecrest, 2013 82
5.1 Visualization of clusters via clustergram 92
5.2 Spatial patterns of clusters in the Charlotte–Gastonia–Rock Hill metropolitan statistical area in 2010 95
5.3 Spatial patterns of clusters in Raleigh-Durham-Cary combined statistical area in 2010 96
5.4 The basic version of Piedmont model 97
6.1 Bay Area homeownership rates, 1950–2010, by county 115
6.2 California house construction, 1975–2010 119
7.1 "Vancouverism" from above 130
7.2 Ratio of jobs to labour force, Vancouver and suburbs, 1971–2006 141
8.1 Commuter train station in the town of Deux-Montagnes, 2016 154
8.2 Saint-Joseph-du-Lac, Lower Laurentians, 2016 154
8.3 Laurentian Boulevard in the City of Laval, 2016 158
8.4 A new residential area in Sainte-Marthe-sur-le-Lac, 2016 165
9.1 New housing near South Farmingdale by the Southern State Parkway 173

viii Figures, Maps, and Tables

9.2 African-American construction workers during the construction of Levittown, NY 176
9.3 Change in manufacturing employment in New York City and its suburbs, 1947–2012 184
9.4 Long Islanders demonstrating for permanent residence for Salvadoran immigrants in Washington, DC, 1995 191
10.1 Phoenix region population growth, 1900–2010 204
10.2 Maryvale 208
10.3 Avondale 215
11.1 Poverty rate in Black and White suburbs of Atlanta, 2000–2014 240
12.1 Portrait of Mill Woods today 246
14.1 Foreclosure Center, Palmdale, California 295
14.2 Postsuburban landscape in Southern California 297
16.1 Halifax suburbs 331
16.2 The commercial centre of Cornell in Markham 332
16.3 Barrie suburbs 333
16.4 Airdrie suburbs 334
16.5 Langley Township suburban enclaves 335
16.6 Medium-density dwellings in Surrey 336

Maps

2.1 The political organization of the Greater Toronto Area to 1998 25
2.2 Patterns of immigrant settlement in the Toronto Area, by census tract, 1965–1971 27
2.3 Population growth in the Greater Toronto Area, by census tract, 2006–2011 32
2.4 Patterns of immigrant settlement in the Toronto Area, 2001–2006 33
4.1 Miami–Dade County 69
4.2 Westward urban growth in Southeast Florida, by zip code area, 1990–2002 73
6.1 The many Bay Areas 111
6.2 Protected areas on the San Francisco Peninsula 117
6.3 New housing starts in the Bay Area, by city, 2000–2010 120
7.1 Sprawl as infestation in Vancouver, 1940s and 1950s 134
8.1 The Montreal metropolitan area and its main divisions 151
8.2 French reported as mother tongue in the Montreal Census Metropolitan Area, 2011 164

Figures, Maps, and Tables   ix

10.1 Geography of Phoenix's post–civil rights suburbs and postwar suburban city 210
10.2 The geography of foreclosures in Maricopa County, January 2004–May 2014 216
11.1 Distribution of population, City of Atlanta, 1924 225
11.2 Map of Atlanta areas "occupied chiefly by Negroes," 1959 230
11.3 Overview map of select Atlanta neighbourhoods 231
11.4 Metro Atlanta population, 1970 234
11.5 Black suburbs in Atlanta in 1990 and 2010 238
12.1 The Alberta Capital Region with future growth plans 248
12.2 Papaschase First Nations Reserve, 1899 253
13.1 Cicero and Berwyn in Greater Chicago 272
14.1 Overview map of the Los Angeles urban region 292
15.1 Jacques Gréber, General Report on the Plan for the National Capital, 1950 310
15.2 Overview map of Ottawa's administrative divisions 311
15.3 Public Administration Industrial Sector in Ottawa-Gatineau 312
15.4 OC Transpo bus stations in Ottawa, 2015 317
15.5 Average individual income in Ottawa-Gatineau, 2012 320
15.6 Typology of neighbourhoods in Ottawa-Gatineau, 2006 321
16.1 Overview and reference map of Halifax and other midsized Canadian cities 329

Tables

2.1 Social and housing characteristics in the Toronto region, 2006 31
4.1 Demographic comparison among the three counties of southeast Florida's consolidated Metropolitan Statistical Area, 2018 74
4.2 Selected characteristics of the City of Miami and six selected "suburbs" in Miami-Dade County 79
5.1 Z-score means across clusters 93
5.2 Populations by cluster 94
5.3 Distribution of clusters 94
5.4 Neighbourhood succession from 1980 to 2010 99
5.5 Neighbourhood growth from 1980 to 2010 101
7.1 Geographies of population change, Metropolitan Vancouver, 1971–2011 138
7.2 Population shares, Metropolitan Vancouver, 1961–2011 139
7.3 Geographies of employment change, Metropolitan Vancouver, 1971–2006 140

7.4 Employment shares, Metropolitan Vancouver, 1971–2006 140
8.1 Population distribution in the Montreal region by area 152
8.2 Demographic growth in suburban municipalities within the Montreal census metropolitan area 152
8.3 Proportion of the population in Quebec having knowledge only of French, 2011 162
9.1 Change in young adults as share of total population, New York, 2000–2010 179
9.2 Manufacturing jobs, regional share of manufacturing jobs, and location quotients for employment, New York 185
10.1 Demographic and housing market characteristics of Phoenix's post–civil rights suburbs and postwar suburban city, 2010 211
10.2 Racial stratification and segregation in post–civil rights suburbs and in the postwar suburban city, 2012 212
13.1 Statistics for Cicero and Berwyn 273
16.1 Population of areas discussed, Canada (2006 and 2011 census) 337
16.2 Population characteristics of areas discussed, Canada (2006 and 2011 census) 337
16.3 Characteristics of households in areas studied, Canada (2006 census) 338
16.4 Density and housing characteristics in areas studied, Canada 339

Preface

This book emanates from a Major Collaborative Research Initiative funded by the Canadian Social Sciences and Humanities Research Council (SSHRC). Managed from York University (Toronto) by Roger Keil, "Global Suburbanisms" is the first major research project to take stock of worldwide suburbanization trends. It is an expansive and multifaceted project involving more than 50 researchers, dozens of students, numerous universities, various non-academic collaborating institutions, and a dedicated and critically important staff at the centre of it all.

The overall project is organized along various themes and multiple world regional clusters. The book before you concentrates on North America and results from the collaborative efforts of about two dozen contributors. North America may be seen as the cradle of suburbanization, but for anyone paying any attention to the scholarly record over the last couple of decades, it should be clear that the suburban landscape has been anything but static. This volume documents suburban trends across the continent in cities large and small, with diverse economic and social histories, and in different regional contexts.

The key question in each chapter is how suburbs have evolved in recent decades, from the ultra-American postwar stereotype that so dominated traditional views: middle class, White, conservative, residential, spacious, sparse, green, and with single-family homes. Suburbs are traditionally defined in relation to central cities and, to be sure, form integral parts of metropolitan regions at large. This means that suburbs, or suburbanization processes, must be understood in the context of wider metropolitan developments: from the foreclosure crisis that peaked around 2010 to the alleged "great inversion" (return to central cities) of recent years. This volume provides a wealth of information and insights regarding North American suburbs and suburbanization, but also regarding metropolitan change at large.

Five of the 17 chapters have been previously published, in somewhat different form, in refereed journals. The chapter on Vancouver appeared in the journal *Cities*, while the contributions on Toronto, Miami, the Piedmont Region, and the San Francisco Bay Area first appeared, in slightly altered form, in a special issue of *Environment and Planning A*. The other 12 chapters are original.

This volume owes much to the key individuals in charge of the Global Suburbanisms project: Roger Keil for his intellectual and organizational leadership, and Sara Macdonald (who has since moved on) for steadfastly managing the productivity of a globally dispersed herd of researchers. At the University of Toronto Press, Acquisitions Editor Doug Hildebrand and his successor Jodi Lewchuk, Associate Managing Editor Robin Studniberg, and Editorial Assistant Kiley Venables all lent their generous professional support to guide this volume to publication.

THE LIFE OF NORTH AMERICAN SUBURBS

Imagined Utopias and Transitional Spaces

1 Introduction: Elusive Suburbia

JAN NIJMAN

This volume, by a group of highly recognized scholars and urban specialists, investigates the nature of suburbs and suburbanization in present-day North America. We have known, or at least suspected, for some time that the stereotypical notion of the suburb that emerged in the 1950s has been diverging from metropolitan realities. The early postwar "sitcom" suburb, singularly dominated by White, middle class families, in spacious and green environs with single-family homes, was short-lived and soon evolved into diversified forms in expanding and increasingly complex metropolitan configurations. We also know that many metropolitan areas have continued to expand outwards while amalgamating with cities in the region, and the notion of the polycentric urban region has become widely accepted among scholars and policymakers. The concepts of edge cities and edgeless cities have been added to the lexicon. Still, the terms suburb, suburbia, suburbanism, and suburbanization have stuck in scholarly or professional jargon as well as in colloquial discourse – they are increasingly difficult to define but the labels persist. Fishman (1987) and others may have announced the end of the common suburb three decades ago, but it is not so clear what followed in its place. The notion of suburbia, no matter how commonplace, has become ever more complex. What is the nature of current processes of suburbanization, how can we best understand present-day suburbs, and how do they fit in larger, often dynamic, metropolitan constellations?

The chapters in this book seek to clarify the meaning of suburbanization today in 15 North American metropolitan areas, from relatively small cities to large conurbations in different regions across the continent. The book contains eight US and six Canadian case studies, as well as a chapter on Mexico City. The latter stands out in some obvious ways but its inclusion serves well to put (especially) the US "model" in perspective, both in regard to the material forms of suburbia and in terms of popular narratives. The national contexts of suburbanism matter, but so do the local and (international) regional scales. The papers are written against the backdrop of the idea of the traditional postwar North American suburb and they concentrate on changes in the past few decades. North America may be considered the birthplace of the modern suburb, but the phenomenon of suburbanization soon went global. If we find ourselves at a historical moment of planetary urbanization, as some suggest, then most of this urban growth is occurring in "suburban" environs. We are also in the midst of intense debates about the portability of (Western) urban theory and, more generally, about the importance of a comprehensive comparative urbanism. In that sense, this collection of papers is relevant not just to North America. If the suitability of conventional (Western) notions of suburbanism to other parts of the world is in question, it is well worth investigating the continued relevance of the notion to North America itself.

North American Suburbanization in Historical Perspective

Since the chapters in this book concentrate on developments in the last couple of decades, it should be useful to set the stage with a brief review of the history of North American suburbanization.[1] The term suburb (of 14th-century English origins) acquired currency during the 18th century – in England as well as in the United States. In pre-industrial times, suburbs were viewed as undesirable and shady places on the edge of town; marginal neighbourhoods with a mix of the poor and people with licentious habits. The word "urbane," instead, referred to sophistication, elegance, and high-class. The elites occupied the centre of these compact pre-industrial cities that mixed residential and economic functions (trade, services).

This arrangement came to an end in the 19th century in the wake of the Industrial Revolution. Cities became sites of industrial production, often with detrimental environmental effects, and they grew increasingly dense. According to authors like Jackson (1985), Fishman (1987), Hayden (2003), and others, this resulted in a growing interest of the elites in new housing on the urban periphery: home as a refuge from work, as a source of happiness and goodness. Upper class status became associated with mansions on large estates in a quiet, lush, suburban environment, while the city centre was increasingly perceived as a scene of congestion, pollution, crime, and crowded working class residential areas. Suburbanization proceeded faster in the United States than in England or elsewhere in Europe because industrialization was more vigorous and sustained in the United States, and as such fuelled a more significant response by way of suburbanization.

The new suburbia of the mid-19th century in places such as West Philadelphia, says Fishman (1987, p. 21), represented a "collective assertion of class wealth and privilege." It was based on exclusion and segregation. If bourgeois demand for grand suburban living drove the process, this cultural impetus was soon accompanied by economic motives: the transformation of agricultural lands just outside the city into residential building plots was by definition a lucrative business.

There was another, cultural reason that suburbanization became such a salient expression in the American landscape. The individualized nuclear family was very much an American institution (closely related to the "American Dream") and demanded a single-family home – which was easier to realize in the spacious suburbs than in the city centre. Hayden (2003) observes that "unlike any other affluent civilization, Americans have idealized the house and yard rather than the model neighbourhood or the ideal town" (pp. 5–6).

If the 19th-century "invention" of the suburb reflected bourgeois imaginaries of Utopia, in reality the process of suburbanization quickly assumed broader significance and a more complicated spatiality. First, escapism from inner city chaos applied not only to residential preferences but also to work. The (re)location of industrial activity to the edges of these still-compact cities – even more so than the suburbanization of residential functions – often required newly built infrastructures (canals, roads, sewers, etc.) and relied on the ownership classes garnering local or state government funding (Walker, 1981; Walker & Lewis, 2001; also see Taylor, 1915).

The number and kinds of residential suburbs increased, too (Douglass, 1925; Warner, 1962). Since the 1890s, the introduction of the electric streetcar in cities across the United States, from Portland to Miami, pushed suburbanization along, allowing middle and lower middle classes to follow the elite out of the central city. The land development and real estate industry became more organized and proactive and began to target potential first-time home buyers. So-called "Why Pay Rent" campaigns from around the turn of the century promoted suburban living to middle and working class households. Not unlike the fate met by home buyers in the early 21st century, many were lured into homeownership they could barely afford, struggling "up a down escalator" entranced with dreams of economic security, saddled with debt, and confused by a false sense of social mobility (Edel, Sclar, & Luria, 1984; Immergluck, 2009).

Increasingly, it seems, demand for suburban living was stimulated on the supply side. Between 1870 and 1920, developers enlarged their area of operations, took a broader view of the urban, and began to promote urban peripheries, often working in partnership with transit owners, utility companies, and local government. The building boom of the "Roaring Twenties" accelerated the creation of suburbs, made possible by the rise of the powerful real estate and construction lobby, in conjunction with new federal regulations that helped subsidize "private development of residential and commercial property on a national basis – largely though tax, banking, and insurance systems ... " (Hayden, 2003, p. 4).

It was also made possible, of course, by the introduction of the automobile, which rapidly increased spatial mobility and allowed access to a greater number of potential suburban designations. Prior to the Second World War, cars were a luxury item out of reach for the bulk of the working classes. In the 1910s and 1920s some new suburbs were designed specifically for the use of automobiles and they were emphatically upscale, with big lots, winding roads, lush vegetation, and no sidewalks. Such suburbs were exclusive, to be sure, and very different in appearance and composition from a range of other suburbs that had already formed by this time.

The exclusivity of the suburb, the former bourgeois Utopia, was under pressure from the beginning. Indeed, it has been argued that "prewar suburbs were as socially diverse as the cities that they surrounded, and it is doubtful whether the city–suburban dichotomy was very significant" (Harris & Lewis, 2001, p. 263). In many big cities, suburbs had formed with a strong working class identity, such as South Gate in southern Los Angeles (Harris, 1992; Nicolaides, 2004). On the eve of the Second World War, a new urban form had emerged due to decentralization of not just a broader part of the urban population but also of industries, services, and retail activities. The process was conditioned by changing technological, economic, and regulatory conditions.

Suburbanization after the war took on such massive proportions that it fundamentally altered the urban order. From 1950 to 1980, the suburban population of the United States tripled; by 1970, more people lived in suburbs than in either central cities or the countryside. Suburbanization was not new; but sometimes enough of a quantitative change implies a qualitative transformation. The United States had become a "suburban nation" (Duany, Plater-Zyberk, & Speck, 2000). Muller (1981) referred to suburbia as "the essence of the late-twentieth century American City."

Reacting to these landmark changes, the 1960 census adopted the category of Metropolitan Statistical Area with a "central city" versus commuting hinterland or "suburbs" (Champion, 2001; U.S. Bureau of the Budget, 1964). It provided, for the first time, an official definition of the suburb and it did so in opposition to the central city, thereby forging a dichotomy that corresponded to traditional imaginaries but that in reality had never been so clear-cut.

The dichotomy was reinforced since the early 1960s as central cities in many parts of the United States declined, economically and socially, as a result of deindustrialization (loss of jobs) and selective outmigration (suburbanization). Inner city decay, thus, had the effect of reinforcing earlier idealistic visions of the suburb. The suburb was thought to be everything the city was not: clean, green, spacious, safe, quiet, harmonious, predictable, and homogeneous. In the process, the deeply American and ideologically inspired notion of the suburb was revitalized. There is no doubt that this imaginary of the suburb played an important role in increasing its desirability for many of those (working and middle classes) who had previously been excluded. Writing about suburbanization in the mid-20th century, Hanlon, Short, and Vicino (2010, p. 6) observe that the suburbs were now part of an American Dream for all:

> There are two mythic journeys in the US. The first ... was the trek to the West, ending in California. The second, the archetypal journey of the mid-20th century, was from the city to the suburbs ... It was a quest signifying acculturation, Americanization, and ultimately success. In this second mythic American journey, the family car replaced the covered wagon, and the single-family home displaced the family homestead as iconic representations.

The renewed suburban imaginary was ingrained in the American psyche on television in a number of wildly popular 1950s sitcoms (Sharpe & Wallock, 1994). The typical suburb was portrayed as the peaceful and comfortable home of White middle class families with traditional gender stereotypes. This imaginary very much articulated desires and choices of individual American households, and at a time that demand for housing was significantly up (unfulfilled demand in the wake of the Great Depression followed by the Second World War; and the baby boom).

The process of suburbanization was, more than ever, driven and facilitated by corporate interests and government intervention. Levittown, the archetypal 1950s' suburb, was the well-documented result of combined private enterprise and new kinds of government policies. The Levitt family planned the Long Island subdivision at the scale of a town but did not include any of the necessary services such as garbage collection, schools, or roads – these responsibilities were passed on to government and were financed through tax dollars. It was a new kind of business and it was made possible through a new regulatory finance environment (Hayden, 2001).

Suburbanization, one might say, had become the business of an extremely powerful industrial conglomerate that employed (and helped generate) the American suburban imaginary to full effect. It included huge corporations such as General Motors (which offered a helping hand in the demise of the electric streetcar) and General Electric (which had embarked on the mass production of household appliances for single-family homes); local "growth machines" (Molotch, 1976) consisting of developers, builders, and banks; local governments that provided conducive zoning and building regulatory frameworks, and sometimes direct subsidies; and, last but not least, a federal government that was central to the financing of homeownership, the construction of highways, and that in various ways espoused suburban ideologies.

Besides the massive acceleration of population shifts to the suburbs (along with shopping malls, hospitals, schools, and other service and retail activities), there was a significant increase in the suburbanization of office work (Mozingo, 2011). The suburbs were considered more representative, more easily traversed, more predictable, and less risky, and better for business. It was a trend that gathered momentum over the decades and resulted in the proliferation of suburban office parks and corporate campuses.

To some, the transformation of the United States into a suburban nation actually signalled the end of the suburban ideal. In 1987, Fishman (1987) argued that "the suburb since 1945 has lost its traditional meaning and function as a satellite of the central city. Where peripheral communities had once excluded industry and large-scale commerce, the suburb now becomes the heartland of the most rapidly expanding elements of the late 20th century economy ... As both core and periphery are swallowed up in seemingly endless multi-centered regions, where can one find suburbia?" (p. 29).

Since the 1960s, architectural critics had begun to depict suburbs as lowbrow, boring, and banal. The monotonous, mass-produced, subdivisions of the postwar years certainly were a long way from the carefully designed elite suburban mansions of the 19th century. More importantly, suburban culture as a whole came to be regarded as uninteresting, conservative, and spiritless. It is not hard to discern elitist undertones in such critiques. Examples include "Jane Jacobs's (1961) picture of her own idyllically bohemian Lower Manhattan neighbourhood in *The Death and Life of Great American Cities* and the wild anger at suburban piggery that pervades James Howard Kunstler's (1993) *The Geography of Nowhere*" (Seal, 2003).

We are now perhaps past such clichéd critiques of the suburbs, but it is important to note that the aesthetic devaluation of suburbs among critics coincided with increased access to suburbs by lower income strata of the population, even if it concerned mainly "drive until you qualify" types of suburbs (Paumgarten, 2007). A home in the suburbs was still sold to the masses as a privileged place in the sun along with all the traditional narratives of the past (Knox, 2008). In other words, there emerged a clear disparity of (changing) discourses about the suburb.

Towards the end of the 20th century, the notion of the suburb had lost its coherence, both in material and in discursive terms. Suburbs now came in many stripes, not least because of a new wave of metropolitan growth and expansion since the mid-1980s. The chapters in this book illustrate this growing diversity of suburbs, their variable expressions in different metropolitan areas and regions, their dynamic and transitory existence and character, and – especially – the complex and contentious relations between suburbs and central cities.

Overview of This Volume

The book is organized into four parts that challenge, in different ways, the traditional understanding of North American suburbia. The first section, "Questioning North American Suburbia," starts off by questioning basic definitions and stereotypes and points to the difficulty of delineating a particular North American experience. It does so through successive investigations into the contrasting geographical experiences of Toronto, Mexico City, Miami, and the Piedmont urban region of North Carolina. As such, it lays out a broad canvas of North American suburban worlds.

In the opening chapter of part one, Richard Harris argues that Toronto does not – and has never – conformed to traditional notions of suburbia. First, the suburban ideal in Toronto has always contained strains of urbanity while having to compete with the more established urbane ideals of living in the central city. Second, Toronto's 20th-century suburbs have varied considerably in terms of density and social diversity. And third, Toronto's suburbia never forged a consensus among its critics: it was neither universally celebrated (in earlier times) nor universally condemned (when fashionable, in later times) for suburbia's allegedly intrinsic cultural blandness. The historical and social origins of the stereotypes recounted by Harris are not always clear even if they sound familiar, but the suggestion is that Toronto is different in part because it is a *Canadian* city. The implication is, of course, that conventional notions about suburbia have a particular US bias.

At the southern end of the geographic spectrum, Mexico City represents another intriguing case to highlight the variegated and splintered nature of North American suburbia. Mexico is as proximate to the United States as is Canada and it, too, shares urban areas that straddle the political border. Liette Gilbert, in her compelling essay on Mexico City, shows that it at once invokes suburban imageries that have travelled intact across the border from the north *and* it displays the most striking negations of those ideals. Mexico City's massive suburban growth of the 20th century has been largely contemporaneous with trends in the United States. While some earlier suburban areas emerged according to ideal-typical US-inspired designs, and some are still home to upper middle income households today, most of Mexico City's "ubiquitous peripheries" are informal settlements (*colonias populares*) inhabited by poor recent migrants. The common label of suburb, says Gilbert, seems ill-suited to these parts, and she instead prefers the notion of "subaltern suburbanism."

Miami, the first US city covered in this volume, is the eighth largest metropolitan area in the nation. Arguably, it is something of an outlier given its recent history, the geographic limits of urban expansion between the Atlantic and the Everglades, and its profound ethnic makeover at the time that postwar suburbanization peaked in North America. Nijman and Clery note that in the case of Miami the city–suburb distinction does not correspond to prototypes and it is equally difficult to find typical suburbs as conventionally perceived. The chapter concentrates on a comparative review of six areas that, on the face of it, could qualify as suburbs, outside the older urban cores along the waterfront. They find that while suburban features can be detected across the metropolis, they almost always combine with characteristics that are emphatically *not* considered suburban, such as high densities or concentrated economic activity. The case of Miami suggests suburbia is a complex and multidimensional notion that finds expression across the metropolitan area, and it questions the continued relevance of the suburb as a discrete and bounded *spatial* entity.

The fourth and final chapter in part one continues the search for suburbia in a very different geographic setting: the North Carolina

Piedmont region (including Charlotte, Raleigh, and Durham-Chapel Hill). This polycentric urban region has grown fast in recent times and is often depicted as economically strong and an attractive area to live. Based on a detailed socioeconomic and spatial analysis of neighbourhood change, Fang Wei and Paul Knox observe strong trends towards increasing socioeconomic diversity and persistent segregation. They do find evidence for the (continuing) presence of more or less classical "sitcom suburbs" dominated by middle class White households, living in single-family homes that represent about one third of the suburban population of the region. But they also observe increasingly segregated suburbs dominated by minorities near city centres, as well as lower income and aging suburbs further out. Wei and Knox summarize their findings in a so-called "Piedmont model," made up of concentric zones, sectors, and clusters, not unlike the classical urban ecology models. They also conclude that if the broad tenets of such models are held up at least in some respects, many suburban areas have undergone processes of succession in various directions and there is overall strong evidence for increased spatial segmentation and polarization. The authors do not claim to be able to generalize from the Piedmont region to the rest of the nation or North America at large – but from the first four chapters in this book, they come the closest to identifying suburbia in its traditional forms.

Part two focuses on the changing political economies of suburbanization. In each of the four chapters, suburbanization and the transformations of suburbia are understood as integral parts of overall metropolitan growth and (re-)development, in turn a function of prevailing forces of economic restructuring and political realignments at national and metropolitan levels. In chapter 6, Richard Walker and Alex Schafran provide a sweeping overview of the Bay Area's enormous growth in past decades, which reset the scale and regional configuration of this metropolitan area in very substantial ways. The tremendous wealth created in Silicon Valley was a powerful engine behind these changes, from polarizing labour markets to some of the most expensive housing in the country, from shifting commuting patterns to municipal incorporations and exclusionary zoning practices. The Bay Area is enormous in size and hugely complex in its spatial ordering, with major central cities in San Francisco, San Jose, and Oakland, a range of edge cities and office parks in the valley and beyond, countless residential areas in between, and often without clear boundaries between peri-urban and rural spaces. Most of the area, say Walker and Schafran, carries suburban features, and "sprawl has been the name of the game." Affluence allowed many people access to single-family homes and cars, and the region became a paradise for developers and banks. But excessive pricing has also caused widespread hardship and necessitated suburbanization of lower income households to distant areas where 2-hour commutes are no exception. Suburbanization, the authors argue, runs on finance and financial opportunities. In the Bay Area, it has resulted in a richly patterned mosaic at multiple scales, where simple city–suburb dichotomies have long dissipated.

In their chapter on Vancouver, Siemiatycki, Peck, and Wyly dissect a rather different – in some ways apparently more Canadian – political economy that has since around 1980 given way to a sub/urban design combining "density, livability, and sustainability." Or, at least, such are the claims of "Vancouverism." This city stood out and became a pace setter in dense, sky-scraping, amenity-rich, residential development in the central city along with compact, transit-oriented development in more peripheral communities. The authors describe Vancouverism as "a contradiction inscribed into the metropolitan landscape – a centralization of suburban modes of development, cultures, aesthetics, and lifestyles into symbolic built forms evocative of urban ways of life in other times and other places." The downtown area, they argue, exudes the suburban logics of privatism and separation. In a way, then, this city is said to have internalized suburbanism, at least in certain respects: a high-density vertical variation of suburbia.

If Vancouverism implied centralized growth and densification, this certainly did not apply to Montreal in the last decade or so. In this Canadian city, sprawl dominated, with population growth of suburban areas being about four times higher than in the central city. Indeed, Hamel and Poitras observe that "the model of the residential suburb that took shape in the United States during the years following the Second World War remains an everyday reality for households in the Montreal region." Montreal, say the authors, has in some ways been on a similar course as other rustbelt cities in the Northeast, even if the central city retained a more dynamic economy and relatively high densities. Further, as in many US metros, some of Montreal's suburbs evolved from bedroom communities to "ethnoburbs." Interestingly, migration out of the central city has been biased to French speakers, and linguistic/cultural identity has been more salient than race. Residential suburbanization and sprawl have been accompanied with the decentralization of cultural amenities and the emergence of edge cities. If metropolitan Montreal has in recent decades undergone comparable political-economic changes as many other older North American cities, these changes have been mediated by its particular linguistic context: the central city has become more English and less French, suburban movers have tended to choose their new residences on the basis of language/culture, and new immigrants have tended to opt for areas of their own ethnic concentration or predominantly English-speaking areas.

In chapter 9, James DeFilippis and Christopher Niedt emphasize the role of economic restructuring and political machinations that have produced ongoing processes of spatial concentration and deconcentration, and a constant sorting of populations and economic activities across the New York metro region. It is hard not to think of New York as being in the vanguard of the transformations across North America's metropolitan landscape in recent decades (and before), against the backdrop of the more or less constant influx of foreign migrants and the global context of New York's economy. The authors sketch a compelling picture of power and place in these transformations, one that in some ways resembles the developments in the Bay Area as presented by Walker and Schafran. Two trends in New York deserve special mention. The first relates to an intriguing demographic shift of highly educated young people to Manhattan and other central cities (Newark, Stamford) and away from more suburban areas. Their departure is a cause of economic concern for suburban communities and civic and business leaders. That concern, however, takes on an additional dimension because these youths are mainly White, while newcomers to the suburbs are in large part ethnic minorities and immigrants. The authors point out that transit-oriented development projects are often aimed at keeping these cohorts in the suburbs. The second trend pertains to the emergence of vibrant social movements in the suburbs of the New York region, particularly in areas with concentrations of immigrants (a development with a parallel in Los Angeles, as we'll see below). Social justice, equity, and inclusion are high on the agenda of these movements, and they are inspired by opposite realities on the ground, especially in regards to housing, work, and health care. Their organization is challenged in the face of political, ethnic, and spatial fragmentation but – sometimes – they are able to forge solidarity across different groups and suburban areas. New York's suburban social movements have become an important part of the metropolitan political landscape.

Part three of the book concentrates on the role of race and ethnicity in the (re)making of North American suburbia. Three of the four chapters focus on the United States where, seemingly more than anywhere else in the Western world, history, institutions, and government policies have converged to make race and ethnicity central factors in the evolution of cities and especially in processes of suburbanization. In chapter 10, Deirdre Pfeiffer provides an excellent analysis of Phoenix, comparing suburbanization trends before and after the 1968 *Fair Housing Act*. Phoenix makes for an interesting case because so much of its population growth and suburbanization happened after the Second World War (and particularly after 1968, when the population quadrupled) and it lacks the weighty history of most sunbelt cities in the US South. The author compares pre- and post-1968 suburbs in terms of racial composition and socioeconomic characteristics. She finds that, while notable disparities remain, post–civil rights suburbs are substantially less segregated and less unequal than the older suburbs. Pfeiffer concludes that post–civil rights suburbs are "far more racially equitable" than older suburbs. Yet she warns that it remains to be seen whether the post–civil rights suburbs reflect a "fundamentally transformed suburban geography or just their relative immaturity."

In chapter 11, Katherine Hankins and Steven R. Holloway turn their attention to Atlanta, the city on which *Ebony* magazine in 1971 famously bestowed the title of "black mecca." It was a narrative to be repeated for decades to come and to be critically deconstructed and reappraised only more recently (Hobson, 2017). If Atlanta has a strong African American middle and upper middle class, this group is also overrepresented among the poor – according to some measures, Atlanta is the most unequal metropolitan area in the nation. The chapter dovetails nicely with that on Phoenix. While the latter tracks minority-majority suburbs over time (before and after 1968), the Atlanta chapter analyses racial patterns across space, between inner and outer suburbs. Hankins and Holloway show how, in light of the racist and restrictive covenants delimiting White neighbourhoods, Blacks "quietly acquired" land at the outskirts of the city "jumping over the whites." The share of Blacks living in the central city dropped significantly, while the Black (outer) suburbs grew virtually nonstop since around 1970. This trend only intensified in later years: between 1990 and 2010, nearly two thirds of Black residents from the central city moved to the suburbs. Generally speaking, there remain major disparities between Whites and Blacks across the metro area and these were compounded by the financial crisis of 2007–9. Subsequently, the geographic disparities between Whites and Blacks have been sharpened by the "great inversion" (Ehrenhalt, 2013). The authors observe that for majority-White suburbs, poverty rates are higher in the outer suburban neighbourhoods than in the inner suburban neighbourhoods – opposite the pattern in majority-Black suburbs. They conclude that if there is (still) any validity to the notion of Atlanta as a Black mecca, it has become a distinctly suburban affair.

The following chapter shifts attention to the Chicago suburbs of Berwyn and Cicero. These old inner-ring suburbs, says David Wilson, have witnessed the emergence of a "new immigrant segregationist politics." Challenges of economic restructuring and the need to foster "suburban downtown" redevelopment have coincided with the influx of international migrants, especially from Latin America. Both suburbs now have a large Latino majority, while local political and civic power remains largely in the hands of the old elite. The new service-oriented and consumer-oriented economy in these areas needs low-paid (and not necessarily highly skilled) immigrant labour, and local business and political leaders have engaged in apparently socially progressive and inclusive narratives about the direction of change. It is, however, a contradictory posture because the downtown revitalization efforts exude nostalgia for past times with stable White all-American prototypical suburban communities. And they are inspired by recent upscale urban revival trends that are associated with "carefully cultivated, vibrant, aesthetic terrains." While the immigrants are a necessary part of the labour requirements, their own economic standing and cultural identities are felt to be a poor fit with the intended appeal of this new economy. As such, the local elite is caught in contradictory economic-political strategies, while the immigrant populations eke out a living on highly ambiguous terrain, at once welcomed and excluded. Change has come fast these suburbs, and their present economic and social constellation seems far from stable.

The final chapter of this section, and one of the most original contributions to this volume, directs attention to the Edmonton suburb of Mill Woods. The Edmonton region is where Canadian Aboriginal peoples were historically settled, but most were pushed out over the years. *Amiskwaciy* (from the chapter title, meaning *Beaver Hills House*) was the Cree name for the late-18th-century Fort Edmonton. Mill Woods was not planned until the early 1970s and was initially predominantly White. Today, it is an ethnically diverse suburb with about 30% of its 80,000 people being recent immigrants, mainly from a variety of Asian countries. Only about 6% of the current population is Aboriginal. Their incomes are well below average, unemployment rates are high, and birth rates are high, too. There are recurring reports and complaints about discrimination, accompanied with "simmering tensions." The authors make two intriguing observations that are of particular importance. First, they posit that dominant White culture in Canada invokes a binary divide between city and country, or civilization and frontier. Low-density outer suburbs tend to occupy an uneasy and ambiguous space in this binary. Edmonton, say the authors, is considered a relatively insignificant city in Canada, not very "urban" and sometimes derided as downright "ugly," and Mill Woods is "a suburb that epitomizes this abject status." Second, the authors point out that the presence of the Indigenous population has been erased from prevailing discourses about the city that today evolve around the suburb as a growing "immigrant community." On the one hand, earlier histories and identities of the area are negated along with the place of Aboriginal peoples in it; on the other hand, Mill Woods is still portrayed as an active settler environment, somewhere between urban civilization and the frontier, with today's settlers being mainly foreign immigrants. It all points to the apparently ceaseless reinvention of suburbs and of suburban narratives. The authors note that "the case of Mill Woods suggests that researchers [should] pay more attention to the temporalities and historical legacies" in the ongoing transformation of suburbia.

The final part of the book concentrates on suburbia's governance challenges and suburbs as politically contested spaces. In chapter 14, Roger Keil and Derek Brunelle point to early 20th-century Los Angeles as a national forerunner in governance trends of market-oriented deregulation, decentralization, and privatism. The early proliferation of gated communities in Los Angeles epitomized these developments. They also argue that the concurrent splintering, sociospatial polarization, and resulting disparities have in more recent decades invoked political counter movements, producing a new terrain for social action. These grassroots movements have "pushed the agenda of their struggle beyond the boundaries of their disparate locations ... and created a regional alliance." They have concentrated their efforts mainly on environmental, transit, and labour issues, leading to new state-based policy initiatives as well as social justice victories throughout the metropolitan area. An early example is the 1996 court victory of the Bus Riders Union, which successfully pushed for more equitable public transit investments across the region to improve access for populations in less advantaged areas. It is in the sense of these politico-spatial contradictions, the authors argue, that Los Angeles represents one of the earliest "postsuburban" metropolitan landscapes in North America. Similar or comparable trends are observed by Niedt and DeFilippis in New York (chapter 9), as noted above, but they do not apply everywhere across the North American metropolitan landscape.

In chapter 15, on Ottawa, Caroline Andrew and Angela Franovic document the growing lack of access to public transit among vulnerable suburban populations, particularly the elderly and low-income immigrants. Ottawa is a relatively small metro of about 1.3 million people that includes Gatineau, on the Quebec side of the Ottawa River. As Canada's national capital, Ottawa has a prominent public sector, with employment concentrated in the old urban core. Transit is heavily oriented towards access to and from the centrally located government offices, and it is considered by many as the most extensive public transportation system in the nation. It is certainly highly developed if compared to US metropolitan areas. But the system has not kept up with rapid suburbanization that has jumped over the Greenbelt area at relatively large distances from the central city. Gentrification of the centre has been accompanied with shifts of lower income groups to the outer suburbs, which is also where new (foreign) immigrants have settled in selected areas. Further, very few of these vulnerable groups are employed in the government sector, and many in these groups work at irregular hours and live in areas that are poorly served by public transit. If compared to Los Angeles and most other US cities, the problems in Ottawa seem less severe (and they are more recent) but in the Canadian context they have given rise to similar kinds of social grassroots movements and political organizations. And they have fairly successfully engaged local government, resulting in transit extension and reduced fares on selected routes.

The final chapter in this volume, by Jill Grant, underscores some of the differences and commonalities among US and Canadian suburbs. The chapter is entitled "Contested Spaces: Suburban Development in Halifax and Other Mid-sized Canadian Cities" and focuses on an important shift in Canadian urban planning paradigms from conventional all-American suburban stereotypes towards "urban intensification." Inspired by notions of new urbanism, sustainable development, and smart cities, the first signs of this shift date back to the late 1970s. It implied growing attention to mixed land use and housing types, diversity, densification, and walkability. The chapter considers the impact of this new kind of planning on six different suburban environments in metropolitan areas across Canada. If the new planning principles have been practically universal across the country, the actual impact on suburbia has not. The reason, says Grant, lies in the variable resistance to these principles from metro area to metro area, and from suburb to suburb – and the resistance is attributed mainly to local developers and political leaders. A good number of Canadian suburbs still embody what many consider the "Canadian dream," a very close cousin indeed of the old stereotypical suburban imaginary in the United States. Those suburbs that have resisted new planning ideals appear to adhere to typical patterns in the United States: they have low densities, are predominantly residential with single-family homes, and are quite homogeneous in terms of demography, class, and ethnicity. Where new planning ideas did have a notable impact, suburbs have assumed a more "urban" character, and in those cases the difference with the United States is, generally speaking, substantial. This chapter reflects the (political) culture differences in yet another way. The "contestation" that is examined here is between planners and prevailing planning ideas on the one hand, and local decision makers on the other hand. This is a rather different view of contested suburbs than that presented in the chapter on Los Angeles (or New York, or Ottawa). Further, Grant argues that local decision makers tend to resist urban intensification planning because it conflicts with local traditions and with the preferences of residents. The author refers to "self-sorting processes" where people *choose* where to live, rendering local developers or politicians as actors responding to popular wishes. It is a perspective not likely to be encountered in any of the preceding chapters that focus on the United States.

The contributions to this volume obviously differ not only in their geographic and thematic focuses, but also in method, even if they are all case studies. Some are quantitative, others more discursive; some focus on entire metropolitan areas, others concentrate on selected suburbs. The thematic organization of the book into its four parts helps to sort the key debates but it is to some degree arbitrary, since there are various different threads throughout the volume. The Epilogue will provide a more explicit comparison of the different chapters and lay out some key observations and conclusions about North American suburbia in the early 21st century.

NOTE

1 This historical overview is in part based on Nijman and Clery (2015).

REFERENCES

Champion, T. (2001). Urbanization, suburbanization, counterurbanization and reurbanization. In R. Paddison (Ed.), *Handbook of urban studies* (pp. 143–61). London: Sage.

Douglass, H. (1925). *The suburban trend*. New York, NY: Century.

Duany, A., Plater-Zyberk, E., & Speck, J. (2000). *Suburban nation: The rise of sprawl and the decline of the American dream*. New York, NY: North Point Press.

Edel, M., Sclar, E.D., & Luria, D. (1984). *Shaky palaces: Homeownership and social mobility in Boston's suburbanization*. New York, NY: Columbia University Press.

Ehrenhalt, A. (2013). *The great inversion and the future of the American city*. New York: Vintage.

Fishman, R. (1987). Bourgeois utopias: Visions of suburbia. Reprinted in S. S. Fainstein & S. Campbell (Eds.), *Readings in urban theory* (2nd ed., pp. 21–31, 2002). Oxford: Blackwell.

Hanlon, B., Short, J.R., & Vicino, T.J. (2010). *Cities and suburbs: New metropolitan realities in the United States.* Oxford: Routledge.

Harris, R. (1992). The unplanned blue-collar suburb in its heyday, 1900–1940. In D. Janelle (Ed.), *Geographical snapshots of North America* (pp. 94–99). New York, NY: The Guilford Press.

Harris, R., & Lewis, R. (2001). The geography of North American cities and suburbs, 1900–1950. *Journal of Urban History, 27*(3), 262–92.

Hayden, D. (2001). Revisiting the sitcom suburbs. *Landlines* 13/2. Washington, DC: Lincoln Institute of Land Policy.

Hayden, D. (2003). *Building suburbia: Green fields and urban growth, 1820–2000.* New York, NY: Pantheon Books.

Hobson, M. (2017). *The legend of the Black Mecca: Politics and class in the making of Atlanta.* The University of North Carolina Press.

Immergluck, D. (2009). *Foreclosed: High-risk lending, deregulation, and the undermining of America's mortgage market.* Ithaca, NY: Cornell University Press.

Jackson, K. (1985). *Crabgrass frontier: The suburbanization of the United States.* New York, NY: Oxford University Press.

Jacobs, J. (1961). *The death and life of great American cities.* New York, NY: The Modern Library.

Knox, P.L. (2008). *Metroburbia, USA.* New Brunswick, NJ: Rutgers University Press.

Kunstler, J.H. (1993). *The geography of nowhere: The rise and decline of America's man-made landscape.* New York, NY: Touchstone.

Molotch, H. (1976). The city as a growth machine: Toward a political economy of place. *The American Journal of Sociology, 82*(2), 309–32.

Mozingo, L. (2011). *Pastoral capitalism: A history of suburban corporate landscapes.* Cambridge, MA: MIT Press.

Muller, P.O. (1981). *Contemporary suburban America.* Englewood Cliffs, NJ.

Nicolaides, B.M. (2004). The neighborhood politics of class in a working class suburb of Los Angeles, 1920–1940. *Journal of American History, 91,* 750–810.

Nijman, J., & Clery, T. (2015). The United States: Suburban imaginaries and metropolitan realities. In P. Hamel & R. Keil (Eds.), *Suburban governance: A global view* (pp. 57–79). University of Toronto Press.

Paumgarten, N. (2007, April 16). There and back again: The soul of the commuter. *The New Yorker.*

Seal, C. (2003). We built this suburb. *The Yale Review of Books.* New Haven, CT. Retrieved from http://yalereviewofbooks.com/monograph-we-built-this-suburb-by-carey-seal/

Sharpe, W., & Wallock, L. (1994). Bold new city or built-up 'burb? Redefining contemporary suburbia. *American Quarterly, 46*(1), 1–30.

Taylor, G. (1915). *Satellite cities: A study of industrial suburbs.* New York, NY: Appleton.

U.S. Bureau of the Budget. (1964). *Standard metropolitan statistical areas.* Washington, DC: Government Printing Office.

Walker, R. (1981). A theory of suburbanization: Capitalism and the construction of urban space in the United States. In M. Dear & A.J. Scott (Eds.), *Urbanization and urban planning in capitalist society* (pp. 383-429). New York, NY: Methuen.

Walker, R., & Lewis, R.D. (2001). Beyond the crabgrass frontier: Industry and the spread of North American cities, 1850–1950. *Journal of Historical Geography, 27*(1), 3–19.

Warner, S.B. Jr. (1962). *Streetcar suburbs: The process of growth in Boston, 1870–1900.* Cambridge, MA: Harvard University Press.

PART 1

Questioning North American Suburbia

2 Using Toronto to Explore Three Suburban Stereotypes, and Vice Versa

RICHARD HARRIS

Introduction

Suburbs – or whatever we choose to call the areas of settlement that extend between the central city and the exurban fringe – are where most North Americans now live. We have to care about the forces that shape such places, what they are like, and how to judge them. Regrettably, on each count we are blinkered by a cliché or stereotype.

One stereotype is that suburbs embody a popular ideal. This involves owning a single-family home in a low-density area near the urban fringe, removed from commercial activity, and blessed with open space and quiet privacy (Clapson, 2003). Some have argued that this ideal has been shaped by clever marketing and government policy, but its pervasive presence is assumed. A related stereotype is that the ideal corresponds to reality. Moreover, because many cannot afford it, suburbs have supposedly been middle or upper middle class sorts of places.

There is a third cliché, a commentary on the other two. Academics and planners, although not North Americans in general, believe that suburbs and their associated lifestyles are to be deplored (Davison, 2013; Nicolaides, 2006). This position involves a judgment of the suburbs as they are actually lived, and especially their reliance on the automobile. It also responds to clichés about who lives in the suburbs, and why. And so the three suburban stereotypes – about the ideal, the reality, and about how both should be judged – are intertwined. Recognizing these interconnections, this synthetic essay uses the postwar experience of Toronto, Ontario, to explore the importance of all three.

Why Toronto? No city can stand for all: metropolitan areas are evolving in one or other of at least two directions (Ehrenhalt, 2012). Some, typified by Detroit, follow a trend of inner city decline. Others, including New York City, have seen gentrification and inner city redevelopment, showing that the suburban dream is not ubiquitous. Toronto fits the latter pattern. Moreover, with a population of 5.5 million, making it fifth or sixth-ranked in North America (depending on where urban boundaries are drawn), it does so on a significant scale. And finally, containing one of the highest proportions of first-generation immigrants of all major North American metros, it is a quintessential immigrant city.

But why a "synthetic essay"? To speak of suburban ideals, patterns, judgments, and their interrelation involves covering a lot of territory. For Toronto, much has been mapped by others, using statistics, interviews, personal observation, and critical reflection. A rich body of writing supports a complex account of postwar suburbs, but no sustained overview has previously been presented, hence the need for a well-documented synthesis.[1]

In Toronto, as elsewhere, the interplay among stereotypes shifted in the late 1960s, mainly because some residents began to reevaluate the inner city. Toronto's inner neighbourhoods had always remained attractive to a segment of the middle class, which is one reason why Jane Jacobs moved to one of them in 1968. But the late 1960s marked a wider shift. "Whitepainting" appeared, Toronto's first phase of gentrification. Many writers started to become dissatisfied with the simple, binary distinction between city and suburb, and looked for more complex models of urban form, and an additional suburban stereotype began to take form, one that distinguished between inner and outer suburbs, less and more desirable places. That is why I deal in turn with the early postwar decades and then the period since around 1970.

The Early Postwar Decades

Toronto's suburbs had always been physically and socially diverse (R. Harris, 1996). The built environment persists, and so Toronto's early postwar suburbs remained varied. In the 1950s, lower-status settlement extended into the suburbs in most directions, and ethnic change centred on suburban York Township, later the City of York (Murdie, 1969, pp. 84–5, 127). As a homemaker later commented, her west-end suburb was "neither purely WASP nor," she added in deference to a stereotype, "dull" (Strong-Boag, 1991, p. 486).

The Character of Toronto's Postwar Suburbs

From the 1950s, suburban diversity persisted but took new forms. In 1953, the Province of Ontario gave important powers to a new level of government, Metro, that embraced the City of Toronto as well as several suburban municipalities (map 2.1) (Bourne, 2001). Metro built

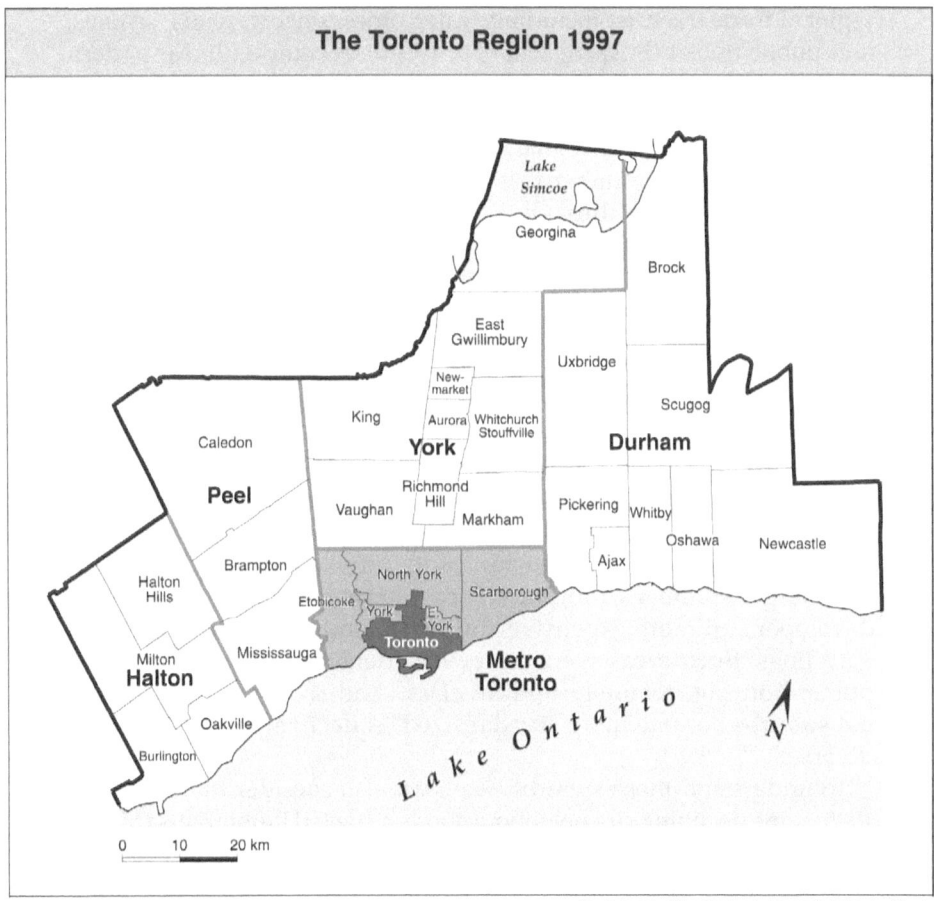

Map 2.1 The political organization of the Greater Toronto Area to 1998. Note that, in 1998, the "Metro" suburbs were amalgamated with the City of Toronto to create an expanded "new" City of Toronto. Source: Larry Bourne. (2001). Designing a metropolitan region: The lessons and lost opportunities of the Toronto experience. In M. Freire and R. Stren (Eds.), *The challenge of urban government: Policies and practices*. Washington, D.C.: World Bank. Used with permission of Byron Moldofsky, Department of Geography, University of Toronto.

regional infrastructure, including water pipes, sewers, roads, schools, and public transit (Frisken, 2007, pp. 130–6). A compact linear pattern of development was favoured, with no satellite towns. In the United States, such restructuring would have been resisted by the suburbs; in Toronto, it was the City that felt it had most to lose. Regardless, as a Canadian province rather than an American state, Ontario had the legal power to carry it through.

Standardized suburbs became common. These soon took a conventional form: neighbourhood units of loops and lollipops focused on schools and framed by an arterial grid. The most ambitious was Don Mills. Other master-planned suburbs followed, notably Erin Mills. But standardized did not mean just tract homes. In the 1960s, federal cost sharing prompted provinces and municipalities to build public housing, often as slabs – high-rise towers. Many were erected in what became known as Metro's suburbs (map 2.1) (Social Planning Council of Metropolitan Toronto, 1979). By the 1970s, when the public housing boom was halted, Metro suburbs contained more than 70% of all assisted-housing units (Frisken, 2007, p. 169). An even larger number of high-rise apartments were built by private developers on more expensive suburban land, adjacent to the subway lines. Such areas were never low density nor, in the case of public housing, remotely middle class. And so, although individual suburbs became more standardized, collectively they remained diverse.

To understand those suburbs, we must also consider the city. After 1945, some declining city neighbourhoods attracted immigrants (Mann, 1961). But so did the suburbs, and there was no inner city stigma, nor middle class flight (map 2.2). Indeed, the challenge was to accommodate redevelopment. Between 1952–62, 1% of the City's land was redeveloped annually (Bourne, 1967, p. 175). Offices were built, and apartments too. Projects aroused NIMBY opposition in downtown neighbourhoods, and by the 1970s residents had elected a reform council to restrain high-density redevelopment.

Slabs sprang up everywhere. Toronto built more in the 1960s than any other North American city except New York, in city and suburbs alike (Keenan, 2013, p. 95). They, too, are a legacy. By 2006, their deterioration prompted the new City of Toronto – formed in 1998 by amalgamating the old City with Metro's suburbs – to mount a "Tower Renewal" program. Tower neighbourhoods are not very walkable (Hess & Farrow, 2011), but they are one reason why Toronto has the highest average densities of any metro region in North America. Unsurprisingly, it has the second-highest per capita transit ridership in North America,

Using Toronto to Explore Three Suburban Stereotypes 27

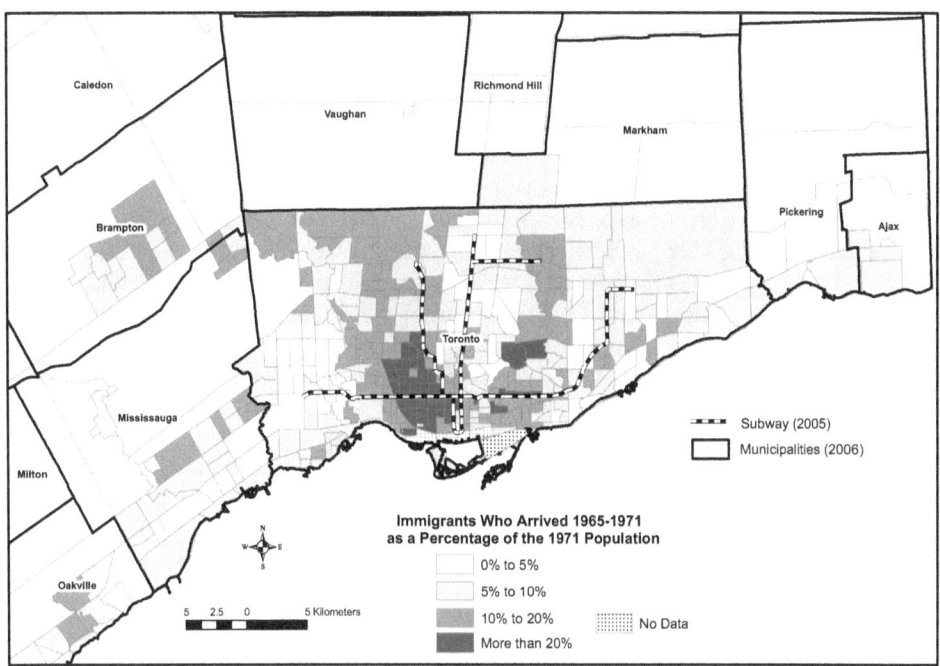

Map 2.2 Patterns of immigrant settlement in the Toronto area, by census tract, 1965–1971. Source: Robert A. Murdie. (2008). Diversity and concentration in Canadian immigration. *Research Bulletin* No. 42, Centre for Urban and Community Studies, University of Toronto. Used with permission of Richard Maaranen, Neighbourhood Change Research Partnership, University of Toronto.

after New York. Among commuters who lived downtown, in 2001 half used transit, while many others walked or biked. Rates declined away from the centre, but of those living 10–15 km away – in what were once Metro's suburbs – a quarter took the bus or subway (Lorinc, 2006, p. 100). These were not stereotypical suburbs.

Contemporary Descriptions and Judgments

Many commentators overlooked or downplayed this suburban diversity, instead focusing on the era's new, stereotypical tract. Even Humphrey Carver chimed in. Carver helped frame the postwar design guidelines of the federal Central (now Canada) Mortgage and Housing Corporation (CMHC), Canada's FHA. Looking back, he (1976,

p. 40) ruefully damned the early tract suburbs: "noble motives seem to produce unexpectedly horrible results," namely "uniformity, conformity ... impersonal, synthetic, exchangeable, temporary."

Intellectuals were beginning to dismiss suburban living, something new to postwar Toronto, and indeed Canada (R. Harris, 2004). Criticism extended to less standardized upper middle class suburbs such as Forest Hill, a Jewish enclave. This point was dramatized in *The Torontonians*, Phyllis Young's best-selling novel about the stresses felt by Karen, a stay-at-home mother in the affluent, WASPy suburb of Rowanwood (Young, 1960/2007). This became was a familiarly critical refrain. But some defended even the blandest place. Ted Relph tackled the criticism that tract housing was generic. Acknowledging that "from the outside" the modern suburb might seem "homogeneous and placeless," he insisted that "from within" they became "closely differentiated into places by the personalization of property, by association with local events and the development of local myths and by being lived in" (Relph, 1976, p. 71). Some experts, then, took the residents' point of view.

And many writers knew that Toronto's suburbs were varied, in ways obvious even to outsiders. In a study of movers' preferences, William Michelson (1977) interviewed households living in houses and apartments in both city and suburban settings. He knew that a binary contrast of city apartments and suburban homes would oversimplify the options. But it was S.D. Clark who undertook the broadest survey. Selecting fifteen widely dispersed districts, Clark found everything from planned suburb to exurban slum (1966, p. 225). He insisted that Toronto's fringe was as physically and socially diverse as could be imagined. Significantly, both Clark's and Michelson's research was funded by CMHC, which must also have recognized suburban variety. Local reporters shared this understanding. When Clark reported his findings, a *Globe and Mail* reporter commented that "it has surely been obvious all along" ("The truth about suburbia," 1960). A suburban stereotype did exist, as did the intellectual's critical response, but Torontonians understood that the suburbs were diverse, and so they varied their judgments accordingly.

The Postwar Suburban Ideal

But what of the *ideal* of suburban living? Even if a place is imperfect, the vision might still be relevant. Michelson asked new residents what had motivated them. Those who bought suburban homes wanted space (Michelson, 1977, pp. 123–124). S.D. Clark (1966, p. 153) made a similar point: he suggested that settlers at Toronto's urban fringe "wanted nothing more than to be left alone," meaning privacy, quiet, and control over affordable living space. He emphasized that people often moved because compelled to: there was only so much city housing, and the suburbs were cheaper. It was those who stayed behind, sacrificing space to enjoy urbanity, who displayed "a strong commitment to a way of life" (Clark, 1966, p. 223). Clearly, many people lived in specific suburbs because they had to. Lower down in the income distribution, the ideal wore so thin as to be almost beyond repair.

More significantly, the ideal was never universal. In 1945, a national survey revealed that three out of ten prospective homebuyers did *not* wish to move to a suburb (Maclean-Hunter Publishing Co., 1945, p. 15). A generation later, Michelson easily found suburbanites in apartments and, compared with their suburban counterparts in homes, they were better-pleased with their access to work, stores, and recreation facilities (Michelson, 1977, pp. 230–231). Conversely, he found buyers of city homes who valued accessibility over interior space or easy parking. In Toronto, city living remained attractive to the middle classes. The suburban ideal always had limits, a fact increasingly apparent since the 1970s.

The Modern Metropolis

Since the 1970s, writers everywhere have increasingly questioned whether it is useful to speak in binary terms about cities and suburbs. The classic suburb, subsidiary to a single urban centre, has been disappearing (Fishman, 1987). Some argue that we should now abandon the term "suburb," and perhaps speak of a "postsuburban" world (Bourne, 1996; Phelps, Wood, & Valler, 2010). Most still use the term, but recognize complexity (e.g. Clapson, 2003; Fishman, 2005; Mikelbank, 2004; Vaughan, Griffiths, Haklay, & Jones, 2009). In particular, many now distinguish "inner" from "outer" versions (Ehrenhalt, 2012; Hanlon, 2010; Kneebone & Bérubé, 2013).

The Shape of Metropolitan Toronto Since 1970

All of these arguments are relevant to Toronto. To an extent that is often overlooked, it is useful to think separately of people and jobs. In terms of employment, although downtown has remained vibrant, most jobs are located in the suburbs, some in polycentric clusters and many in dispersed scatterings. How did these patterns arise? Toronto has long been the centre of a diversified economic region in which blue-collar work played a significant but never dominant role. Since the 1970s, the region has lost manufacturing. Especially around the downtown, however, jobs lost from industry have been more than replaced by white-collar work. Toronto is Canada's financial capital, and the third-ranked financial centre (after New York and Chicago) in North America (Bourne, Britton, & Leslie, 2011, p. 256). Lately, job growth at the centre has gathered pace. Between 2006 and 2011, Toronto's central area actually gained more jobs than the outer suburbs (TD Economics, 2013). Toronto, along with cities such as Los Angeles and Singapore, displays a "re-corporatized downtown" (Florida, 2013).

Beyond the downtown, many suburban jobs have dispersed along highway office strips or in ways that defy generalization (Keil & Young, 2009). Others have coalesced. The major, low-density cluster, with half as many jobs as the downtown, has grown around the international airport (Bourne et al., 2011, p. 245). Planners have encouraged the growth of suburban nodes since 1980s. These edge cities are denser and more tightly clustered than most of their American counterparts (Relph, 2014) but they are still works in progress.

But if the language of downtown, polycentricity, and dispersion is vital to understanding the economic geography of urban areas such as Toronto, residential geographies must be understood in different terms. Here, as in many cities, two patterns are evident. The first is monocentric. The urban area still grows outwards. Newer areas lie close to the urban fringe, where land values encourage lower density development. New housing is rarely built for lower income households, who must live where they can. The average age of the housing stock falls steadily with distance from the central business district (CBD), while the proportion of single-family dwellings, homeownership rate, household size, proportion of unpaid work carried out in the home, and single-person commutes by car all rise (table 2.1).

But if many physical and certain lifestyle characteristics change steadily away from the centre, that is not true in overall social terms. Here, marked distinctions have emerged between gentrifying central neighbourhoods, a declining inner ring of suburbs, and a prosperous outer belt (map 2.1). These contrasts have reshaped local stereotypes about the residential ideals of local residents, about suburban realities, and about the judgments appropriate to each. To make sense of these stereotypes, we must consider the causes and character of these recent sociogeographical shifts.

The Ideal, Reality, and Judgment of Urbanity

In Toronto, there were always some people who viewed inner city neighbourhoods as desirable, and recent job growth downtown has reinforced

Table 2.1 Social and housing characteristics in the Toronto region, 2006

Distance zone (kilometres)	Population ('000s)	First-generation immigrants (%)	Visible minorities (%)	Average household size	Single-family dwellings (%)	Owner-occupied dwellings (%)
0–5	384	48	36	2.0	6	40
5–10	645	51	32	2.4	27	51
10–20	1,456	64	50	2.7	35	61
20–30	1,361	63	56	3.3	53	82
30–40	804	47	38	3.2	61	86
40+	412	22	8	2.9	74	86
Toronto region	5,064	55	43	2.8	42	68

Source: Markus Moos & Ann Kramer. (2012). *Atlas of Suburbanisms*. Waterloo: University of Waterloo (from census).

a new wave of enthusiasm for downtown living. Gentrification gathered momentum in the 1970s. Continuing steadily, it has concentrated in neighbourhoods adjacent to downtown (Walks & Maaranen, 2008, p. 5). Simultaneously, a condominium boom emerged in the 1980s, lapsed, and surged back in the 2000s. In 1981, condos accounted for 2% of housing units in the metropolitan area; it is now 18% (Rosen & Walks, 2013). Currently, the metropolitan area experiences population growth mainly in two areas, the centre and the fringe (map 2.3) (Toronto Community Foundation, 2012, p. 17). Condo dwellers are mostly singles or childless couples who, in the preceding generation, would have rented (Kern, 2010). Now most own, and these include retirees and empty nesters. The condo boom has pushed up homeownership rates across the region from 58 to 68 percent, from 1996 to 2006. Like settlers in raw suburbs, these new homeowners – young and old – plan to stay, and so complain about inadequate services (McGinn, 2013). Until the 1970s, the suburban dream was never the only residential ideal in Toronto, but it was dominant. Today, it must compete.

The growth of CBD employment, and the number of people committed to living close by, has passed a critical threshold. Downtown is now a self-sufficient world: many suburbanites experience the city when they travel to work, but city dwellers need never see the suburbs. The councillor for a downtown ward "once proudly told a reporter he had never visited the suburbs around Toronto" (McMahon, 2012). For a growing number of influential downtowners the suburbs have become a largely irrelevant *terra incognita*.

Map 2.3 Population growth in the Greater Toronto Area, by census tract, 2006–2011. Source: Toronto Community Foundation (2012). *Toronto's Vital Signs*. City of Toronto, City Planning, Policy Analysis & Research.

A Shifting Suburban Stereotype

The growing ideal and reality of downtown urbanity has reinforced the old, negative stereotypes about the suburbs as dull. Local community and ethnic newspapers may cast individual suburbs in a favourable light, but academics, public intellectuals, and reporters at the major dailies see things differently. In the early 2000s, local planners came under the influence of "creative city" ideas, and in 2007, Richard Florida's arrival in town tipped the scales (Lehrer & Wieditz, 2009). This line of thinking has reinforced city planners' emphasis on downtown.

The inner city contains the overwhelming concentration of artists, artists' studios, and cultural institutions (Silver, 2010, p. 33). Cultural workers who do not occupy converted loft workspaces live within a bicycle or streetcar ride of downtown (Silver, 2010, p. 28). Statistics Canada (2011, Table A5) has reported the numbers of persons aged 25 to 44 who, between 2001–6, moved between the new City and the outer suburbs. As suburbanization continued, the ratio favoured out-movers over in-movers by 14:5. But for creative-class occupations such as

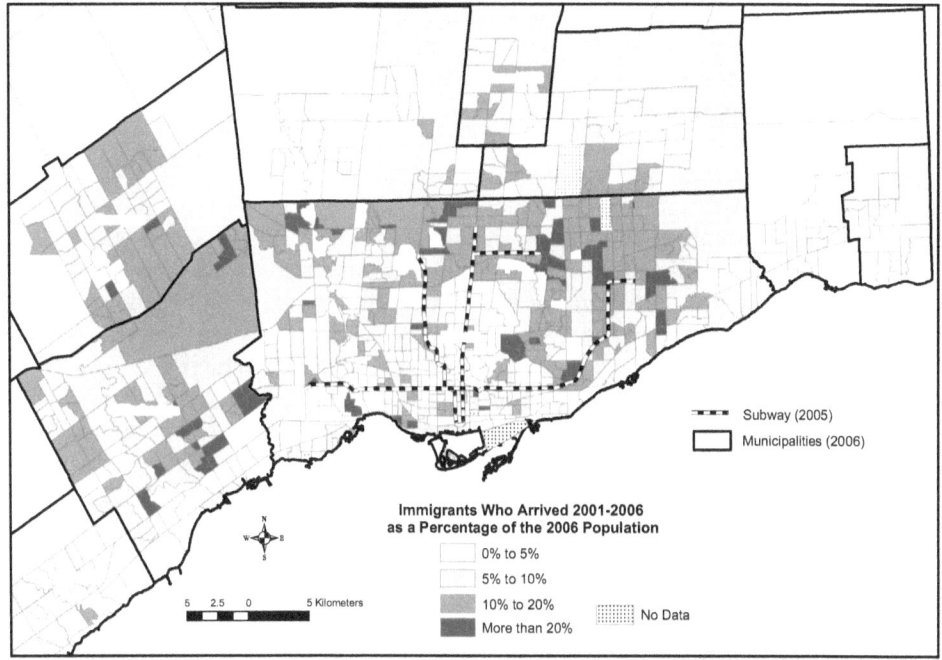

Map 2.4 Patterns of immigrant settlement in the Toronto area, 2001–2006. Source: Robert A. Murdie. (2008). Diversity and concentration in Canadian Immigration. *Research Bulletin* No.42, Centre for Urban and Community Studies, University of Toronto. Used with permission of Richard Maaranen, Neighbourhood Change Research Partnership, University of Toronto.

writers and public-relations professionals (8:12), university professors (5:13), and creative or performing artists (5:14), the ratios were reversed. Such groups embody, articulate, and reinforce a negative stereotype of the suburbs (Silver & Clark, 2015). In that context, many people still think in terms of a city-suburb binary (Fiedler, 2011).

At the same time, gentrification has helped to overthrow one aspect of the old stereotype – that of the whitebread suburb. Toronto is one of the most multicultural places in the world; half of its residents are immigrants, with representatives from every continent and almost every country (Hiebert, 2012; Murdie, 2008; Qadeer, Angrawal, & Lovell, 2010). Until the 1970s, many immigrants settled in inner districts, but gentrification has redirected them elsewhere (Murdie & Teixeira, 2011). In 2006, newcomers comprised almost two thirds of the residents in Toronto's most populated suburban zones (map 2.4; table 2.1; Toronto Star, 2013). In the United States, the "ethnoburb" is seen as a noteworthy recent development (Li, 2009). In Toronto, the term is unnecessary: immigrant settlement is so taken for granted that the presence or dominance of ethnic minorities has become part of the local suburban mythology.

Suburban Realities and Stereotypes Evolve

If some elements of the suburban stereotype have been revised while others have persisted, still others have fragmented in the face of a growing contrast between the inner and outer rings. This bifurcation now shapes politics at the municipal, provincial, and federal levels.

In 1970, most census tracts in Metro's suburbs, which are now the inner suburbs, were middling in income. Since then, most have declined (Hulchanski, 2010). Today, the incidence of poverty is higher there than in the old city or the outer suburbs, as the inner suburbs receive lower income immigrants (Murdie, 2008; Qadeer et al., 2010). They include enclaves rather than ghettos, and some serve immigrants well (Hiebert, 2012; Walks & Bourne, 2006). But in many, the residents of deteriorating high-rises have poor access to public transit and well-paid employment. As a result, in a metro with one of the lowest crime rates in Canada, the inner suburbs contain most of Toronto's pockets of drug-related, violent criminal activity.

Income differences are apparent in commercial landscapes. The old city boasts homeopathists, yoga instructors, bicycle dealerships, and theatres at rates that range from three to ten times those in the older suburbs; the latter stand out for their auto-body shops, bowling alleys, and bingo halls. Both have coffee shops, but while one has Starbucks and boutique operations, the other has Tim Hortons, the definitive Canadian chain.[2] In transitional areas these landscapes can be strikingly juxtaposed.

City-suburban contrasts are equally apparent politically. The 1970s saw the emergence of "polarized patterns of political debate and social commentary" between the old city and Metro's suburbs (Social Planning Council of Metropolitan Toronto, 1979, p. 5). It has since hardened, partly through fallout from divisive amalgamation debates in 1998 (Boudreau, 1999). It is reflected in voting patterns in civic, provincial, and federal elections (Taylor, 2011; Walks, 2006, 2008). In 2010 Rob Ford, the successful mayoral candidate, won almost all polls in the inner suburbs and almost none within the old city.

The downward trend of the inner suburbs has been understood since the late 1970s by social agencies, planners, and researchers, and was general knowledge by the 1990s. Some writers have worked to counter the negative stereotype, celebrating the effective immigrant enclaves, notably Thorncliffe Park (Saunders, 2010). In his walking guide to Toronto, Shawn Micallef (2010, p. 206) describes the Kingston-Galloway area of Scarborough thus: "Just like in downtown Toronto, very different environments can co-exist close together ... undermining the cinematic ... image of blighted neighborhoods that go on for miles." But the effort is forced. Even Micallef (2010, p. 222) concedes that "with the influx of new Canadians and less affluent residents, a pedestrian culture has been grafted onto an automobile landscape, and sometimes it's a hard fit."

And so a negative stereotype has solidified. If the early postwar suburbs were seen as bland, inappropriate settings for crime (A. Harris, 2010, p. 272), they are now seen as epicentres of social problems. This is a change acknowledged by the location of "priority areas" designated by the new city and reinforced by the media (Lindgren, 2009). Aware of stigma, residents often disguise their home address when applying for jobs (Zaami, 2012).

The reality and stereotype of the inner suburbs sets this zone apart not only from the inner neighbourhoods, but also the outer suburbs. If Toronto residents and reporters have a generalized mental map of the city's geography, it speaks of these three zones, and the same model is routinely used by researchers (Bourne, 2001; Keenan, 2013, pp. 120–1; Relph, 2014; Taylor & Burchfield, 2010, p. 33). It is no longer plausible to speak, without qualification, of "the suburbs."

Of course, some features are shared by the inner and outer rings. Immigrants, most of whom conform to what the Canadian census refers to as "visible minorities," characterize both (table 2.1) (Toronto Star, 2013). Indeed, the proportion of visible minorities is highest in parts of the outer ring (table 2.1). The larger ethnic communities are diverse and defy generalization (Salaff & Chan, 2007). Broadly, however, if those with low-to-moderate-incomes settle in the inner ring, the more affluent live beyond (Lo, 2006, pp. 145–6).

What distinguishes the two rings is income and the trappings of homeownership, auto dependency, and sprawl. As everywhere, sprawl has been encouraged by the underpricing of infrastructure (Blais, 2010), and also by the creation of large, competing jurisdictions beyond the new city boundaries (Bourne, Bunce, Taylor, Luka, & Maurer, 2003; Frisken, 2007; Keil & Boudreau, 2005; White, 2016) (map 2.1). Lately, however, provincial policies have favoured intensification within the existing urban envelope. From the 1990s there has been a substantial amount of infill, much being concentrated in edge cities such as the Square One in Mississauga (Burchfield, 2010, pp. 19, 20;

Taylor & Burchfield, 2010, p. 33). Since 2006, development has been shaped by the designation of an outer greenbelt (Bunce, 2010; Sandberg, Wekerle, & Gilbert, 2013). This has slowed and further focused growth, helping to raise land values, house prices, and residential densities.

It is in the outer ring, and beyond, that the stereotypical suburb – as dream and as reality – lives on, albeit in modified form. Densities are higher than in past decades, and Markham contains the largest node of New Urbanist development on the continent (Gordon & Vipond, 2005; Skaburskis, 2006). But in 2001, 96% of outer suburban households owned at least one car, and a similar proportion used it to commute (Lorinc, 2006, p. 107). Automobility is a choice. A survey found that Toronto residents made "affordability" their top criterion in choosing a home. However, city dwellers ranked "ease of walking" second, while outer suburbanites valued "size inside home" (Toronto Public Health, 2012). Here, then, a version of the suburban ideal lives on, with strong appeal for immigrants (Diamond, 2000; Teixeira, 2006). For Jews, it has even won compromises from an Orthodoxy skeptical of material wealth. Significantly, newcomers' preferences are passed on. The children of South Asian immigrants who grew up in outer suburban Brampton now buy homes nearby (Kataure & Walton-Roberts, 2013).

But do even these outer suburbs embody the traditional suburban ideal? Often, practical considerations reign: ease of commute or affordability. People trade the cost of housing against the journey to work: they drive until they qualify. The bastion of the ideal may be exurbia, where incomers strongly prefer low-density living and a "natural" environment, itself an artifact (Sandberg et al., 2013). The irony is that as exurbanites embrace nature they destroy it or deny it to others in the name of conservation (Cadieux & Taylor, 2013). Notably, exurbia is White, containing fewer immigrants than the inner or the outer suburbs (table 2.1; Bollman, 2010). Indeed, many exurbanites have left the city "to get away from the visible minorities" (Barrett, 1994, p. 95). Even those moving into exurbia, then, have mixed motives.

And of course the outer suburbs are damned. Their residents display higher rates of obesity (Toronto Public Health, 2012, p. 27). How much is due to the suburban elements of their lifestyle – as opposed to the sedentary nature of their work or their eating habits – is unclear. Regardless, for health, and even more for environmental reasons, suburbia (especially sprawl) has attracted waves of criticism (Sahely, Dudding, & Kennedy, 2003). Local experts are still at war with the suburban ideal and its associated reality. The difference is that they now recognize that their critique is relevant above all to the outer zone of the broad suburban field.

Conclusion

To make sense of Toronto's suburbs we need to use different models and language for its economic and social geography. The former requires a language of centralization, polycentricity, and dispersion; edge and edgeless cities. For Toronto – and arguably elsewhere – however, residential patterns can be interpreted in terms of a new zonal arrangement. For in residential terms, Toronto has never fitted any of the three suburban stereotypes. Its early postwar suburbs were known to be diverse and sometimes judged sympathetically. Although many residents shared a suburban ideal, their motives were often pragmatic: they wanted an affordable home. And always, for a minority, the city beckoned. Lately, these observations have become doubly true. Downtown acquired a new image – and reality – of urbanity. The vision of city living attracts families with children, as well as the young and hip. It is city folk who exhibit the strongest preference for a particular built environment and lifestyle; suburbanites buy what they can afford.

All of this has become familiar in local discourse. As Christopher Hume (2008), the *Toronto Star's* long-standing urban affairs columnist, says, "Having remade the face of North America, the tide now seems to be turning against the suburbs." Of course, not everyone wants to live downtown, and fewer can afford to. But in Toronto, those who choose the suburbs are self-conscious. Suburbs – the ideal and the reality – are not the only game in town, and everyone knows it. Everyone also knows that no stereotype can cover all suburbs. The old package persists for the outer fringes: affluent familism. But there are new clichés about the inner suburbs: poor, underserved by transit, struggling, and problematic. This understanding, once confined to social workers, policy wonks, and public housing tenants, is now a political brand: the ex-mayor's critics and supporters alike spoke of Ford Country. In Toronto, then, stereotypes about the suburbs have been eroded from the inside and are, as well, being challenged from without.

No urban area has ever conformed perfectly to any of the stereotypes that I have discussed, still less all three. But in Toronto the balance has been different from, and shifted more than, the average. There are few American metros where downtown living poses such a real and symbolic alternative to the 'burbs; where the bifurcation of inner and outer suburban rings has been so marked; and where assessments of the suburbs have had to become so nuanced. In these respects, Toronto exemplifies a certain type of North American city.

What has made Toronto unusual? A cultural and financial capital, Toronto's downtown is a major employment hub; White flight was never an issue; the blight of downtown freeways has been moderate.

In 1953, Ontario created metropolitan government, and so a generation of suburban growth took an unusually diversified form. Finally, many peoples' aspirations for the suburban ideal have been redirected into "cottage country," beyond the exurban fringe (Luka, 2010). There are few major North American cities where such a high proportion of residents have ready, if intermittent, access to low-density living in a rural setting. Cottages make city living more tolerable, and perhaps attractive. In these ways, Toronto is untypical. That is why its experience is so useful. It forcefully probes some sweeping assumptions that academics have brought to their interpretation of urban areas across the North American continent. It is unclear whether many North American cities will follow its example. Regardless, its experience demonstrates that, within a North American context and beyond the confines of New York, alternative visions, realities, and judgments can thrive.

So should we dispense with stereotypes, clichés, and the like? In one sense the question is idle: we have always relied on them and, in one form or another, we will continue to do so. A more constructive response is to put them to use, as I have attempted to do here. In the process it becomes clear that if Toronto's experience challenges those stereotypes, it can also be illuminated by them, if only as a point of reference.

ACKNOWLEDGMENTS

I would like to thank Larry Bourne, Robert Lewis, Jan Nijman, Ted Relph, Danielle Robinson, Dan Silver, Zack Taylor, Carol Town, Alan Walks, and Richard White for helpful comments on an earlier draft. Thanks to Byron Moldofsky for cartographic assistance. I would like to offer a special, lifetime thank you to Larry Bourne for professional advice since 1975.

NOTES

1 A longer version with more local references may be found in *Environment and Planning* 46 (2015).
2 Thanks to Dan Silver for making available unpublished calculations based on census data, and elsewhere for advice on the interpretation of cultural activity.

REFERENCES

Barrett, S. (1994). *Paradise: Class, commuters and ethnicity in rural Ontario.* Toronto: University of Toronto Press.
Blais, P. (2010). *Perverse cities: Hidden subsidies, wonky policies, and urban sprawl.* Vancouver: UBC Press.
Bollman, R. (2010). Canadians living beside a city: A profile. In K. Beesley (Ed.), *The rural-urban fringe in Canada: Conflict and controversy* (pp. 23–54). Brandon, MB: Rural Development Institute, Brandon University.
Boudreau, J.-A. (1999). Megacity Toronto: Struggles over differing aspects of middle-class politics. *International Journal of Urban and Regional Research, 23*(4), 771–81.
Bourne, L. (1996). Reinventing the suburbs: Old myths and new realities. *Progress in Planning, 46*(3), 163–84.
Bourne, L. (2001). Designing a metropolitan region: The lessons and lost opportunities of the Toronto experience. In M. Freire & R. Stren (Eds.), *The challenge of urban government: Policies and practices* (pp. 27–46). Washington, DC: World Bank.
Bourne, L.S. (1967). *Private redevelopment of the central city: Spatial processes of structural change in the city of Toronto* (Research Paper No.112). Chicago, IL: Department of Geography, University of Chicago.
Bourne, L.S., Britton, J.N.H., & Leslie, D. (2011). The greater Toronto region: The challenges of economic restructuring, social diversity and globalization. In L.S. Bourne, T. Hutton, R. G. Shearmur, & J. Simmons (Eds.), *Canadian urban regions: Trajectories of growth and change* (pp. 236–68). Toronto: Oxford University Press.
Bourne, L.S., Bunce, M., Taylor, L., Luka, N., & Maurer, J. (2003). Contested ground: The dynamics of peri-urban growth in the Toronto region. *Canadian Journal of Regional Science, XXVI,* 251–70.
Bunce, M. (2010). Entrenching the countryside in the rural-urban fringe: The role of public policy in the Toronto region. In K. Beesley (Ed.), *The rural-urban fringe in Canada: Conflict and controversy* (pp. 328–36). Brandon, MB: Rural Development Institute, Brandon University.
Burchfield, M. (2010). *Implementing residential intensification target: Lessons from research on intensification rates in Toronto.* Toronto: Neptis.
Cadieux, K.V., & Taylor, L. (Eds.). (2013). *Landscape and the ideology of nature in exurbia.* New York, NY: Routledge.
Carver, H. (1976). Building the suburbs: A planner's reflections. *City Magazine, 3*(7), 40–5.
Clapson, M. (2003). *Suburban century: Social change and urban growth in England and the USA.* Oxford: Berg.

Clark, S.D. (1966). *The suburban society*. Toronto: University of Toronto Press.

Davison, G. (2013). The suburban idea and its enemies. *Journal of Urban History, 39*, 829–47.

Diamond, E. (2000). *And I will dwell in their midst: Orthodox Jews in suburbia*. Chapel Hill: University of North Carolina Press.

Ehrenhalt, A. (2012). *The great inversion and the future of the American city*. New York, NY: Knopf.

Fiedler, R. (2011). The representational challenge of the in-between. In D. Young, P. Wood, & R. Keil (Eds.), *In-between infrastructure: Urban connectivity in an age of vulnerability*. Praxis (e)press, 67–85

Fishman, R. (1987). *Bourgeois utopias: The rise and fall of suburbia*. New York, NY: Basic Books.

Fishman, R. (2005). The fifth migration. *Journal of the American Planning Association, 71*(4), 357–66.

Florida, R. (2013, March 27). Interview with Sam Bass Warner and Andrew Whittemore. *CityLab*. https://www.citylab.com/life/2013/03/why-our-cities-look-and-work-way-they-do/4965/

Frisken, F. (2007). *The public metropolis: The political dynamics of urban expansion in the Toronto region, 1924–2003*. Toronto: Canadian Scholars' Press.

Gordon, D., & Vipond, S. (2005). Gross density and New Urbanism: Comparing conventional and new urbanist suburbs in Markham, Ontario. *Journal of the American Planning Association, 71*(1), 41–54.

Hanlon, B. (2010). *Once the American dream: Inner-ring suburbs of the United States*. Philadelphia, PA: Temple University Press.

Harris, A. (2010). *Imagining Toronto*. Toronto: Mansfield Press.

Harris, R. (1996). *Unplanned suburbs: Toronto's American tragedy, 1900–1950*. Baltimore, MD: Johns Hopkins University Press.

Harris, R. (2004). *Creeping conformity: How Canada became suburban, 1900–1960*. Toronto: University of Toronto Press.

Hess, P. M., & Farrow, J. (2011). *Walkability in Toronto's high-rise neighborhoods*. Toronto: University of Toronto Cities Centre.

Hiebert, D. (2012). *A new social order? The social geography of visible minority and religious groups in Montreal, Toronto and Vancouver in 2031*. Ottawa, ON: Citizenship and Immigration.

Hulchanski, J.D. (2010). *The three cities within Toronto: Income polarization among Toronto's neighborhoods, 1970–2005*. Toronto: University of Toronto Cities Centre.

Hume, C. (2008, March 3). Downtown density will prevail over slums of suburbia. *Toronto Star*.

Kataure, V., & Walton-Roberts, M. (2013). The housing preferences and location choices of second-generation South Asians living in ethnic enclaves. *South Asian Diaspora, 5*(1), 57–76.

Keenan, E. (2013). *Some great idea. Good neighborhoods, crazy politics and the invention of Toronto.* Toronto: Coach House Press.

Keil, R., & Boudreau, J.-A. (2005). Is there regionalism after municipal amalgamation in Toronto. *City, 9*(1), 9–22.

Keil, R., & Young, D. (2009). Fringe explosions. Risk and vulnerability in Canada's new in-between urban landscape. *Canadian Geographer, 53*(4), 488–99.

Kern, L. (2010). *Sex and the revitalized city: Gender, condominium development, and urban citizenship.* Vancouver: UBC Press.

Kneebone, E., & Bérubé, A. (2013). *Confronting suburban poverty in America.* Washington, DC: Brookings.

Lehrer, U., & Wieditz, T. (2009). Condominium development and gentrification: The relationship between policies, building activities and socio-economic development in Toronto. *Canadian Journal of Urban Research, 18*(1), 82–103.

Li, W. (2009). *Ethnoburbs: The new ethnic community in urban America.* Honolulu: University of Hawaii Press.

Lindgren, A. (2009). News, geography and disadvantage: Mapping newspaper coverage of higher-needs neighborhoods in Toronto, Canada. *Canadian Journal of Urban Research, 18*(1), 76–97.

Lo, L. (2006). Suburban living and indoor shopping: The production of the contemporary Chinese landscape in Toronto. In W. Li (Ed.), *From urban enclave to ethnic suburb: New Asian communities in the Pacific rim countries* (pp. 134–54). Honolulu: University of Hawai'i Press.

Lorinc, J. (2006). *The new city: How the crisis in Canada's urban centres is reshaping the nation.* Toronto: Penguin.

Luka, N. (2010). Of McMansions, timeshare cottages, and zebra mussels: Dispatches from the second home settings of central Ontario. In Ken Beesley (Ed.), *The rural-urban fringe in Canada: Conflict and controversy* (pp. 199–220). Brandon, MB: Rural Development Institute, Brandon University.

Maclean-Hunter Publishing Co. (1945). *The housing plans of Canadians.* Toronto: Maclean-Hunter Co.

Mann, W.E. (1961). The social system of the slum: The lower ward, Toronto. In S.D. Clark (Ed.), *Urbanism and the changing Canadian society.* Toronto: University of Toronto Press, 39–69.

McGinn, D. (2013, March 2). Buildings in search of a community. *Globe and Mail.*

McMahon, T. (2012, August 9). Vertical development: A dense idea. *Toronto Star.*

Micallef, S. (2010). *Stroll: Psychogeographic walking tours of Toronto.* Toronto: Coach House.

Michelson, W. (1977). *Environmental choice, human behavior and residential satisfaction*. New York, NY: Oxford University Press.

Mikelbank, B.A. (2004). A typology of U.S. suburban places. *Housing Policy Debate, 15*(4), 935–64.

Moos, M., & Kramer, A. (2012). *Atlas of suburbanisms*. Waterloo: University of Waterloo. Retrieved 30 April 2013 from http://env-blogs.uwaterloo.ca/atlas/

Murdie, R.A. (1969). *Factorial ecology of metropolitan Toronto, 1951–1961* (Research Paper No. 116). Chicago, IL: Department of Geography, University of Chicago.

Murdie, R.A. (2008). *Diversity and concentration in Canadian immigration* (Research Bulletin No.42). Centre for Urban and Community Studies, University of Toronto.

Murdie, R.A., & Teixeira, C. (2011). The impact of gentrification on ethnic neighbourhoods in Toronto: A case study of Little Portugal. *Urban Studies, 48*(1), 61–83.

Nicolaides, B. (2006). How hell moved from the city to the suburbs, urban scholars and changing perceptions of authentic community. In K. Kruse & T.J. Sugrue (Eds.), *The New Suburban History* (pp. 80–81). Chicago, IL: University of Chicago Press.

Phelps, N., Wood, A.M., & Valler, D.C. (2010). A post-suburban world? An outline of a research agenda. *Environment and Planning A, 42*(2), 366–83.

Qadeer, M., Angrawal, S.K., & Lovell, A. (2010). Evolution of ethnic enclaves in the Toronto metropolitan area, 2001–2006. *International Journal of Migration and Integration, 11*(3), 315–19.

Relph, E. (1976). *Place and placelessness*. London: Pion.

Relph, E. (2014). *Toronto: Transformations in a city and its region*. Philadelphia, PA: University of Pennsylvania Press.

Rosen, G., & Walks, A. (2013). Castles in Toronto's sky: Condo-ism and urban transformation. Unpublished manuscript, University of Toronto, Toronto.

Sahely, H.R., Dudding, S., & Kennedy, C. (2003). Estimating the urban metabolism of Canadian cities: Greater Toronto area case study. *Canadian Journal of Civil Engineering, 30*, 468–83.

Salaff, J., & Chan, P. (2007). Competing interests. Toronto's Chinese immigrant associations and the politics of multiculturalism. *Population, Space and Place, 13*(2), 125–40.

Sandberg, L.A., Wekerle, G., & Gilbert, L. (2013). *The Oak Ridges Moraine battles: Development, sprawl and nature conservation in the Toronto region*. Toronto: University of Toronto Press.

Saunders, D. (2010). *Arrival city: The final migration and our next world*. Toronto: Knopf.

Silver, D. (Ed.). (2010). *From the ground up: Growing Toronto's cultural sector*. Toronto: Martin Prosperity Institute and City of Toronto.
Silver, D., & Clark, T.N. (2015). The power of scenes. Quantities of amenities and qualities of places. *Cultural Studies, 29*(3), 425–49.
Skaburskis, A. (2006). New urbanism and sprawl: A Toronto case study. *Journal of Planning Education and Research, 25*(3), 233–48.
Social Planning Council of Metropolitan Toronto. (1979). *Metro's suburbs in transition. Part 1: Evolution and overview*. Toronto: The Council.
Strong-Boag, V. (1991). Home dreams: Women and the suburban experiment in Canada, 1945–1960. *Canadian Historical Review, 72*(4), 471–504.
Taylor, Z. (2011, May). *Who elected Rob Ford and why? An ecological analysis of the 2010 Toronto election*. Paper presented at the Annual Meetings of the Canadian Political Science Association, Waterloo, Ontario. Retrieved 10 June 2013 from http://www.cpsa-acsp.ca/papers-2011/Taylor.pdf
Taylor, Z., & Burchfield, M. (Eds.). (2010). *Growing cities: Comparing urban growth patterns and regional Growth policies in Calgary, Toronto and Vancouver*. Toronto: Neptis.
TD Economics. (2013). *Toronto – A return to the core*. Toronto: TD Economics. Retrieved from https://www.td.com/document/PDF/economics/special/ff0113_toronto.pdf
Teixeira, C. (2006). Residential experiences and the culture of suburbanization: A case study of Portuguese homebuyers in Mississauga. *Housing Studies, 22*(4), 495–521.
Toronto Community Foundation. (2012). *Toronto's vital signs*. Toronto: The Foundation. Retrieved 30 April 2013 from https://torontofoundation.ca/wp-content/uploads/2018/02/TVS12FullReport.pdf
Toronto Public Health. (2012). *The walkable city: Neighborhood design and preferences, travel choices and health*. Toronto: Toronto Public Health. Retrieved 30 April 2013 from https://www.toronto.ca/wp-content/uploads/2017/10/9617-TPH-walkable-city-report.pdf
Toronto Star. (2013). Maps: Where immigrants settled in the GTA. Retrieved 6 June 2013 from http://www.thestar.com/news/immigration/gta_immigration_history.html
"The Truth about suburbia." (1960, July 11). *Globe and Mail*.
Vaughan, L., Griffiths, S., Haklay, M., & Jones, C. E. (2009). Do the suburbs exist? Discovering complexity and specificity in suburban built forms. *Transactions, Institute of British Geographers, 34*(4), 475–88.
Walks, R.A. (2006). The causes of city-suburban political polarization: A Canadian case study. *Annals, Association of American Geographers, 96*(2), 390–414.

Walks, R.A. (2008). Urban form, everyday life, and ideology. Support for privatization in three Toronto neighborhoods. *Environment and Planning A, 40*(2), 258–282.

Walks, R.A., & Bourne, L.S. (2006). Ghettos in Canada's cities? Racial segregation, ethnic enclaves and poverty concentration in Canadian urban areas. *Canadian Geographer, 50*(3), 273–97.

Walks, R.A., & Maaranen, R. (2008). *Neighborhood gentrification and upgrading in Montreal, Toronto and Vancouver (Research Bulletin* No. 43). Centre for Urban and Community Studies, University of Toronto.

White, R. (2016). *Planning Toronto: The planners, the plans, their legacies, 1940–1980.* Vancouver: UBC Press.

Young, P.B. (2007). *The Torontonians.* Montreal and Kingston: McGill-Queen's University Press. (Original work published 1960.)

Zaami, M. (2012). *Experiences of socio-spatial exclusion among Ghanaian immigrant youth in Toronto: A case study of the Jane-Finch neighbourhood* (Unpublished master's thesis). University of Western Ontario, London, Ontario.

3 Mexico City: Elusive Suburbs, Ubiquitous Peripheries

LIETTE GILBERT

Introduction

Metropolitan Mexico City challenges our conventional North American understanding of suburban experiences as upper middle class romantic aspirations or bourgeois utopias. Like its North American counterparts, Mexico City is the product of postwar urban development, industrialization, and capital accumulation but its peripheralization has been predominantly experienced by low-income residents pushed towards urban edges in informal settlements and, in more recent years, middle class populations having access to mortgage financing to secure mass-produced housing. In the metropolitan area of Mexico City, where urban expansion has increasingly covered former lakebeds and volcanic hillsides, peripheral "suburban" developments take many forms.

Yet for most Mexicans, *Suburbia* refers to a chain of relatively low-priced clothing department stores, subsidiary of Walmart (Mexico). What most people north of Mexico refer to as suburbanization is better known in Latin America as peripheralization or periurbanization: a process of metropolitan expansion articulated around urban and rural enmeshing, urban corridor developments, and the rearticulation of urban subcentres (Connolly, 2003). This distinction is important as suburb, with its middle and upper class connotation of the ideal life translates poorly to most peripheral (and often irregular) expansion characterizing metropolitan Mexico City (Heinrichs & Nuissl, 2013).

With a population of more than 20 million, the metropolitan Mexico City Area (hereafter Mexico City) is North America's largest city (INEGI, 2010). Sitting at the centre of the metropolitan area, the Federal District (with a population of 8.8 million living in 16 boroughs) is surrounded by 59 municipalities of the State of Mexico and one municipality in the State of Hidalgo. Urban proliferation in metropolitan Mexico

City has produced a polycentric urban form incorporating the expansion of (pre-Hispanic) towns and rural areas of the Federal District and State of Mexico into a complex metropolitan structure, radiating along major transport corridors towards nearby secondary cities. Mexico City is a metropolitan conurbation often described as "many cities within the city" (Garza, 2000). By virtue of its particular size, history, and geography, Mexico City is a complex metropolitan region that encompasses many forms of urbanization, most of them highlighting the deep contradiction of Mexico's social and political economies. The aphorism of "many cities within the city" could, in the next 25 years, take on a new meaning as megapolitan expansion weaves the surrounding cities of Toluca, Pachuca, Cuernavaca-Cuautla, Puebla-Tlaxcala, and Queretaro into a contiguous urbanized fabric.

Three dominant modes of peripheral housing production characterized metropolitan Mexico: private subdivisions for middle and higher incomes, massive subdivisions of social interest housing for sectors with limited resources but able to access financing programs, and the prevailing so-called informal settlements (or *colonias populares*) established on communal or private lands by people with low income. After tracing a rapid overview of the peripheral expansion of metropolitan Mexico City, I briefly present four cases representing different modes of (sub)urban production in more or less distant relations to traditional prototypical post-1950s North American suburb. Although there are many interesting examples to illustrate the metropolitan expansion of Mexico City, I first selected Ciudad Satélite, probably the development that most closely attempts to reproduce a North American suburb. I then present an unusual case in Mexico, Ciudad Santa Fé, an edge city representing Mexico's global future. These two forms of suburbanization are contrasted with two other types of peripheral developments that are more illustrative of the metropolitan fabric: Ixtapaluca and its mass-produced so-called social-housing development; and Chalco Solidaridad, one of the many irregular or informal urban developments continually extending the margins of the metropolis outwards. The presentation of these four cases draws on the works of Mexican scholars who have written extensively on these particular cases.

Brief Overview of Metropolitan Expansion

Mexico City's population grew most rapidly from 3 million in 1950 to 13 million in 1980. Growth rates slowed down afterwards, adding about 3 million per decade, to reach 18.5 million in 2000, which was short of the apocalyptic projection of 31 million (Brockerhoff, 1999).

Mexico City's rapid urban growth was the result of a national economic modernization scheme focused on import substituting industrialization programs that privileged the concentration of industrial production in the Federal District. Rural migrants, attracted by manufacturing jobs and the increasing demand for cheap domestic services for emergent middle class households, arrived in the capital to immediately face a serious housing shortage, forcing many of them to improvise their homes. Workers expanded the city outward, first settling informally in settlements around the centre of the city in proximity to jobs, transportation, and public services. A 1954 municipal bylaw preventing growth and land subdivision in central areas resulted in the massive expansion of urban development in adjacent areas of the State of Mexico (e.g., in Netzahualcoyotl, Naucalpan, Tlatenpantla, and Ecatepec), which proffered lax planning norms, contrasting greatly with the Federal District's restrictive urban development policies (Davis, 1998; de Alba & Capron, 2010). As a result of demographic growth and weak planning coordination, irregular settlements increasingly grew on the hillsides of the Trans-Mexican Volcanic Belt and in the dried-up lakebeds that characterized the Valley of Mexico. Private and communal (ejidal) lands held generally for agricultural production were informally settled by lower income populations, but some were also illegally developed into elite and upper income estates (Davis, 1998). Authorities allowed the informal takeover of lands because of their own incapacity to regulate growth and provide housing.

National economic growth culminated with the 1968 Olympic Games in Mexico City and the oil boom in the 1970s. Up to the 1980s, land and credit availability for real estate development and relatively low interest rates enabled investments in construction, land-value increases, and the regularization of informal settlements. The 1980s' recession tightened markets and job opportunities; unemployment and lower wages pushed struggling middle class families to look for cheaper (and often informal) housing on the outskirts of the city. The core of the Federal District lost 1 million of people, while the peripheries grew by more than 3 million. A devastating earthquake in 1985 killed more than 10,000 people and destroyed many buildings in the central areas. Despite reconstruction programs and the mobilization of urban and housing organizations to expand and to rehabilitate housing in the Federal District, urban expansion kept spreading out to peripheral agricultural centres north (e.g., in Cuauliltlan and Tecamac) and south of the State of Mexico (e.g., in Xochimilco and Chalco).

The economic climate recovered in the early 1990s but plummeted again in 1994 as Mexico was entering into the North American Free

Trade Agreement and privatizing utilities and services. Mexico's financial crises increased the pressure from international financing institutions (e.g., IMF and World Bank) to guide the housing sector towards free-market conditions. Part of the ubiquitous peripheralization of Mexico City was significantly shaped by the restructuring of the housing financial system, leading to the emergence of a booming mortgage and housing industry and to massive tract developments in the peripheries (Connolly, 2009; Monkkonen, 2011). Seeing the housing-sector market as an important economic driver and as an essential part of the modernizing project for the Mexican economy, the federal government turned to homebuilding companies for the construction of social interest housing. Reform of governmental lending practices led to a shift in the production and acquisition of housing in Mexico from self-built housing to housing built by private-sector homebuilding companies and predominantly financed with mortgages offered by the National Workers' Housing Fund, known as Infonavit (Monkkonen, 2011). Between 1995 and 2012, Infonavit issued between 65% to 75% of the housing loans in the country whose funds are obtained from compulsory dues levied on private-sector workers (Monkkonen, 2011). While financing has enabled qualifying middle-income households to acquire their homes, the poorest households (notably the estimated 60% of people working in the informal economy) remain excluded from such programs. Infonavit is the cornerstone of housing policy whose main function is to "provide subsidised mortgage credit to guarantee effective demand for housing developers" (Connolly, 2009, p. 13).

Benefitting from land banking, housing financing, and international investments, a thriving building industry flourished in the early 2000s. Following the end of the 71-year ruling by the Institutional Revolutionary Party, an estimated 9 million homes were built between 2000 and 2012. Massive high-density subdivisions developments, characterized by thousands of small, identical, relatively low-cost houses, emerged on the outskirts of the city, where land was cheaper but often prone to seasonal flooding. Developers maximized their profits by achieving economies of scale through cheaper peripheral lands, homogeneity of construction, and vertical integration of production, promotion, and sale activities. Quantitatively, the boom in housing production was seen as a success (García Peralta & Hofer, 2006). Housing development was, however, reduced to buildings with very limited infrastructure and nearly no public amenities such as schools, transportation, health clinics, parks, etc. (Connolly, 2009). In many cases, this mass-industrialized housing production infilling many municipal territories added to services and infrastructure systems that were already stretched. Investments in infrastructure and services have not followed housing investments, and as a result, homeowners contend daily with limited access to employment and schools, daily long commutes, higher transportation costs, deficient services, higher costs of services delivery, and environmental pollution. Many households confronted with these daily challenges have had no other choice than to abandon their dream of homeownership, and rates of vacancy in recent developments have reached close to 50% in some municipalities (Fundación CIDOC and SHF, 2011). Metropolitan Mexico City faces serious problems and deficits in the provision of basic services and infrastructure in irregular settlements as well as in mass-produced social interest housing developments. Infrastructural demands have long outpaced fiscal capacities of all levels of governments (Davis, 1998). In an attempt to curb peripheral sprawl increasingly encroaching on ecologically sensitive areas and to address infrastructural and services deficits, growth policy has recently favoured densification and redevelopment of central mixed-use areas. Let us now review some cases of urban expansion to better understand some of the relations, processes, and narratives at play in metropolitan Mexico City.

Ciudad Satélite: The First and Perhaps Only "Suburb"

It is not the actual American city that "travels" from the US to the rest of the planet, but rather an imaginary based on it.

(Capron & de Alba, 2010, p. 160)

Located in Naucalpan on the edge of the State of Mexico (16 kms from the centre of the city), Ciudad Satélite was one of the first suburban neighbourhoods to benefit from the development ban in the Federal District in the late 1950s. Satélite also benefitted from the construction of the Mexico-Querétaro highway (extending north all the way to the United States) during the same period. Part of Satélite was partly built on agricultural lands belonging to ex-president Miguel Alemán Valdés (in office between 1946 to 1952), whose administration led rapid industrialization that has historically benefitted the development of the capital region as well as cozy and enriching relationships with big businesses, developers, and investors. Communal lands were illegally purchased by development companies sponsoring housing subdivisions for the growing middle class and elite populations (Flores Peña, 2013). At the core of the development and implementation of Satélite was an ambitious ecological plan designed to escape the chaos of the Federal District. The close distance to the central areas, the relatively low cost of lands, and the lax planning regulation of the State of Mexico contributed to the implementation of what appears to be the closest ideological and morphological incarnation of a "typical" North American suburb.

The vast residential development was designed by Mario Pani in 1954. Pani, a famous Mexican architect, has been responsible for many social-housing projects (e.g., Nonalco-Tlateloco in the centre of the city and Ciudad Universitaria in the southern part of the city). Pani was well known for his *unidad habitacional* model of a "city within the city," anchoring large residential apartment buildings into a dense fabric of schools, hospitals, stores, cultural and recreational facilities, and green spaces. Pani's master plan for Ciudad Satélite sought to emulate a garden city with a greenbelt separating the new suburban development from the rest of the city. Pani's vision for Satélite was to build a "city outside the city" for those who could afford to distance themselves from the urban chaos of the central city (while remaining connected to the rest of the city with the new highway). Ciudad Satélite provided a distinct "suburban" model of development inspired by the modernism of the 20th century, clearly breaking from Mexico City's urban and architectural traditions (Cordova Gonzáles, 2001).

Innovative aspects of Mario Pani's design for Satélite included an organic vehicular circulation pattern (without crossings and streetlights), superblocks, and the construction of the first commercial centre, Plaza Satélite, in 1968 (then the largest mall in in Latin America). Satélite offered the dream of a single-family home for the middle class, away from the polluted air, traffic congestion, urban density, and noise of the central city. The garden-city-like development offered tranquillity, and its suburban imaginary was directly linked to both home and car ownership. As the first residential development built in the northern edge of the Federal District, Satélite was initially a dormitory suburb. The detached single-family home, the car(s), the shopping centre were sold as a new and different lifestyle to the growing middle class who could afford it. As Capron and de Alba (2010, p. 168) write, "Ciudad Satélite is seen, in general and by its own inhabitants, as an American suburb with American ways of life. It is clear that the initial urban models were not suburban, nor were they solely 'American.' The arrival of this imaginary in Mexico started with the hybridization of progressive urbanism as it came into contact with a society whose political context was different."

But as the periphery urbanized, the accessibility of Satélite was also its demise as the "suburban" island soon became immersed in the growing urbanized fabric. Initially planned for a population with middle

Figure 3.1 Ciudad Satélite, Naucalpan, State of Mexico. Photo courtesy of author.

income, the growing demand transformed Satélite into an upper middle enclave, as the tranquillity, convenience, and homogeneity of suburban living away from the urban "anarchy of other neighborhoods of the city" became unaffordable to the middle class (Tarrés, 1999, p. 426). The colourful Towers of Satélite (designed by Mexican architect Luis Barragán and sculptor Mathias Goeritz in 1957), standing tall in the middle median of one of the busiest highway in the country, remain the proud symbol of modernity for this locality (figure 3.1). Recognized by UNESCO as architectural heritage, the five towers ranging from 30 to 52 metres (allegedly representing the ideal family of two parents and three children) nowadays barely cast a shadow over the urban expansion of the State of Mexico.

Ciudad Santa Fé: From Sand Pit and City Dump to Global City

Santa Fé is certainly the most exclusionary space in the city, embodying the symbolic place of globalized Mexico.

(Moreno Carranco, 2010, p. 24)

Figure 3.2 Ciudad Santa Fé, Federal District. Photo courtesy of author.

Santa Fé is located on the western edge of the Federal District, strategically located between downtown and nearby Toluca, capital of the State of Mexico. The neighbourhood benefitted from the construction of a major (Mexico-Toluca) highway, but Santa Fé has remained an isolated, inaccessible, and exclusive enclave along this major corridor. In the early 1990s, the neighbourhood experienced a radical transformation from sand pits and garbage dumps surrounded by low-income settlements to a corporate and high-scale residential enclave.

With the signing of the North American Free Trade Agreement in 1994, this hilly depreciated area soon became the seat of many multinational corporate headquarters.

Santa Fé, with its stylish corporate towers, large-scale commercial facilities, upmarket gated housing developments, private schools, and universities, stands as the symbol of a globalizing Mexico City asserting its place into the global economy (figure 3.2). Santa Fé became the new branded image of Mexico City, replacing the old industrial city, a "symbol of a new Mexico strategically positioned in the world economy" (Pérez Negrete, 2009, p. 34). Santa Fé, like Ciudad Satélite, marks a clear break with the architectural and urban traditions of Mexico City. The area blatantly imitates the business and residential centres of the global north, drawing directly on architectural and urbanistic influences from Los Angeles, London, and Paris.

The Santa Fé mega-development was raised from the pits and dumps in the 1990s, but it had been in the works since the 1970s as one of five megaprojects conceived by the Federal District's Regent (appointed by the president until the election of the first independent municipal government in 1997 since 1928). Carlos Hank González, appointed regente in 1971 (to 1982), built his political legacy on improving the public and roadway transportation systems, but his dream was to establish a first-world business district, which he referred to as his own "Manhattan" (Moreno Carranco, 2010). The site of Santa Fé was chosen for Hank González's project, given its location as the extension of the main business corridor of Reforma Avenue and as a sparsely populated area. Low-income residents had lived near and worked the putrid garbage dumps and depleted sand pits that characterized the neighbourhood since the 1950s. As often the case in Mexico, the realization of this global (and vertical) suburb project was entangled in a series of economic crises, conflicting political visions, fluctuations of capital availability, and corruption (Moreno Carranco, 2010). Hank González allegedly donated a plot of land on a hill overlooking the garbage dumps to then President José López Portillo, who later donated 20 hectares of the land for the construction of the new campus of the IberoAmericana University (where his daughter was studying).

Santa Fé is one of the rare urban development projects in Mexico to have (but not necessarily to follow) a master plan initially created by Mexican architect Ricardo Legoretta (and collaborators) in the late 1980s and early 1990s. Santa Fé was conceived as an American-style suburban development/edge city inspired by Century City in West Los Angeles (Moreno Carranco, 2010), but the severe economic crisis of 1994–7 halted the project. By 1997, liberal reforms triggered real estate investment, and the construction of Santa Fé resumed, fuelled by the desire to attract capital and to position Mexico on the global scene. According to García Canclini (2000, p. 210), the 650 hectares of Santa Fé redevelopment "is the most radical transformation of land use and urban topography to affect the capital in recent decades: Where there had been precarious immigrant neighbourhoods surrounded by garbage dumps, in less than a decade there has emerged a postmodern architecture of residential and consumer structures built to upscale First World standards."

The project initially rested on the Mexican government's ability to convince many national and transnational corporations to build their headquarters in Santa Fé. More than 150 corporations have their offices in Santa Fé (including DaimlerChrysler, Hewlett Packard, Erickson, GE, IBM, Citibanamex, Televisa, Jose Cuervo and Bimbo, among many others). Expedited construction often proceeded without official permits and with lagging infrastructure and public services. As Moreno Carranco (2010, p. 95) remarks, "Disconnections between plans and realities are an ongoing narrative of Mexico City's urban history."

Like in Ciudad Satélite, the goal of the Santa Fé megaproject was to create an urban enclave that was separated from the rest of the city. Santa Fé's concentration of corporate towers, "American"-style shopping malls (the "new" largest in Latin America), restaurants, hotels, and exclusive gated communities and apartment buildings in Mexico City is splintered by the highway and surrounded by walls and bridges, separating the "global" megaproject from the "local" neighbourhood. Hilly topography, cul de sac roads, and gated streets discourage access and interaction. The original concept of greenbelt (created by inaccessible ravines) served to buffer the exclusive from the precarious. Mexico City is a city with tremendous economic and social inequalities, but Santa Fé embodies "new geographies of acute social contrast" given that it "is the only area in the city in which people with the highest per capita income live territorially adjacent to people with the lowest per capita income in the city" (Moreno Carranco, 2010, pp. 34, 189).

Ixtapaluca: The Mass-Produced Periphery

[T]he real estate production, despite being supported by the government, seems to make no contributions in the planning of city growth.
(Hastings, 2008, p. 26, author's translation)

Ixtapaluca is located on the southeastern edge of the Federal District, nestled between the Mexico-Puebla and the Mexico-Cuautla highways. Like many other areas of the State of Mexico, Ixtapaluca has been dramatically transformed over the last 30 years. Agricultural fields and pastures were taken over by informal settlements in the late 1980s, followed by large housing developments approved in the 1990s. These infilling subdivisions of single-family housing subsidized by the state range from a few houses to more than 20,000-unit developments. Ixtapaluca was one of the prime locations of the accelerated mass-produced housing development that spread throughout the nation from the mid-1990s to the early 2010s. Ixtapaluca's population tripled from

Figure 3.3 San Buenaventura, Ixtapaluca, State of Mexico. Photo courtesy of author.

1990 to 2010 (to reach 467,361 residents), and its housing stock grew from 68,625 houses in 1990 to 94,280 homes in 2005 (García Peralta & Hofer, 2006; Hastings, 2008).

This rapid metropolitan expansion was generated by large social interest housing developments built by a handful of large homebuilding companies (notably Geo Corporation and Ara Consortium) in different areas of the municipality. Large housing tracks were literally built overnight (at a rate of 2,500 homes per day at its peak) with very limited, if any, planning of roads, transportation systems or public services (Corona Benjamin, 2011). Driven exclusively by a market logic, homebuilding companies (the largest trading on the Mexican Stock Exchange) maximize their profits by achieving economies of scale through cheaper peripheral lands (and secured land reserves), homogeneity of construction, and full integration of construction, financing, and promotion/sale activities.

The San Buenaventura development in Ixtapaluca, with its 24,000 units, became the poster child of social interest housing market in Mexico (figure 3.3). This development was built on one of the oldest archeological vestiges of the Valley of Mexico, dating from more than

2,000 years BC, but nevertheless received authorization from national and state authorities (Lázaro, 2000). In the late 1990s, Geo Corporation built some 4,000 units on the former San Buenaventura hacienda lands, and Ara Consurtium added some 20,000 units on the former Canutillo hacienda lands in 2002. San Buenaventura (typical of many other similar developments) is characterized by a monotonous urban morphology where long rows of subdivisions of thousands of identical small single-family homes are separated by streets. These relatively low-cost homes generally range from 45 m^2 to 65 m^2 and sit on lots varying from 48 m^2 to 88 m^2 (Hastings, 2008). This particular development was based on two or three prototypical models of two or three rooms (of one or two floors), with a small patio service in the back and a one-car parking space in the front (Hastings, 2008). Rigid regulation established by the development company binds residents to a homogenous architectural image, and any modification to the house or public areas is considered a civil code violation (Hastings, 2008). However, many residents who are achieving their dream of homeownership modify their dwellings even though it might not be legally or structurally possible. Given the lack of services in the area, some residents have converted their living rooms into a small convenience store or a restaurant as a way to provide services and generate revenues.

Ixtapaluca's municipal authorities have developed a close relationship with homebuilding companies, but in their eagerness to attract investors, they often disregard their own and/or the developers' responsibilities to create or consolidate infrastructure and public services. San Buenaventura mega-housing development, as is often the case of social interest housing subdivisions, suffered from a serious public services deficit (water, electricity, drainage, etc.). Such services are often installed after the homeowners have taken possession of their homes, with the result of significantly increasing living costs. Only three markets, one supermarket, two parks, and about 20 schools were built for more than 23,000 households, not including the surrounding informal/self-built settlements, which have benefitted from the very minimal commercial and education amenities of the development (Hastings, 2008). Adding to the lack of infrastructure, services, and amenities are the great distance faced by resident to commute to work or study or to seek services. The 32 kms separating Ixtapaluca from downtown can take up to 2 to 3 hours (one way). For many residents relocating from surroundings areas of the States of Mexico and Puebla in this outer fringe of the metropolitan area, homeownership, security and tranquillity appear to initially compensate for poor services and long commutes (Linares Zarco, 2009). Peripheral expansion is the result of economic forces and speculative processes, planning policy, and lack of affordable land in the centre, as well as a competitive climate between municipalities of the State of Mexico, but it is also the result of individual motivations and dreams of affordable homeownership, pushing both formal and informal urban expansion further outwards.

Chalco Solidaridad: The Expanding Periphery

[M]any lower income families do not choose informal arrangements as the best alternative but instead it is often the only option for them.
(Iracheta Cenecorta & Smolka, 2000, p. 5)

Chalco Solidaridad is a municipality in the southeastern part of the State of Mexico created in 1994 (out of lands in five adjacent municipalities: Chalco, Chicoloapan, Chimalhuacán, Ixtapaluca, and La Paz). The region, known for its extensive "informal" urbanization, is often referred to as Valle de Chalco. Urbanization in Valle de Chalco has spread partially over the drained lakebed of Chalco Lake, a freshwater lake of the endorheic basin that once covered most of the pre-Hispanic Valley of Mexico. Chalco Lake, once surrounded by many Indigenous communities whose main activities were related to fishing, was partly drained in the 19th century. Emerging agricultural lands were later distributed as communal (ejidal) properties to local communities. The Valle de Chalco area started to change dramatically in the 1980s due to urban expansion over inexpensive and unserviced peripheral lands. Low-income populations from the Federal District and immediate peripheries move outward to the Valle de Chalco and the overall population of the area grew about 20-fold between 1970 and 2010 (Iracheta Cenecorta, 2000). Irregular settlements started spreading from the northern edge of the valley along the highway (to Puebla) to the south, absorbing the existing village of Xico located at the foot of the extinct volcano of the same name (figure 3.4). In 20 years, the population of Chalco Solidaridad reached 357,645 residents, many from various Indigenous populations (Molinar Palma, 2003).

Chalco Solidaridad was once the largest informal settlement in Latin America (Molinar Palma, 2003). Irregular settlements are generally understood as invasions of agricultural lands by poor populations, but so-called "illegal" occupation of land is better explained by a series of interacting factors, including the shortage of affordable land and housing, the absence of financial assistance for low-income households, land speculation, lack of or lax planning and land-use policies, political

58 Liette Gilbert

Figure 3.4 Xico, Chalco Solidaridad, State of Mexico. Photo courtesy of author.

acceptance and inaction, and difficulty for agricultural sector to resist urbanization (Hiernaux Nicolás, Lindón Villoria, & Noyola Rocha, 2000; Iracheta Cenecorta & Smolka, 2000; Molinar Palma, 2003). Thus, "informal" urbanization entails a series of political and cultural institutionalized practices ranging from land transactions (by *ejidararios*, professional intermediaries or developers), unauthorized land development, and regularization of dubious or inexistent property titles, to the post-facto implementation of basic urban services through clientelistic relations (in exchange for political support). For Connolly (2003, p. 19), the "multifaced and confusing legal status" of self-built, low-income irregular settlements is not their most important characteristic; rather, what such settlements usually share is the complete absence of credit for building and purchasing lots. This lack of credit generally translates into incremental and self-built construction according to each household's financial capacity and family needs. In metropolitan Mexico City, irregular settlements represent between 60% to 70% of the urban housing development production (Connolly, 2009; Iracheta Cenecorta & Smolka, 2000; Pérez & Cervantes Borja, 2005). Therefore, self-built urbanization, despite its substandard housing, infrastructure deficits, and tenure insecurity, cannot be seen as marginal but rather is fundamental in securing housing (and family patrimony) for some of the poorest residents of metropolitan Mexico City (Connolly, 2009; Duhau & Jacquin, 2008).

Federal, state, and municipal authorities sought to regularize the large number of irregular settlements starting in the late 1970s. Regularization processes usually entail, at very different paces and degrees and by different agencies, land titles being secured, infrastructure and services being installed, houses being improved, amenities (commercial, educational, recreational, and medical) being provided. Regularization of remote and flood-prone irregular settlements is generally incremental (starting with electricity, road improvements, and water and drainage systems), given limited public resources (linked to insufficient tax base) and the demand on already scarce resources available.

The regularization of Chalco Solidaridad was handled by State of Mexico and entailed the expropriation and indemnification of the ejidal lands by presidential decree and the resale of lots to the settlers. This process of regularization started in 1978 and took about 5 years (Connolly, 2003). Urban and building improvements in Valle de Chalco were financed by a federal poverty programme, Solidaridad (hence the name), in which authorities and associations of residents collaborate to implement services destined to alleviate precarity. With about $160 million (US) between 1989 and 1993, Chalco Solidaridad benefitted "not only from regularization, electrification, water, drainage and some paved roads, but also a civic center and a visit from the Pope" (Connolly, 2003, p. 34). Urban services and clientelism have been tightly linked in areas of irregular settlements, and as Connolly (2003, p. 34) explains: "apart the magnitude of the problem and its visibility from one of the main roads out of Mexico City, a major reason for targeting Chalco was the low vote for the PRI [Institutional Revolutionary Party] in that area in the 1988 presidential elections."

Regularization of irregular settlements remains nevertheless a long and cumbersome bureaucratic process. According to Fundación CIDOC and SHF (2011), a little more than 40% of dwellings have title deeds in the Federal District, while the number is a little less than 40% in the State of Mexico (the national average being 45%). Most families without title deeds are generally interested in securing property titles, but procedures are often too costly, too complicated, and too time-consuming (Fundación CIDOC & SHF, 2011). Basic services are the most important because once a settlement is serviced and tenure is not challenged by eviction, the actual legitimation of property titles becomes less important. Beyond the individual case, tenure legalization is also seen as a strategy of social and economic change (Varley, 2002). The intensification of regularization since the 1970s was tied to the government's need to incorporate irregular settlements into the formal land market, to establish greater local authority over land use and planning control (which often resulted in expropriation and resale, or intense negotiations with community leaders), and to incorporate residents into land tax and service registries (Ward, 1998).

However, despite federal investments and clientelism, Valle de Chalco still has some of the worst housing condition in Mexico City; 80% of the population lives in poverty – of which 60% live in extreme poverty (Enciso, 2013). Moreover, increased regularization of irregular settlements has, according to Iracheta Cenecorta and Smolka (2000), the paradoxical effect of increasing living costs for the most marginalized, who are then forced out to cheaper and more remote settlements. Representing the growing outward periphery of the metropolitan area, Valle de Chalco continues to experience urban squatting (albeit at a much lesser scale than before) from migrants coming from all over the country. Although Mexican authorities have generally limited evictions given its *laissez faire* recognition of social needs – and tacit non-eviction agreements in exchange of political support – about 3,000 occupants were evicted from irregular settlements located in San Francisco Tlaltenco-Liconsa and in Santa Catarina in Valle de Chalco (Salinas, 2014).

Mexico City in North American Perspective

Like its North American counterparts, suburbanization or (more accurately) peripheralization processes in metropolitan Mexico City are the product of postwar urban development, industrialization, and capital accumulation. However, while suburbanization in the United States and in Canada has generally and respectively been enabled by economic mobility of upper classes (Beauregard, 2006) and middle classes (Harris, 2004) starting in the early 19th century, suburbanization in Mexico City is more recent, ongoing, and predominantly experienced by either low-income residents pushed towards urban edges or rural migrants settling informally on the margins of metropolitan area. Low-income households and rural migrants have successively settled in irregular settlements in the eastern part of the valley and in the hills of the Sierra de Guadelupe in the north, incrementally building their own homes with limited resources (Connolly, 2009; Pezzoli, 1998; Ward, 1998). Differentiated neighbourhoods were created not only by incomes, but also by quality of housing stocks, availability of public services and facilities, and disparities in land values (Monkkonen, 2011).

Urban growth in Mexico City has been a contested and contingent process (Davis, 1998). Urban planning never really succeeded at managing growth but rather attempted, with mixed results, to address housing shortages, to manage resources, and to provide basic services through subsidized private developments or regularized self-built settlements (Ward, 1998; Wigle, 2014). Housing policies, both at the national and local levels, have generally focused on increasing housing access rather than controlling metropolitan expansion. Planning restrictions in central areas, lack of political will, and rising real estate and land values pushed populations to the outskirts into irregular settlements or out of the city limits into smaller towns. Thus, urban growth was not only the product of rapid industrialization and demographic growth, it was also the product of competing urban policies and political struggles in and between the capital city (Federal District) and surrounding municipalities of the State of Mexico (Davis, 1998; García Peralta & Hofer, 2006; Ward, 1998). As a result, close to 80% of the metropolitan region's population lives in what Ananya Roy (2011) calls subaltern urbanism – or perhaps more appropriately in this case subaltern suburbanism. Mexico City is the only large North American metropolitan region where irregular/informal urbanization predominates.

Metropolitan Mexico City is indeed "many cities within the city." Mexico City has grown outwards since the 1950s through various waves of informal settlements and formal developments. Although inspired at times by North American sub/urban concepts (as in the cases of Satélite and Santa Fé), Mexico City's urban form has remained predominantly characterized by (formal and informal) precarized urbanization (as in the case of Ixtapaluca and Chalco Solidaridad). Despite the recent increase in housing production for the middle class (facilitated by financing access), the fact that developments are generally chronically underserviced exposed the failure of governments to regulate urban and growth conditions to benefit social development. Still, what seem the most constant and comparable elements across elusive suburbs and ubiquitous peripheries of metropolitan Mexico City – and other North American suburbs – is the desire for homeownership at the cost of long commutes.

REFERENCES

Beauregard, R. (2006). *When America became suburban*. Minneapolis: University of Minnesota Press.

Brockerhoff, M. (1999). Urban growth in developing countries: A review of projections and predictions. *Population and Development Review, 25*(4), 757–78.

Capron, G., & de Alba, M. (2010). Creating the middle-class suburban dream in Mexico City. *Culturales, VI*(11), 159–83.

Connolly, P. (2003). The case of Mexico City, Mexico. Retrieved from https://www.ucl.ac.uk/dpu-projects/Global_Report/pdfs/Mexico.pdf

Connolly, P. (2009). Observing the evolution of irregular settlements: Mexico City's colonias populares, 1990 to 2005. *International Development Planning Review, 31*(1), 1–35.

Cordova González, L.A. (2001). Ciudad Satélite Pani: valoración de la modernidad urbana. *Esencia Espacio, 13,* 45–6.

Corona Benjamin, L. (2011). Two million homes for Mexico. Retrieved from http://www.liviacorona.com/#S7,T598,Two_Million_Homes_for_Mexico

Davis, D. (1998). The social construction of Mexico City: Political conflict and urban development, 1950–1966. *Journal of Urban History, 24*(3), 364–415.

de Alba, M., & Capron, G. (2010). La Publicité immobilière à l'assaut de l'environnement dans une grande ville du sud, Mexico, 1950–2000. *Ecologie & politique, 1*(39), 55–71.

Duhau, E., & Jacquin, C. (2008). Les ensembles de logements géants de Mexico: Nouvelles formes de l'habitat social, cadres de vie et reformulations par les habitants. *Autrepart, 47,* 169–85.

Enciso, A. (2013, November 8). Valle de Chalco fure laboratorio social, pero hoy 80% vive ne probreza: expert. *La Jornada,* 37. Retrieved from http://www.jornada.unam.mx/2013/11/08/sociedad/037n2soc

Flores Peña, S.A. (2013). Are first generation suburbs in Mexico City shrinking? The case of Naucalpan. In M. Hibbard, R. Freestone, & T. Øivin Sager (Eds.), *Dialogues in urban and regional planning volume 5* (pp. 115–40). New York, NY: Routledge.

Fundación CIDOC (Centro de Investigación y Documentación de la Casa) and SHF (Sociedad Hipotecaria Federal). (2011). Current Housing Situation in Mexico 2011. http://www.shf.gob.mx/English/Press/Publications/Documents/EAVM%20INGLES%202011.pdf.

García Canclini, N. (2000). From national capital to global capital: Urban change in Mexico City (Trans. P. Liffman). *Public Culture, 12*(1), 207–13.

García Peralta, B., & Hofer, A. (2006). Housing for the working class on the periphery of Mexico City: A new version of gated communities. *Social Justice, 33*(3), 129–41.

Garza, G. (Ed.). (2000). *La Ciudad de México en el fin del segunda milenio.* Mexico City: El Colegio de México, Centro de Estudios Demográficos y de Desarollo Urbano. Gobierno del Distrito Federal.

Harris, R. (2004). *Creeping conformity: How Canada became suburban*. Toronto: University of Toronto Press.

Hastings, I. (2008). El problema cualitativo en la producción del habitat popular en la Ciudad de México: Análisis cualitativo de la vivienda popular. *Informes de la Construcción, 60*(511), 25–40.

Heinrichs, D., & Nuissl, H. (2013). Latin America at the urban margin: Sociospatial fragmentation and authoritarian governance. In R. Keil (Ed.), *Suburban constellations: Governance, land, and infrastructure in the 21st Century* (pp. 170–5). Berlin: Jovis Verlag GmbH.

Hiernaux Nicolás, D., Lindón Villoria, A., & Noyola Rocha, J. (2000). *La construcción social de un territorio emergente: El Valle de Chalco*. Zinacantepec. Estado de México: El Colegio Mexiquense.

INEGI, Instituto Nacional de Estadistica y Geografia. (2010). Mexico en cifras. Retrieved from http://www.inegi.org.mx/

Iracheta Cenecorta, A. (2000). La urbanización metropolitan descapitalizada: El Valle de Chalco. In D. Hiernaux Nicolás, A. Lindón Villoria, & J. Noyola Rocha (Eds.), *La construcción social de un territorio emergente: El Valle de Chalco* (pp. 167–204). Zinacantepec, Estado de México: El Colegio Mexiquense.

Iracheta Cenecorta, A., & Smolka, M.O. (2000). *Access to service land for the urban poor: The regularization paradox in Mexico*. Mexico City: CEPAL. Retrieved from https://pdfs.semanticscholar.org/0924/6a3a427bdf3d1cc79def9b6cbd7fe5c1eaba.pdf

Lázaro, J. (2000, December 9). Fraccionarán sobre vestigios. *El Universal*. Retrieved from http://archivo.eluniversal.com.mx/estados/24571.html

Linares Zarco, J. (2009). *La imagen urbana, México en el siglo XXI: Entre la crisis y la transición urbana*. Mexico City: Miguel Angel Porrúa.

Molinar Palma, P. (2003). Valle de Chalco Solidaridad: Reflexiones sobre las nuevas formas de asentamientos urbanos. *Clio Nueva Epoca, 2*(29), 103–18.

Monkkonen, P. (2011). Do Mexican cities sprawl? Housing-finance reform and changing patterns of urban growth. *Urban Geography, 32*(3), 406–23.

Moreno Carranco, M. (2010). *The socio/spatial production of the global: Mexico City reinvented through the Santa Fé Urban megaproject*. Saarbrücken: VDM Publishers.

Pérez Negrete, M. (2009). Santa Fé: A "global enclave" in Mexico City. *Journal of Place Management and Development, 2*(1), 33–40.

Pérez, E.M., & Cervantes Borja, J.F. (Eds.). (2005). *La producción de vivienda del sector provado y su problemática en el municipio de Ixtapaluca*. México DF: Plaza y Valdés.

Pezzoli, K. (1998). *Human settlements and planning for ecological sustainability: The case of Mexico City*. Cambridge, MA: MIT Press.

Roy, A. (2011). Slumdog cities: Rethinking subaltern urbanism. *International Journal of Urban and Regional Research, 35*(2), 223–38.

Salinas, J. (2014, May 20). Desalojan a 3 mil personas de dos asentamientos irregularies en Valle de Chalco. *La Jornada en línea*. Retrieved from www.jornada.unam.mx

Tarrés, M.L. (1999). Vida de familia. Prácticas privadas y discursos públicos entre las clases medias de Ciudad Satélite. *Estudios Sociológicos, 17*(50), 419–39.

Varley, A. (2002). Public or private: Debating the meaning of tenure legalization. *International Journal of Urban and Regional Research, 26*(3), 449–61.

Ward, P.M. (1998). *Mexico City* (2nd ed.). New York, NY: John Wiley & Sons.

Wigle, J. (2014). The "graying" of "green" zones: Spatial governance and irregular settlement in Xochimilco, Mexico city. *International Journal of Urban and Regional Research, 38*(20), 573–89.

4 Searching for Suburbia in Metropolitan Miami

JAN NIJMAN AND TOM CLERY

Introduction

Between 1960 and 2018 the combined populations of Miami–Dade County and Broward County grew from 1.3 million to about 4.6 million people. To the north, Palm Beach County reached a population of nearly 1.9 million, its urban area steadily creeping south to meet with Broward's advancing perimeter. Accordingly, in 2004 the US Census recognized a new consolidated metropolitan statistical area covering the tri-county region. With more than 6 million people, it is the nation's seventh-largest metropolitan region and Florida's demographic centre of gravity. It is estimated that nearly 2 million may be added by 2030. In addition to the three counties, the region now comprises more than one hundred municipalities with their own elected governments, police forces, public works, planning departments, and so on.

Greater Miami has in the past half century expanded and gobbled up space at a rapid pace, in a wetland environment poorly suited for urban growth. Urban sprawl, for that reason alone (and others), has been a contentious matter in southeast Florida. In that sense, "suburbanization" here has been obvious and for all to see. But the social geography and history of Greater Miami is quite unusual: it is a comparatively young city that started out as a seasonal beach resort and without a single strong urban centre; it has experienced, since the 1960s (supposedly the high tide of suburbanization in North America), an extreme demographic makeover due to massive foreign immigration; and it sits hemmed in between Everglades National Park and Biscayne Bay, limiting expansion either east or west. The resulting spatial configuration appears unusual.

The main question in this chapter is whether suburbanization in these parts followed the same logic as elsewhere in North America and how the Miami case reflects on entrenched conceptual notions of the American suburb. These notions were born largely from the particular (sub)urban constellations that prevailed in the first couple of decades following the Second World War. They coloured our understanding of what came before – that is, earlier phases of suburbanization were often overlooked; and what came after – that is, the persistence of notions of the typical 1950s after its retreat. Theory about suburbanization has, therefore, been overshadowed by a-historicism and it has carried a penchant for dichotomies (suburb versus central city; White versus Black; rich versus poor, etc.).

Suburbanization in North America tends to be understood as urban growth from the centre outwards, where emergent suburbs stand in contrast to the central city in terms of peripheral location, low density, exclusive residential functions, and newness (e.g., Harris, 1992; Jackson, 1985; Knox, 2008; Nijman & Clery, 2015; Walker & Lewis, 2001). Greater Miami may be a rather exceptional place in some regards but it is one that is "good to think" – it makes for an interesting case study precisely because it allows a different angle on standard concepts and compels us to reconsider established notions in a different light. What is suburbia like in a city that joined the North American urban system so late; a city that experienced neither 19th-century industrialization nor the early stages of suburbs as "bourgeois utopias" (Fishman, 1987); a city where native Whites, elsewhere the main entrants of postwar suburbs, were reduced to an overall minority almost overnight?

Historical and Geographical Context

Miami is of recent vintage, its history as a city of any significance beginning as late as the early 20th century (Nijman, 2011). The railways did not reach here until 1896, the same year that the city of Miami was incorporated. In 1880 the recorded population of present-day Miami–Dade and Broward counties was a scant 257; by 1920 southeast Florida counted 40,000 people, at which time Los Angeles passed the 1 million mark (figure 4.1). Miami was simply too far and too disconnected, without a major river or port, and without a hinterland of any importance.

In the beginning, Miami's evolving spatial order took shape in two different ways. First, as a beach resort of growing appeal, mainly to the northeastern United States, the city's early spatial organization was along the waterfront. The initial settlement of the city was on the coast near the Miami River; from there, expansion went either north or south and stayed close to Biscayne Bay. Thus, early urban Southeast Florida

Searching for Suburbia in Metropolitan Miami 67

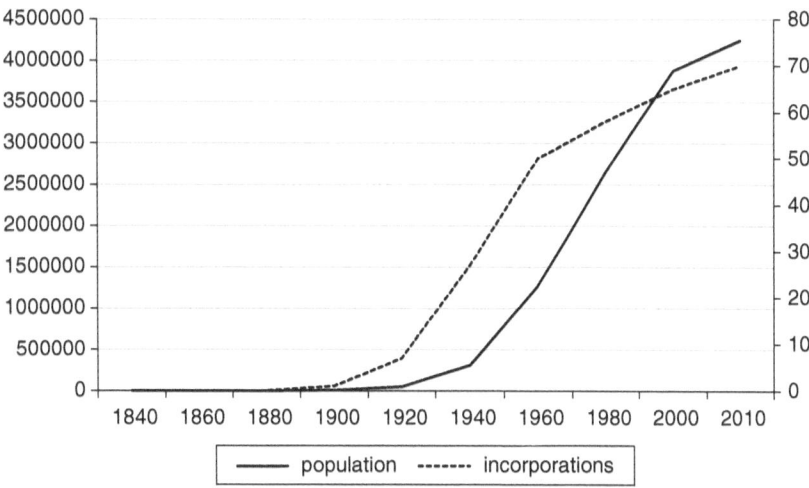

Figure 4.1 The combined population (left axis) and municipal incorporations (right axis) of Miami–Dade County and Broward County. Sources: US Census; Miami–Dade and Broward County websites.

was not so much centred but rather developed linearly and it did so, in that first quarter of the 20th century, more or less simultaneously. For example, Fort Lauderdale, to the north, was incorporated in 1911, Miami Beach in 1915, and Coral Gables, to the south, followed in 1925.

Second, the early development of Southeast Florida was driven by the desire to drain the Everglades and make way for agriculture. The wetlands were, in those days, considered useless and unhealthy, and the financial prospects of reclamation seemed considerable – the subtropical climate providing for year-round growing seasons and for crops, such as citrus, that adhered to warm weather. This means that, at the same time, several farming communities emerged further away from the coast, where land was cheaper. Homestead, its name indicative of its agricultural origins, is the second-oldest city (1913) in the area, about 25 miles southwest from the City of Miami. Neighbouring Florida City, another agricultural settlement, was incorporated a year later. Today, it is the last town before driving off the mainland to the Florida Keys.

Thus, Southeast Florida from the beginning lacked a clear centre. Along the coast, downtown Miami on the river's north bank soon had to compete with Miami Beach, Coral Gables, and Fort Lauderdale for upscale tourists and a growing residential (winter) population; further inland, quiet agricultural towns developed according to their own logic, a long way from the conspicuous glamour of the coastal resorts.

Suburbs at this time did not exist; at least they were not generally thought of in that manner.

Most urban growth, in part owing to the advent of air conditioning, took place after the Second World War, in terms of infill development along the coast and gradual westward expansion. It is at this time that the region first experienced suburbanization as it is generally understood – as a physical process: urban expansion from existing centres outwards, with the new suburbs depending in various ways on the centre for employment and high-level services. But population growth, which was explosive from about 1960, was in large part fuelled by foreign immigration. The influx of Cubans, later joined by other Latin American or Caribbean arrivals, transformed the city (Nijman, 2011; Portes & Stepick, 1993). Within two decades the large non-Hispanic White majority became a minority in Miami–Dade County. The influence on intraurban migration and the shaping of the suburbs in terms of ethnicity, race, and class was enormous (Nijman, 2007a).

There is one other particularity of Miami's geography that needs mentioning at this point. Miami–Dade is the only county in the United States that has two national parks within its borders: Biscayne National Park (most of which covers the bay) to the east and Everglades National Park to the west. It means that, over time, urban expansion became notably constrained, forcing relatively high densities and the formation of an elongated north–south metropolis. The emergence in the early 21st century of a tri-county conurbation from Miami to Palm Beach reflected as much (map 4.1).

Suburbs in Local Discourse

Arguably, local notions of suburbia (anywhere) are the strongest when they are juxtaposed to an identifiable and established urban core. But downtown Miami is actually small, compared with the downtowns of other major North American cities, and the central City (municipality) of Miami is quite large and highly diverse compared with other cities in the metro area. Most locals would probably have hard time drawing its municipal boundaries. The City of Miami has also lacked, throughout most of its history, a tradition of planning, established and recognizable designs, or famous public spaces. One reason lies in the city's extreme transience and the constant turnover of local leadership. Architectural historian Jean-Francois LeJeune speaks of a "city without a memory." In the early postwar years, Miami was "the consenting victim of resort-city development patterns combined with a boom-and-bust psychology

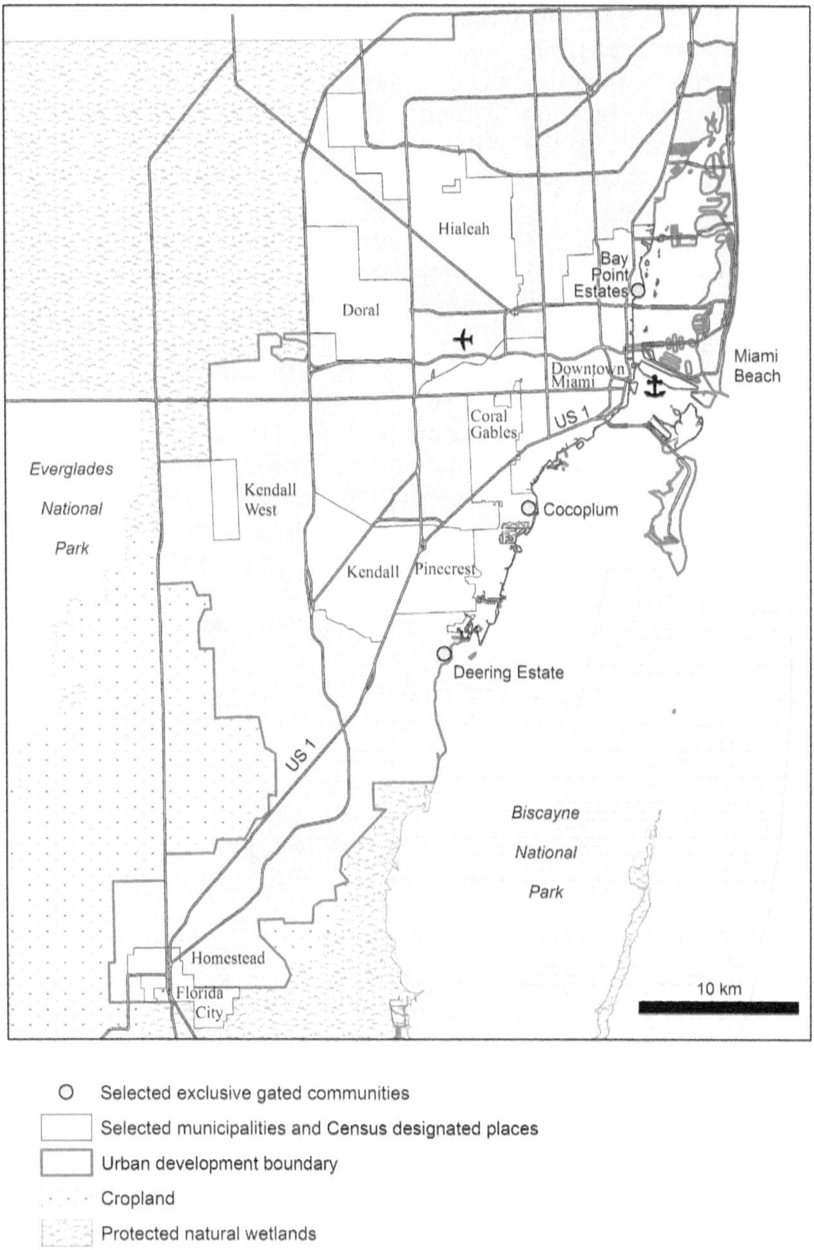

Searching for Suburbia in Metropolitan Miami   69

Map 4.1 Miami–Dade County. Map by Chris Hanson, University of Miami.

that prevented any long-term vision of productive city development and infrastructure" (2009, p. 36).

This propensity is illustrated in the location of Miami's city hall – which was never built as a city hall in the first place (it is a reconstructed hangar) – away from the centre, southward on the bay. It invokes the city's identity at large: decentred, detached, and with a sense of impermanence (Nijman, 2011, p. 195).

Usage of the term "suburb" in Southeast Florida was rare until after the Second World War. The one major exception, Coral Gables, stood in contrast to the city of Miami and most other parts of the area in that it was designed from scratch. The chief planner and architect was George Merrick, whose statue graces the front lawn of Coral Gables' city hall (which, in contrast to that of the City of Miami, does have a central location). Coral Gables was a prime example of suburban design in the 1920s that was geared to the automobile, evidenced in the playfully winding roads, lush vegetation, and the absence of sidewalks. These were exclusively single-family homes, and the prevailing architecture was Mediterranean Revival. The lots were big, and prices were high. Coral Gables was – and many parts still are today – quiet, spacious, and green; a different world from Miami's gritty downtown. Yet from the start, Coral Gables also distanced itself from the city of Miami in that it was more than a suburb. Downtown Coral Gables, centred around Miracle Mile, soon featured clusters of office employment, retail activity, and other services. One could argue that this part of Coral Gables turned into an edge city avant-la-lettre. This crossbreed of sub/urb was captured in Merrick's designation of Coral Gables, which it still carries today: the "City Beautiful." The official website (2013) emphasized these mixed qualities of the urban and suburban:

> Incorporating secluded residential enclaves and commercial areas inspired by the architectural style of the Mediterranean, Merrick envisioned a City that would offer every amenity to its residents and at the same time would become a center for international business.

Since the Second World War, the term suburb gained currency in Greater Miami as it did elsewhere. The western stretches of Kendall, about 90 blocks south and 100-plus blocks west of downtown, are often considered terra firma of Miami's suburbia, with their predictable designs, large scale, and homogeneous (lower) middle-class composition. But beyond stereotypical Kendall, definitions of suburbia are vague, inconsistent, and subjective. In colloquial usage, the term appears to connote (1) distance from the older and more established eastern core of the urban area; (2) relatively recent origins; (3) little or no high rise; (4) homogeneous design; and (5) affordability in middle-class terms.

Thus, today, Coral Gables tends not to be regarded as a suburb, and the same is true for most areas along the waterfront as long as they are not too far out north or south; many of which, it should be noted, are secluded, gated, and pricy communities, from Bay Point Estates on Miami's upper east side, to Cocoplum and Deering Bay in the south (see map 4.1). In other cities, newly developed areas on the urban perimeter sometimes carry prestige and valuation far beyond those of the old centre, but in Greater Miami, where the first established developments were on the bay, successive expansion tends to find itself at increasing distance from the water and is therefore considered less desirable and less prestigious. It is the pricing of real estate, generally with a steep gradient from east to west, that is essential to local perceptions and mental maps of Greater Miami's urban landscape. Given the wealth that is concentrated for miles along and near the waterfront, the more recent suburbs to the west are by definition valued less. It attests to the enormous gap between the rich and the masses in Southeast Florida that these same suburbs are still aspired to by large numbers of households of the region's middle classes.

But Miami's suburbanism is more complicated still: while the older communities on the seaboard generally have a vintage and status and proximity to the core of the urban region that precludes the label of suburb, on the inside they do have a suburban appearance. They are spacious and green and dominated by single-family homes. The western suburbs, on the other hand, while at a greater distance and of more recent origins, are often characterized by mixed use with, sometimes, considerable economic activity. It is precisely because of the leisure- and amenity-oriented nature and strict zoning policies of the waterfront communities that industrial activity would find itself further west, in the "suburbs." The telling contrast between coastal Pinecrest and inland Doral will be elaborated below. Finally, from the point of view of the self-indulgent hip crowds on Miami Beach, all these distinctions seem unnecessarily complicated: to them, all of mainland Miami is suburbia.

The Formation of Miami's Suburbia

Miami's urban growth shifted from the historic core towards the west, south, and north. The old core, as said, evolved in linear fashion along the waterfront, with the earliest growth centres in the cities of Miami and Fort Lauderdale. Most congruous urban growth was focused initially on the rest of the waterfront. Then, from the 1950s onwards, *suburbanization* took shape, geographically speaking, at two scales.

First, westward urban expansion was fuelled by relative congestion and high land values along the coast. This was in part a matter of spillover of previous residents who used to live in the urban core but, more so, it concerned new arrivals in the region who could not afford proximity to the ocean breezes. Over time, the (sub)urban boundary crawled further west, and a pattern emerged in which the historical origin of many suburban neighbourhoods can be predicted on the basis of distance from Biscayne Bay. This is reflected in map 4.2, which shows population growth rates across the entire three-county area from 1990 to 2002 and which indicates that over time, growth became concentrated in the western parts of the region. Homestead, some 25 miles southwest of downtown Miami and not long ago considered a remote old farming community, had become the fastest growing city in the entire State of Florida by 2007, and most of its residential developments were emphatically suburban.

Much of Greater Miami's expansion since the 1970s sat on what were once the quiet eastern stretches of the Everglades. As suburbanization progressed, South Florida's precarious ecology set off alarm bells on repeated occasions, especially in Miami–Dade. In 1983 it resulted in the promulgation of the urban development boundary (UDB): a Miami–Dade macro-zoning instrument that separates urban/suburban development from rural/ natural land uses. To the west, urban uses are prohibited; to the east, they are encouraged. The result is clearly visible to anyone with a window seat on an airplane about to land at Miami International Airport: a sharp western border to a sea of monotonous red roofs signalling the residential suburbs. In southern Miami–Dade County, the UDB curls east and then north, up against Biscayne National Park (see map 4.1).

In the south the UDB separates recent residential developments from farmlands. The boundary is more hotly disputed here because, until recently, agricultural activities dominated both sides. In the 1990s and early 2000s landowners on the line's "urban" side cashed in by selling property to large developers, who in turn made huge profits creating residential subdivisions. Presently, land on the urban side of the UDB has a market value about ten times that on the other side. Predictably, farmers on the "green side" have often lobbied for the UDB to shift so they too can reap the profits.

The second way to geographically conceive of the suburbanization process in Southeast Florida is at a larger regional scale. Since the 1960s there has been a steady relocation of urban residents from Miami–Dade

Searching for Suburbia in Metropolitan Miami 73

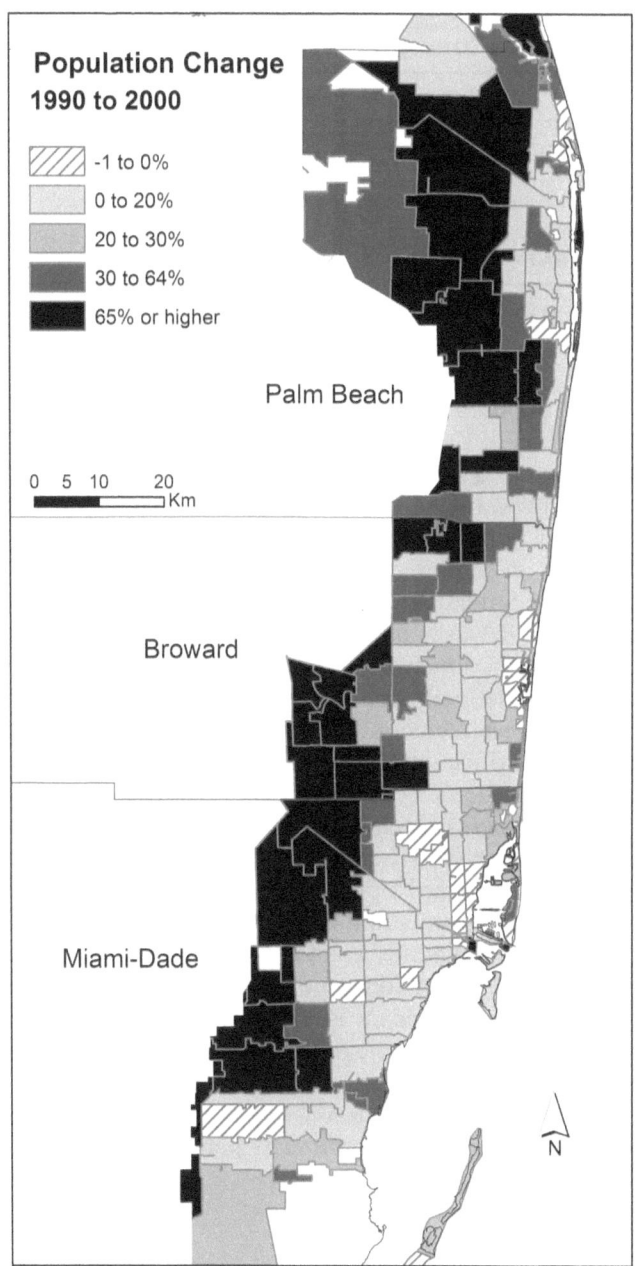

Map 4.2 Westward urban growth in southeast Florida, by zip code area, 1990–2002. Source: Broward County, Office of Urban Planning and Redevelopment, Planning Services Division: Broward-by-the-Numbers, March 2004, number 20.

74 Jan Nijman and Tom Clery

Table 4.1 Demographic comparison among the three counties of Southeast Florida's consolidated Metropolitan Statistical Area, 2018

	Miami–Dade	Broward	Palm Beach
Total population	2,761,581	1,951,260	1,485,941
Foreign born (%)	52.9	33.1	24.6
Hispanic (%)	69.1	30.4	22.9
Black or African American (%)	17.9	30.9	19.7
Non-Hispanic White (%)	13.0	35.6	54.1

Source: US Census Quick Facts, States and Counties.

County to the suburbs of Broward County, much of this in reaction to the rapid influx of Cubans and other foreigners in parts of Miami–Dade. Thus, there is a specific pattern to intraregional migration: most newcomers in Broward hail from Miami–Dade while most migrants to Palm Beach County come from Broward (Broward County, 2004). South Florida is in some ways like a reverse-flow funnel, sucking in large numbers of foreign migrants from the south and spreading successive northbound waves of movers. Commuter patterns underscore this urban–suburban dimension across the region. In 2013 the US Census reported that the daily commuter flow from Broward to Miami–Dade was the biggest intercounty flow in Florida at 125,000, with 71,000 going the opposite direction (McKenzie, 2013).

In recent years, it should be noted, Broward has become more diverse, with growing numbers especially of Haitians and Jamaicans, and Hispanics who have started to drift from Miami–Dade into Broward. The Hispanic share of Broward's population increased from 17% in 2000 to 30.4% in 2018. Broward's non-Hispanic Whites no longer hold the absolute majority, but Latinos are not nearly as dominant there as in Miami–Dade, where they make up over two thirds of the population. The share of African Americans is higher than in Miami–Dade. Table 4.1 summarizes some key demographic statistics for all three counties in the region. It underscores their continuing differences: from Miami–Dade to Broward to Palm Beach, the landscape becomes more suburban, at least as suburbia is conventionally perceived in ethnic terms.

City versus Suburb in Southeast Florida

In North America, central cities are traditionally clusters of economic activity, with a range of services that are accessible to the wider region. People living in the central city often also work there, while another part of the workforce commutes into the central city from the suburbs. Typically, central cities have a higher daytime population than nighttime population, and the opposite is true for the conventional suburb. Even with the development of edge cities and edgeless cities in recent decades, central cities generally continue to exercise a degree of dominance in their respective metropolitan areas.

As we have already noted, Miami's origins as a resort town and its late growth into a major city during postindustrial times help explain why the central city is not quite as dominant here and why some of the "suburbs" have attracted substantial economic activity. According to the US Census, in the United States as a whole, of the workforce in metro areas living in principal cities, about 20% works outside that principal city: that is, in the suburbs of that metro area. For the City of Miami, the corresponding figure is 35% – indicative of the relatively high level of economic decentralization in Southeast Florida.

In its early history as a resort town, the largest sectors of the urban economy were retail, tourism, and construction. Manufacturing was always relatively small here, and producer services and finance did not take off until the late 1970s (Nijman, 2011). When the office sector gathered momentum, the City of Miami captured only a part, and much of it was established elsewhere. Coral Gables, in particular, is home to a sizable cluster of banks and other financial firms. Census data for 2006 to 2010 show that Coral Gables' increase of daytime population (+62.2%) actually exceeded that of the City of Miami (+54.6%), testifying to the pronounced edge city qualities of the Gables.

Perhaps more importantly, the office sector in Southeast Florida is not so much concentrated in edge cities but rather spread around the metropolitan landscape in "edgeless" fashion. Lang (2003) designated Miami the most decentred office city in the nation, the ultimate "edgeless city," with no less than 72.1% of all office space in edgeless places and in sharp contrast to core-dominated metropolitan areas such as New York or Chicago.

The dispersal of economic activity, particularly in Miami–Dade County, is not the only cause of fuzzy city–suburb distinctions – the other has to do with population densities across the metropolitan area. Densities outside the central cities here are much higher than in comparable sunbelt cities: census data indicate that in 2010 Greater Miami's "suburbs" were three times as dense as those of Phoenix and four times as dense as those of Atlanta. The ratio of density in central city versus suburbs in Greater Miami is relatively low at 1.26, compared with 1.52 in Phoenix and 2.02 in Atlanta. This can be explained in part by the employment opportunities outside Miami's central city, but it is also related to geographical constraints on urban expansion due to the UDB, which has, especially in more recent years, stimulated infill development across the metro area.

Finally, in as far as Greater Miami has adhered to conventional characterizations of the suburbs as more affluent, more White, and with fewer minorities, this is changing quickly. Figure 4.2 shows trends of convergence among the central city and suburbs in Miami–Dade and Broward counties in terms of poverty rates and population shares of Blacks and Hispanics.

The convergence between central city and suburbs in Greater Miami is indicative of change in the suburbs but reflects change in the central city as well. Urban growth to the west has reached its environmental limits, and the remote suburban developments are of little interest to young high-income residents. Miami's downtown area since around 2000 has been the scene of massive new construction, mainly of condos targeted at precisely that audience. Downtown Miami's population jumped from 39,000 in 2000 to 72,000 in 2011 (Goodkin Consulting, 2011). The decadal growth rate of the central city went from 2.8% from 1990 to 2000 to 29.1% from 2000 to 2010; elsewhere in the metro area, decadal growth rates dropped from 8.8% to 5.6% (US Census). The downtown area became one of the epicentres of the national real estate bubble that burst in 2008 with significant surplus inventory, but by 2013 the market was moving again.

The "renovation" and revival of downtown has had two important consequences. First, the attraction of high-income, highly educated, big-spending, and often White new urbanites has reinforced the trend of the central city (well, at least large parts of downtown) becoming more prosperous, thereby furthering the *convergence* of the central city compared with the suburbs shown in figure 4.2. However, the notable densification of downtown has resulted in renewed *divergence* in terms of density between city and suburbs. Seen from the condo towers on the waterfront, the rest of the metro area is starting to look increasingly suburban. But the lifestyles of new downtown dwellers are not altogether urban either: as in a number of other metropolitan areas in the United States, we are witnessing a kind of suburbanization of the city, where the new "returning" residents want it all: spacious accommodation; dense, trendy, and lively neighbourhoods; a parking space; nearby amenities and entertainment; good schools for the kids; greenery; private enclosures with security; and, yes, a certain socioeconomic homogeneity of the residential areas (Nijman & Clery, 2015). The new condo towers in downtown Miami, in a sense, are like vertical suburbs.

Searching for Suburbia in Metropolitan Miami 77

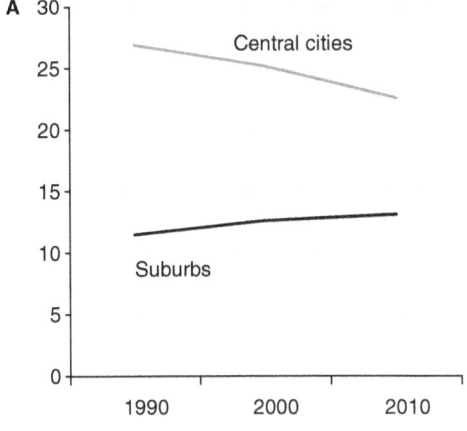

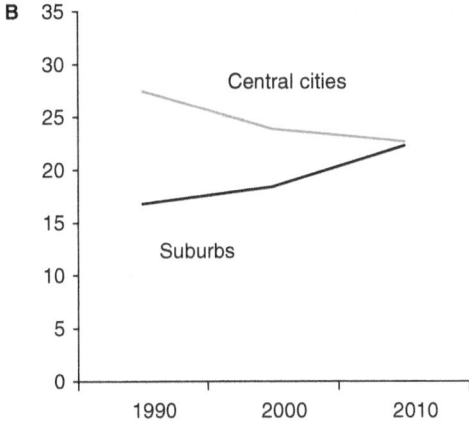

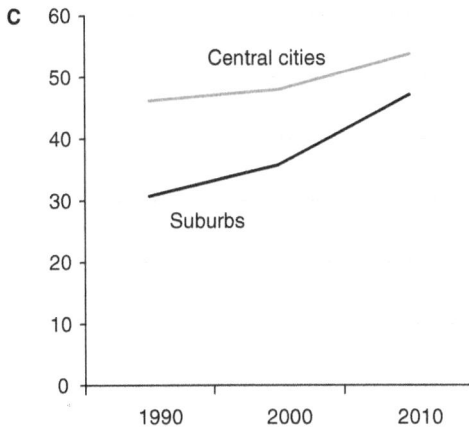

Figure 4.2 Central cities versus suburbs in Miami–Dade and Broward counties, 1990, 2000, and 2010: (A) The percentage of the population living in poverty as a share of the total population; (B) Black population as a share of the total population; (C) Hispanic population as a share of the total population. Source: US Census.

Elusive Suburbia

Aggregate statistics of "suburbia" in Southeast Florida (if one were able to delineate it) would likely conceal as much as they reveal. This section provides short profiles of six selected areas across the Miami–Dade County area. The areas are reviewed comparatively in terms of their suburban qualities and in juxtaposition to the City of Miami. They are chosen from different parts of the county, at varying distances from downtown, and with different historical origins. This is, then, an exercise in comparative urbanism (Nijman, 2007b; 2015) at the level of the suburb inside a single metropolitan area. Hialeah, Pinecrest, Florida City, and Doral are all municipalities, while Kendall and Kendall West are census designated places (CDPs), and as such form part of unincorporated Miami–Dade County (map 4.1).

Table 4.2 lists various demographic, economic, and social indicators that are derived from the dichotomous nature of traditional concepts of the suburbs rooted in the 1950s prototype. The suburb stood in contrast to the central city in terms of socioeconomic status, race or ethnicity, density, and function: the typical suburb was more affluent and more expensive, more White, less dense, and more residential. Furthermore, suburbanism was understood to increase with distance from the central city.

The table underscores the eclectic nature of Greater Miami's urban landscape and the difficulty of identifying a common suburbia. The contrasts between different areas are often stunning. Among some areas, particular similarities can be found, but then they are widely different on other scores. Some areas seem suburban in one way but not at all in another. For example, Doral may seem a typical suburb in that it is of recent vintage, relatively far from the city centre and not very dense; but it has relatively more daily in-commuters than the central city itself, and its share of Whites is less. In another example, Kendall West is also of recent origins and even further out, but densities are nearly as high as in the central city and home values substantially lower than in the central city. What follows is a more detailed, yet succinct, profile of each of the selected "suburbs."

Hialeah is the sixth-biggest city in the State of Florida and its incorporation dates back to 1925. In the 1920s it covered only the area closest to downtown and it was very much a part of the tourist and spectacle-oriented culture of Miami. It was home to a highly rated horse-race park, a greyhound racetrack, and a jai-alai stadium. In the 1930s parts of the area turned into a working-class type of suburb but also attracted considerable small manufacturing. From 1960 onward Hialeah became

Table 4.2 Selected characteristics of the City of Miami and six selected "suburbs" in Miami–Dade county

	Population, 2010	Daytime population change, 2006–10 (%)	Population density per square miles, 2010	Median home value, 2007–11 ($)	Median household income, 2007–11 ($)	Persons in poverty, 2007–11 (%)	Black, 2010 (%)	Hispanic, 2010 (%)	Non-Hispanic White, 2010 (%)
City of Miami	399,497	54.6	12,140	257,500	30,270	27.7	19.2	70.0	11.9
Doral	45,709	108.6	3,316	340,100	340,100	8.3	2.5	79.5	14.6
Hialeah	224,669	−9.5	10,534	197,800	197,800	21.6	2.7	94.7	4.2
Pinecrest	18,223	−15.4	2,464	755,600	755,600	5.5	2.0	41.3	50.4
Florida City	11,245	−3.8	1,896	174,800	174,800	45.2	52.4	42.4	5.6
Kendall	75,371	4.8	4,687	321,500	321,500	9.4	4.4	63.7	28.4
Kendall West	36,154	−39.6	10,280	215,500	215,500	12.8	3.2	88.3	8.2

Source: US Census; see also map 4.1.
Note: Kendall and Kendall West are census designated places; the others are municipalities.

an important destination for large numbers of Cuban migrants, and within a few decades they had displaced the original inhabitants. By 2010 this city of more than 230,000 people was about 74% Cuban and 95% Hispanic and became predominantly known as a lower middle (working) class area. In more recent times, the area has lost manufacturing and has become more and more residential. There is now a net outflow of daily commuters. Hialeah is said to be the densest American city of its size not to feature a single skyscraper. The city's website typifies the area's cultural character as follows: "Predominantly Hispanic, Hialeah residents have assimilated their cultural heritage and traditions into a hard-working, diverse community proud of its ethnicity, as well as its family oriented neighborhoods." (https://www.hialeahfl.gov/469/History-of-Hialeah).

Hialeah's home values are about 20% below the county average, and 21% of the people live below the poverty line. The share of Blacks is less than 3%, compared with 19% countywide. If Hialeah can at all be considered a suburb, it should be qualified as one that is very large (certainly in the context of Greater Miami), at close proximity to the core of the metropolis, with high density, and one that is virtually exclusively Hispanic. In 2009 *Forbes* listed Hialeah as one the "most boring" cities in the country (a rather questionable characterization altogether and one that smacks of old-fashioned suburban prejudice), overlooking the area's remarkable idiosyncrasies.

Florida City, the southernmost municipality in the state, lies more than 25 miles away from downtown Miami and is one of the smallest cities in the region at just over 11,000 people. It was one of the earliest incorporations in the county, dating from 1915, but it developed in relative isolation. This was a small farming community (800 inhabitants by 1922), and its main relation to the central city was in terms of transportation of produce, supplies, and general urban services. For most things Florida City depended on the larger neighbouring farming town of Homestead. Socially and culturally, Florida City was a long way from the urban core and remains so to this day. It became part of the contiguous urban region through amalgamation. In the early 20th century most incorporated areas in Southeast Florida had strictly segregated Black neighbourhoods. Over time, Florida City's Black community expanded and came to dominate the town. Blacks still comprise nearly half the population, and this is one of the few areas in the county where Hispanics are not an absolute majority. Another contrast to Hialeah lies in the very low density here (figure 4.3). The economy is weak, and many commute to Homestead or elsewhere. Poverty rates are among the highest in the state, and home values are well below the county average.

Figure 4.3 Street view of Florida City, 2013. Photo courtesy of authors.

Pinecrest lies entirely east of the US1, near the bay, and about 7 miles south of downtown Miami. It became a municipality only in 1996, more or less according to the standard fiscal logic of incorporation in the United States (Peck, 2014): the area had developed a considerable tax base but saw relatively little in return in terms of county services; secession followed suit. The official website of The Village [!] of Pinecrest states that "during the 1950s and 1960s the area flourished with the development and construction of ranch-style homes on acre lots which laid the foundation for the community's rural and lushly landscaped residential character. Rapid growth and local issues during the 1990s inspired a movement led by residents ... to incorporate the area" (http://www.pinecrest-fl.gov/index.aspx?page=47).

Pinecrest is an affluent and expensive area in an attractive location with abundant greenery. Real estate values are more than three times the county average, as are household incomes. Densities are very low, only about one fifth of those in the City of Miami. Statistically, Pinecrest and Florida City have two things in common: very low density (figure 4.4) and a relatively low share of Hispanics of around 42%. But in all other respects, these two places could not be further apart: Pinecrest has virtually no Black population and it is among very few places in Southeast Florida that by 2010 still had an absolute majority of non-Hispanic Whites (barely, at 50.4%; by 2018, the number had dropped to 43.3%).

Figure 4.4 Street view of Pinecrest, 2013. Photo courtesy of authors.

The City of **Doral** came into existence in 2003 and this, too, was a case of fiscal logic. The city's website is plain and clear:

> Incorporation began in earnest in 1995 with the realization that residents were paying a very high price for services received; they wanted more services at a reasonable price... . Doral had been classified as a "donor community," meaning that the taxes paid were more than the cost of operations... . In January 2003, following a seven-year battle, 85% of the voters in Doral voted in favor of incorporation. At long last, they had their own new city with a local government and more service for their tax dollars. (www.cityofdoral.com)

Doral is located in between Miami International Airport and the UDB. It is about 12 miles from downtown Miami and it seems even further because the large grounds of the airport that separate it from the metropolitan core. The entire area was swamp until the 1950s when the first structure was built, a country club with a golf course that would acquire national stature – today it is known as the Doral Golf Resort and it hosts a PGA tournament every year.

Residential developments followed, but so did a host of logistics-related economic activities, many connected to the airport. The city website boasts more than 3,000 transportation/warehousing companies and about being home to the biggest "tile district" in the United States. Doral's daytime population is more than twice as large as its nighttime population, by far the highest percentage increase of daytime population (108.6%) of any municipality in Southeast Florida, including the City of Miami (54.6%). Home values and incomes in Doral are considerably higher than the county average but below those of Pinecrest. Like Pinecrest, there are virtually no Blacks here, but Doral in 2010 had fewer non-Hispanic Whites and a much higher share of Hispanics (79.5%). Doral is in some ways a prototypical suburb that formed around a country club far away from the central city, yet it is also a major economic hub. The city website describes the area as "the premier place to live, work and play, our many assets provide for a superior quality of life in an *urban center* known for its commerce" [italics added].

Kendall often features as the poster child of suburbia in Greater Miami. Development here did not take off until the early 1970s when Kendall Drive (SW 88 St) was constructed, a major traffic artery running due west from the US1. It was the speed with which this residential area grew and expanded to the west and southwest, and the accompanying homogeneous neighbourhoods in design and demography, that made Kendall stand out and that invoked its prototypical suburban qualities. It is a huge area (bigger than any of the other "suburbs" discussed in this section), covering about 60 blocks north to south and 100 blocks east to west, where it sits against the UDB. By 1990 it was home to no less than 200,000 people and it constituted an "urban realm by itself" (Muller, 1991, p. 66). The area was initially anchored on Dadeland, the shopping centre and office complex at the US1 and Kendall Drive on its eastern edge. Dadeland could be considered an edge city; in the early 1990s it was already being described as South Florida's "most mature suburban downtown" (page 66). In local discourse, Kendall still tends to comprise the entire area described above but it presently holds several CDPs. Kendall CDP itself is now confined to the oldest and most easterly part, in relative vicinity to Dadeland (map 4.1).

Kendall West is another CDP, a smaller suburb of more recent origins to the west and north of Kendall proper. The differences between the two areas are instructive and call into question the conventional North American distinction between inner-ring and outer-ring suburbs (Hanlon, 2010). Kendall West's population density is twice that of Kendall, indicating that the region ran out of space over time but also that later suburban developments were increasingly aimed at lower income households, with smaller lot sizes. In other words, in later periods, density increases outward. Median home values in Kendall (inner suburb) are nearly 50% higher than in Kendall West (outer suburb). The Kendall "inner-ring" suburb's share of non-Hispanic Whites is double the county average and three-and-a-half times that of Kendall West, where Hispanics account for nearly 90% of the population. Incomes in Kendall West are lower, and poverty rates are higher. Kendall West is almost entirely residential; its daytime population change, at −40%, is the second-highest negative value of any municipality or CDP in Miami–Dade County (after Palmetto Estates, a CDP in the southwest of the county). The average commute from Kendall West is nearly 40 minutes, by far the highest of the areas discussed here. Kendall, however, has a net inflow (though modest), of commuters, reflecting the significance of the "downtown" Dadeland area but also its more mixed urban uses and scattered "edgeless" office parks, along with commercial strips, medical care facilities, schools, and so on (another explanation why population densities are lower here). In some ways, then, the more recent and further outward Kendall West makes older Kendall look urban – but in other respects it is the reverse.

Conclusion

The Miami case highlights three theoretically interrelated issues. First, the ahistorical perspective of much work on suburbanization (resting on static representation of the 1950s suburb) obscures different successive stages of the process, with different outcomes. Thus, for example, a suburb from the 1920s is a very different place from an early 21st-century suburb, and the processes of suburbanization that gave rise to it were equally dissimilar. Second, acknowledgment of earlier trends of suburbanization raises the question as to how new suburbs relate to older "suburbs." This, in turn, means that the dichotomous thinking that pervades much work on suburbanization (suburb versus central city, rich versus poor, White versus Black, sparse versus dense, residential versus mixed functions) is out of sync with evolving metropolitan constellations. Third, ahistorical conceptualizations of suburbanization are often accompanied by a kind of geographic fetishism: the idea that there is an identifiable suburban space; that there is actually such a common thing as a spatially delineated coherent suburb. If this were at all the case in the 1950s, it is increasingly not anymore. Instead, that suburban composite has come apart and is selectively scattered across the metropolitan landscape. Geographically, suburban traits increasingly combine with features that are traditionally urban, such as high densities or concentrated economic activity. The Greater Miami case highlights suburbanism as a complex and multidimensional notion that in the modern metropolis can manifest itself partially and in different ways in various places.

The one area in Miami–Dade County that in local parlance has been emphatically suburban is Kendall. That happened relatively late as it gathered force in the 1970s (at least a decade after the proliferation of the prototypical "sitcom" suburbs elsewhere in the nation), and it did not last very long. Kendall today is an edge city with lots of edgelessness to add and it has passed the torch to newer "suburbs" like Kendall West. The Kendall "suburbs" occupied the local headlines for some time, and their discursive dominance seems to have overshadowed what is actually a much more varied, complicated, and eclectic metropolitan landscape.

At a larger scale, and for a short time, suburbanization adhered a bit more to generalized North American patterns. This applied to commuter patterns between Broward and Miami–Dade but also to "white flight" in the form of substantial relocation of middle-class Whites from Miami–Dade County to Broward County in reaction to the even more substantial influx of Cubans and other foreign migrants in Miami–Dade. This "suburban" movement gathered momentum from the 1970s as well, with anxiety expressed on bumper stickers such as "Will the Last American to Leave Please Bring the Flag." Many of the inter-county movers settled in Broward's westerns suburbs that are thus, too, becoming increasingly Hispanic. There are now only a few areas where Hispanics are not in an absolute majority. Florida City and Pinecrest, in very different ways as we have seen, are oddities in that they have held on for so long to sizable non-Hispanic (relative) majorities.

If suburbs are in fact shifting parts of a dynamic metropolitan configuration, so is the central city. The City (municipality) of Miami, at the core of this metro area, was for years listed as the poorest city of its size in the United States, but downtown has in recent years experienced a revival. By 2018, the City of Miami was the fastest growing municipality in the metro area, underscoring the significance of the "great inversion" (Ehrenhalt, 2012). This should not be mistaken for a return to old city–suburb dichotomies. The revival is highly concentrated in the downtown area and leaves much of the central city (sparsely built up and relatively poor) unaffected. Moreover, the prevailing lifestyle of the new downtown apartment dweller, as in various other cities in the United States and elsewhere, presents a peculiar mix of alleged urban and suburban qualities that involve spacious homes, nearby cultural amenities, parking spaces, greenery, and "gated" condominiums with relatively homogeneous populations. Traits that used to be considered typically suburban.

REFERENCES

Broward County. (2004, October). Domestic migration patterns. Office of Urban Planning and Redevelopment *Broward-by-the-Numbers* number 27.
Ehrenhalt, A. (2012). *The great inversion and the future of the American city*. New York, NY: Knopf.
Fishman, R. (1987). Bourgeois utopias: Visions of suburbia. In S. Fainstein & S. Campbell (Eds.), *Readings in urban theory* (2nd ed., pp. 21–31, 2002), Oxford: Blackwell.
Goodkin Consulting. (2011, September). Population and demographic profile. Downtown Development Authority District and Adjacent Areas of Influence. Report prepared for the Downtown Development Authority. Retrieved from http://dpanther.fiu.edu/sobek/FIGO000254/00001
Hanlon, B. (2010). *Once the American dream: Inner-ring suburbs of the metropolitan United States*. Philadelphia, PA: Temple University Press.
Harris, R. (1992). The unplanned blue-collar suburb in its heyday, 1900–1940. In D. Janelle (Ed.), *Geographical snapshots of North America* (pp. 94–9). New York, NY: Guilford Press.
Jackson, K. (1985). *Crabgrass frontier: The suburbanization of the United States*. New York, NY: Oxford University Press.
Knox, P. (2008). *Metroburbia, USA*. New Brunswick, NJ: Rutgers University Press.
Lang, R.E. (2003). *Beyond edge city: Office sprawl in South Florida*. Washington, DC: Brookings Institution.
Lejeune, J. (2009). City without memory: Planning the spectacle of Greater Miami. In A.T. Shulman (Ed.), *Miami, modern metropolis* (pp. 34–59). Miami Beach, FL: Bass Museum of Art.
McKenzie, B. (2013). *County-to-county commuting flows 2006–2010* (Working Paper). U.S. Census. Retrieved from https://www.census.gov/library/working-papers/2013/acs/2013-McKenzie.html
Muller, P. (1991). The urban geography of South Florida. In T. Boswell (Ed.), *South Florida: The winds of changes* (pp. 63–9). Miami, FL: Association of American Geographers.

Nijman, J. (2007a). Locals, exiles, and cosmopolitans: A theoretical argument about identity and place in Miami. *Tijdschrift voor Economische en Sociale Geografie,* 98(2), 167–78.

Nijman, J. (2007b). Comparative urbanism. *Urban Geography,* 28(1), 1–6.

Nijman, J. (2011). *Miami: Mistress of the Americas.* Philadelphia: University of Pennsylvania Press.

Nijman, J. (2015). "The theoretical imperative of comparative urbanism." *Regional Studies,* 49/1: 183–6.

Nijman, J., & Clery, T. (2015). The United States: Suburban imaginaries and metropolitan realities. In P. Hamel & R. Keil (Eds.), *Suburban governance: A global view* (pp. 57–79). Toronto: University of Toronto Press.

Peck, J. (2014). Chicago school suburbanism. In P. Hamel & R Keil (Eds.), *Suburban governance: A global view* (pp. 130–52). Toronto: University of Toronto Press.

Portes, A., & Stepick, A. (1993). *City on the edge: The transformation of Miami.* Berkeley: University of California Press.

Walker, R., & Lewis, R.D. (2001). Beyond the crabgrass frontier: Industry and the spread of North American cities, 1850–1950. *Journal of Historical Geography,* 27(1), 3–19.

5 Spatial Transformations in the Suburbs of the North Carolina Piedmont Region

FANG WEI AND PAUL KNOX

Introduction

Until the middle of the 20th century, the social ecologies of metropolitan America could be conceptualized accurately by the textbook models of Alonso (1964), Burgess (1925), and Hoyt (1939) in terms of processes of congregation, segregation, bid-rent, and sequent occupancy – all of which pivoted tightly around a dominant central business district and transportation hub. Since the middle decades of the 20th century, however, remarkable changes have transformed metropolitan America. In spite of these considerable changes, only passing attention has been given to how these changes have been transcribed into the settlement patterns of cities and metropolitan areas. Have the stereotypical patterns of the Chicago School and of Murdie's famous model (1969) shown signs of giving way to a more complex social spatial structure? Or should Los Angeles be considered the paradigmatic postmodern American city? All these questions remind us to reevaluate the contemporary importance of traditional theories and to test our long-standing ideas about how cities grow and change.

In this chapter, a set of cluster analysis and GIS-based spatial analyses have been developed to capture the spatiotemporal patterns of neighbourhood change in the North Carolina Piedmont region based on an analysis of decennial census tract data between 1980 and 2010.[1] This chapter focuses on (1) the spatial patterns of neighbourhood distribution; (2) neighbourhood transformation over the past three decades; and (3) the spatial patterns of neighbourhood transformation over time. We examine the spatiotemporal patterns of high-resolution changes of socioeconomic development and show how macro-level socioeconomic changes have reshaped neighbourhoods and altered the spatial structure of the metropolis.

Socioeconomic Change in the Piedmont Metropolitan Region

The Piedmont metropolitan region, including Charlotte–Gastonia–Rock Hill (CGR), Raleigh-Cary (RC), and Durham-Chapel Hill (DC) metropolitan statistical areas (MSAs), refers mainly to the hilly plateau between coastal plains and mountains in North Carolina. The fast-growing development in the North Carolina Piedmont region is somewhat typical of the United States. For more than a hundred years this vast rural area that formerly produced cotton and tobacco has developed textile, processing, and other related industries. The Piedmont has been the most populated region in North Carolina (NC) for nearly a century (Meade, 2008) and for decades has been the top metropolitan area, showing continued significant growth, strong competitiveness, a strong economy, and one of the best living areas in the state (Frey, 2005, 2010; Hughes, 1990; Katz, 2010; Meade, 2008).

The Charlotte–Gastonia–Rock Hill MSA is anchored by the city of Charlotte. Hanchett (1998) has described how the city's early development involved a "sorting out" along racial and class lines. Initially driven by economics, an original antebellum "salt and pepper" ecology of spatially intermixed African American and White populations gave way to a "patchwork quilt" of racial segregation in the 1880s. A booming textile economy at the turn of the 20th century produced new wealth that quickly found expression in streetcar suburbs that were sharply segregated through restrictive covenants. After the Great Depression, the city's social ecology changed again in response to modernization and the advent of the automobile, developing – like many other North American cities – a sectoral pattern in terms of income and race. By the mid-20th century, the basic layout of modern Charlotte had been formed, with wealthy and upper middle class White families dominating the south and southeast of the city, while the north and west sides were dominated by the more modest homes of the city's large African American and working-class White populations.

Raleigh's sociospatial development followed a similar chronological pattern, but with a different geography. Here the locus of affluent White neighbourhoods was in the north of the city, while poor African American neighbourhoods were concentrated in the south. While Charlotte and Raleigh both presented a clear demarcation between rich White and poor African American populations, with each occupying one end of the city, Durham developed a distinctively different social ecology. African American business thrived in Durham, a unique phenomenon in the early South. The hub of African American businesses was Parrish Street, widely known as "Black Wall Street," adjacent to the town's tobacco warehouses. The early streetcar suburbs in Durham, in contrast to those in Charlotte and Raleigh, were largely established to serve African American communities such as Trinity Park, Morehead Hills, Club Boulevard, and Needmore (Turner, 2002).

Since the mid-1970s, the population of the Piedmont region has grown rapidly (Berube, Katz, & Lang, 2006; The Brookings Institution, 2000). The population of Charlotte–Gastonia–Rock Hill MSA was 829,824 in 1980, and by 2010, it reached 1,758,038. The population of Raleigh-Durham-Cary combined statistical area (RDC CSA) reached 1,634,847 by 2010 from just over 635,131 in 1980. Central city Charlotte, for example, topped the American core-cities list with a 70% population growth rate (Landis, 2009). Migrants and immigrants, drawn first by manufacturing jobs, relocated from the deindustrializing northeast and then – and in much greater numbers – by the growth of "new economy" jobs in banking, advanced business services, digital technologies, and biotechnology, contributed to rapid growth.

One aspect of this growth in the Piedmont has been changing family structure, including a growth in the numbers of married couples with children, single-person households, and senior households. The Piedmont metro areas also became an immigration gateway in the 1990s, resulting in a marked increase in foreign-born populations, especially Hispanics and Asians (Singer, 2004; Smith & Furuseth, 2004). Meanwhile, racial segregation and its attendant inequalities have persisted within the Piedmont metro areas, despite an overall increase in affluence (The Brookings Institution, 2000).

Another aspect of this growth has included changing employment structure. As those regions have grown, the spatial locations of some sources of employment have become more decentralized, such as in Charlotte MSA; some may have experienced small changes in the spatial locations of employment, such as in Durham MSA (Kneebon, 2009). With respect to job creation and job sprawl, Weitz & Crawford (2012) used 2001–6 data to show that Raleigh MSA and Charlotte MSA gained in job creation but decreased in job accessibility. In another study, Stoll (2005) showed that Charlotte and Raleigh MSAs experienced relatively higher job sprawl but lower job mismatch for African Americans among 300 metropolitan areas based on year 2000 data.

An important factor in the growth of the Piedmont metros has been the Research Triangle, anchored by the University of North Carolina (Chapel Hill), Duke University (Durham), and North Carolina State University (Raleigh). The Triangle has been listed among the nation's top high-tech regions in terms of labour-force quality (Koo, 2005), top competence, and top population gains since the 1990s (Landis, 2009). Centred on Research Triangle Park, this area has fostered the growth of the region's new economic industries. Just as in other metropolitan regions, these employers have sought new settings well away from congested central city areas. The result has been the emergence of a "metroburban" metropolitan form (Knox, 2008), with a polycentric structure that incorporates urban realms and corridors, "edge cities," "edgeless cites," "exurbs," "micropolitan" centres, and "boomburbs."

The Piedmont was historically a rural agricultural region but has now urbanized so much that it is swarming with the largest and fastest growing cities. Now what characterizes the Piedmont region, especially the cities of Charlotte, Raleigh, and Durham, are fast-growing, vibrant job centres for financing and hi-tech, top population growth, and racial diversification.

Data and Methods

The primary source of the data used in this study at the census tract level from 1980 to 2010 is derived from the Longitudinal Tract Data Base (LTDB) and prepared by Spatial Structures in the Social Sciences (S4). To track neighbourhood changes directly over time, we use the LTDB data that have been standardized to 2010 boundaries. Central cities of 1980 are identified based on the indicator in the Neighborhood Change Database (NCDB) produced by the Urban Institute and GeoLytics. The boundaries of the 2010 census used in GIS analysis are from the National Historical Geographic Information System (NHGIS). The central city boundaries are those of 1980, and therefore some suburban tracts in 1980 might have become central city tracts in 2010. This will affect the numbers and proportions of clusters in 2010 to a certain extent. However, the focus of this study is the spatial pattern of neighbourhood change, and the boundaries of central cities will not influence the patterns of spatial distribution and transformation.

Census tracts with populations lower than 200 in each census year have been excluded from the analysis to avoid estimations based on a small number of data. After excluding these tracts and tracts with missing data, the pooled data include a total of 2,874 tracts (632 tracts in 1980, 740 tracts in 1990, 749 tracts in 2000, and 753 tracts in 2010). All variables for each tract were standardized as z-scores relative to all the other tracts in the same census year. The major advantage of our study is that it allows direct comparisons of the relative importance and spatial organization of each major tract type from one census year to another. When analysing the spatial changes of socioeconomic distribution from 1980 to 2010, only those tracts that have specific typologies in both census years (629 tracts both for 1980 and 2010) were included in the analysis.

In seeking to delineate the sociospatial transformation of these Piedmont metros, we have pooled the standardized tract-level data from 1980 to 2010 for the three MSAs and employed a k-means cluster analysis to

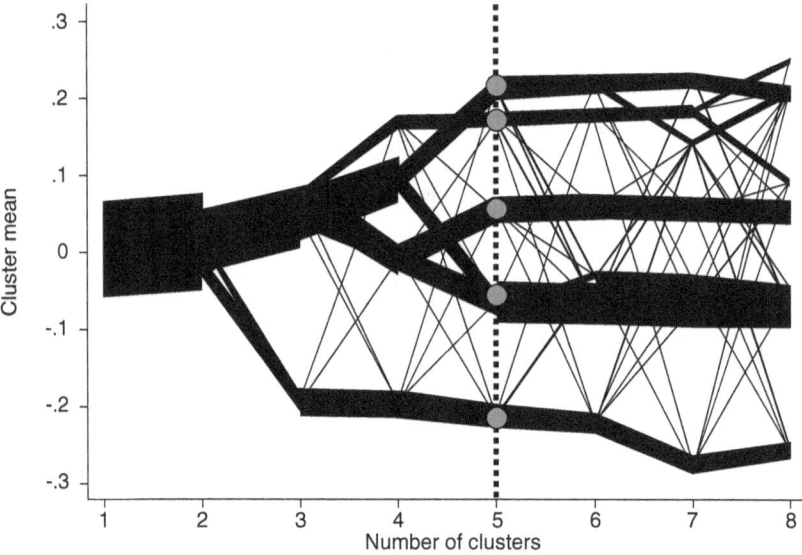

Figure 5.1 Visualization of clusters via clustergram.

develop the overall typologies. K-means is a method of cluster analysis that partitions N observations into k clusters. In this process, each observation belongs to the cluster with the nearest mean. To identify the number of clusters that are relatively stable, we relied on a data-visualization technique – clustergrams – to guide the choice of the number of clusters. The clustergram is used to examine how the members of these clusters are formed as the number of clusters increases. The width of the line segments indicates the number of observations that are assigned to a cluster. Ten variables related to demographics, socioeconomic status, and housing characteristics were selected based on the literature of neighbourhood typologies (Hanlon, 2009; Kitchen & Williams, 2009; Mikelbank, 2004, 2011; Morenoff & Tienda, 1997). To determine just how the spatial patterns of Piedmont metros have changed since 1980, we have drawn on tract-level decennial census data on this standard set of ten socioeconomic variables. We first divided the pooled data into two to eight clusters using k-means algorithms. The three to five group solutions have the relatively larger values based on the Calinski/Harabasz pseudo-F test. Then, a clustergram indicates the relative stability of the five-cluster choices (figure 5.1). Thus, our analysis is based on the fivefold classification of census tracts in the Piedmont metros.

Table 5.1 Z-score means across clusters

Variable	Middle class	Lower/ aging	Black/ poor	Upper income	Immigrant/ renter
Demographic					
Percentage of persons age 17 years and under	0.29	0.10	0.51	0.49	−1.09
Percentage of persons age 60 years and over	0.04	0.73	0.17	−0.40	−0.33
Percentage of persons of Black race, not Hispanic origin	−0.22	−0.14	2.42	−0.69	−0.03
Percentage of foreign born	−0.38	−0.61	−0.24	0.35	1.07
Socioeconomic status					
Percentage of persons with at least a four-year college degree	−0.25	−0.91	−0.93	1.27	0.61
Percent unemployed	−0.32	0.27	1.77	−0.59	−0.07
Percentage of manufacturing employees (by industries)	−0.17	1.46	−0.20	−0.34	−0.77
Median household income	0.11	−0.62	−1.23	1.53	−0.44
Housing characteristics					
Percentage of owner-occupied housing units	0.58	0.19	−1.11	0.65	−1.24
Percentage of vacant housing units	−0.33	0.10	0.54	−0.14	0.18

Table 5.1 lists the means of the ten variables for the five clusters. A significant group of tracts are dominated by *middle class* households – the classic demographic of America's "sitcom suburbs" (Hayden, 2003): family-oriented, White, and relatively stable home-owning households. In comparison, *lower/aging* tracts have lower homeownership and median household incomes with a relatively higher proportion of seniors. *Black/poor* tracts are characterized by the highest percentages of African American populations and the lowest median household incomes. We describe the fourth group of tracts as *upper income*: households with significantly higher than average median household incomes, as well as the highest proportion of persons with a higher education. *Immigrant/renter* describes those tracts with the lowest homeownership and the highest proportions of foreign-born populations. It should be emphasized that these results reflect the dominant general patterns among the pooled tract data for the period 1980 to 2010.

Table 5.2 Populations by cluster

Year	Total tracts	Middle class	Lower/ aging	Black/ poor	Upper income	Immigrant/ renter
1980	632	342,756	313,787	186,017	224,807	233,931
1990	740	532,238	472,646	209,501	292,937	401,728
2000	749	813,804	580,484	220,320	434,422	502,557
2010	753	1,259,841	575,685	301,266	617,502	636,792

Table 5.3 Distribution (tracts %) of clusters

Location	Year	Total tracts	Middle class (%)	Lower/ aging (%)	Black/ poor (%)	Upper income (%)	Immigrant/ renter (%)
Suburbs	1980	509	38.90	22.00	6.09	21.02	11.98
Suburbs	1990	618	39.00	23.14	4.37	18.77	14.72
Suburbs	2000	627	36.52	24.88	4.31	20.41	13.88
Suburbs	2010	631	37.40	20.44	7.45	19.02	15.69
Cities	1980	123	13.01	5.69	30.08	16.26	34.96
Cities	1990	122	9.02	2.46	32.79	10.66	45.08
Cities	2000	122	8.20	3.28	31.15	9.02	48.36
Cities	2010	122	5.74	4.92	31.97	12.30	45.08

Changing Spatial Patterns of Neighbourhood Distribution

This section revolves around the question of what spatial patterns of neighbourhood distribution may be discerned in the Piedmont region. Given the nature and extent of changes in metropolitan form and in social and demographic structure, it is reasonable to expect that the remarkably consistent social ecology of mid-20th-century North American cities – zones, sectors, and clusters – has evolved in significant ways. The contemporary urban social fabric might be fragmented at the fine-grained level but integrated at the macro level (Marcińczak & Sagan, 2011). Thus, we expected a consistent pattern in the spatial expression of neighbourhood distribution.

Plotting the spatial distribution of each tract type over the three decades reveals some interesting patterns in segmentation, diversification, and evolution of different socioecological settings in the North Carolina Piedmont region. Tables 5.2 and 5.3 summarize the populations and distributions of clusters. Given the overall growth of MSAs over the period, the general trend for most tract types is towards an increase

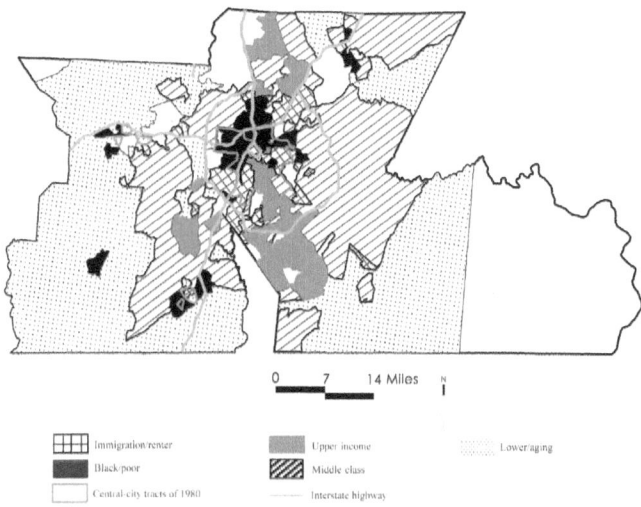

Figure 5.2 Spatial patterns of clusters in the Charlotte–Gastonia–Rock Hill meropolitan statistical area in 2010.

in aggregate populations and tract numbers, but the percentages of tracts vary by tract types. Figures 5.2 and 5.3 show the spatial patterns of the five types of neighbourhoods in CGR MSA and RDC CSA in 2010. Generally speaking, *immigrant/renter* and *Black/poor* tracts are generally located near urban cores. *Middle class* socioecologies form rings in the suburbs of the region. The *upper income* neighbourhoods occupy one or several sectors. The *lower/aging* tracts are located generally in outer suburbs and exurbs. In many cases, *immigrant/renter* tracts occupy the sectors between *Black/poor* and *upper income* neighbourhoods. Using the extended urban region – the North Carolina Piedmont – as the basis for the empirical analysis, a spatial model (Piedmont model) is developed and illustrated in figure 5.4.

In contrast to our expectations, cities are still organized with "zones," "sectors," and "clusters" as described in the classic urban models. The first point to make here is that the urban social patterns of Piedmont cities still bear some resemblance to the concentric zone model. A broad observation is that the *middle class* dominance in suburbs, and the *immigrant/renter* and *Black/poor* dominance in city centres have been maintained numerically and spatially throughout these decades. In the Piedmont model, neighbourhoods of immigrant/renter and Black/poor are generally located in rings near urban cores, with the middle class in suburban rings.

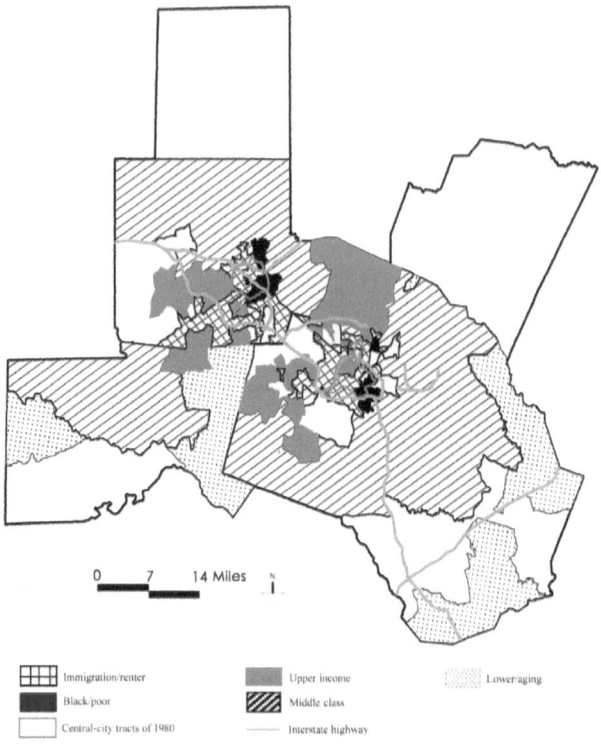

Figure 5.3 Spatial patterns of clusters in Raleigh-Durham-Cary combined statistical area in 2010.

At a more detailed scale, much of the sociospatial differentiation in the Piedmont region exhibits a sectoral pattern. As in most of metropolitan America, and indeed as established in the early development of the Piedmont metros, most affluent White and low-income African American neighbourhoods occupy one or several sectors; they are not only spatially segregated but located at some distance from each other. Both Charlotte and Raleigh have traditional affluent regions, barely changed except for accretion, that date back to the streetcar era. The exception, as in its own early development, is Durham, where *Black/poor* tracts are in propinquity to *upper income* White tracts. But in recent decades the upper income tracts have been separated from *Black/poor* tracts. This finding is consistent with Hoyt's sector model to a certain extent, where residential functions tend to grow in wedge-shaped patterns with sectors of low-income households and high-income households.

Spatial Transformations in the Piedmont Region 97

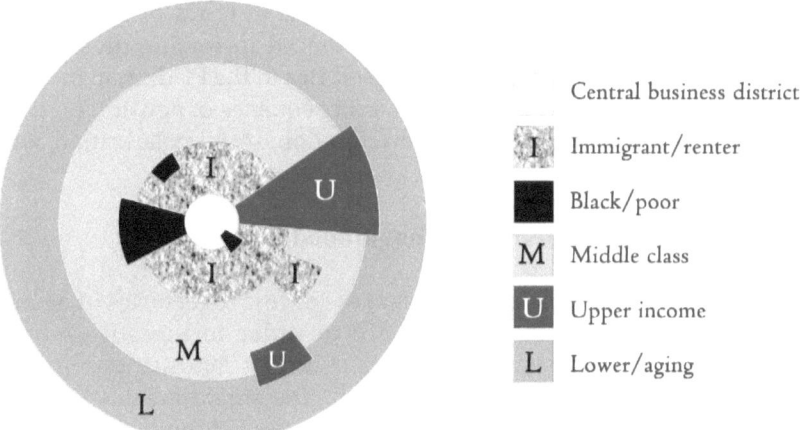

Figure 5.4 The basic version of Piedmont model.

In addition to zones and sectors, the Piedmont model also incorporates a clustered pattern for some social groups. For example, many new houses are developed on the edges of cities due to decentralization of retail, people, and jobs. In suburbs and exurbs, we see clusters of upper income neighbourhoods. Due to urban regeneration and gentrification, clusters of upper income neighbourhoods can also be found in central cities. Some other types of neighbourhoods also show signs of clusters. Most *immigrant/renter* tracts in the Piedmont region, for instance, are located near university areas. The North Carolina Piedmont region is home to a large number of colleges and universities (The Brookings Institution, 2000). These characteristics suggest that these areas have a higher education rate but lower homeownership rate. The finding that universities may attract well-educated immigrants and renters is consistent with certain aspects of the multiple nuclei model of Harris and Ullman.

Another broad observation, then, is that the sociospatial ecology of the Piedmont metros is characterized by a juxtaposition of a relatively stable structure with a growing fragmentation and diversification at a finer grained level. Since the 1980s, the Piedmont metros have become substantially postsuburban (Phelps, Wood, & Valler, 2010). Despite the dominance of the middle class in suburbia, the Piedmont region has shown an increasingly diverse and fragmented mosaic. Spatial patterns have become more diversified due to sociospatial restructuring, but the basic, traditional patterns and relative locations of neighbourhood types have remained largely stable over time. In other words, the

Piedmont region has largely remained stable in terms of overall patterns of distribution but has also experienced increasing diversity at the micro scale. Thus, sociospatial trajectories in the Piedmont are both an extension of historical trends and a consequence of new forces, such as new streams of migration and immigration, social polarization, and gentrification.

Changing Spatial Patterns: Neighbourhood Succession

In this section, we look in more detail at specific trajectories of neighbourhood change from 1980 to 2010 in order to investigate how socioeconomic changes are reflected in urban landscapes. Several further observations emerge. Stability is the single greatest dimension of metropolitan change; while for those neighbourhoods that have changed their attributes, their succession and growth reveals some interesting patterns. In this study, neighbourhood upgrading and downgrading are identified based on changes in neighbourhood income.

First, we focus on the succession process, defined as the sequences of neighbourhood change where typologies of neighbourhoods come to occupy a territory formerly dominated by another typology. Specifically, we examine the different evolutionary trajectories of ecological change across the Piedmont metro region. Table 5.4 illustrates the types of neighbourhood succession from 1980 to 2010. In looking at the overall succession patterns of the five types of tracts in the entire Piedmont region, we found that succession patterns differ by clusters:

(1) *Immigrant/renter* and *Black/poor* neighbourhoods: A few *immigrant/renter* and *Black/poor* neighbourhoods have been upgraded into *middle class* or *upper income* neighbourhoods, especially in suburbs. Several *immigrant/renter* neighbourhoods in inner cities that are adjacent to *upper income* neighbourhoods have been gentrified into *upper income* neighbourhoods, especially in the Charlotte MSA. In addition, transitions between *Black/poor* and *immigrant/renter* neighbourhoods occurred both in central cities and suburbs of 1980.
(2) *Upper income* neighbourhoods: The fringes of sectors of *upper income* neighbourhoods, especially those close to inner-ring suburbs, have generally been superseded by *middle class* tracts and, to a lesser extent, *immigrant/renter* tracts. However, those *upper income* neighbourhoods located in central cities are relatively

Table 5.4 Neighbourhood succession from 1980 to 2010

	Types of neighbourhood change[a]	Total tracts	Suburbs	Cities	Suburbs (%)	Cities (%)
Middle class succession	11	122	120	2	98.36	1.64
	12	18	17	1	94.44	5.56
	13	20	12	8	60.00	40.00
	14	30	30	0	100.00	0.00
	15	23	18	5	78.26	21.74
Lower/aging succession	21	35	35	0	100.00	0.00
	22	69	66	3	95.65	4.35
	23	6	6	0	100.00	0.00
	24	2	1	1	50.00	50.00
	25	7	4	3	57.14	42.86
Black/poor succession	31	5	5	0	100.00	0.00
	32	2	2	0	100.00	0.00
	33	47	18	29	38.30	61.70
	34	4	4	0	100.00	0.00
	35	10	2	8	20.00	80.00
Upper income succession	41	38	33	5	86.84	13.16
	42	3	1	2	33.33	66.67
	43	0	0	0	-	-
	44	57	45	12	78.95	21.05
	45	29	28	1	96.55	3.45
Immigrant/renter succession	51	10	10	0	100.00	0.00
	52	3	3	0	100.00	0.00
	53	5	3	2	60.00	40.00
	54	16	14	2	87.50	12.50
	55	68	30	38	44.12	55.88

[a] Each two-digit number represents a neighbourhood type in each census year from 1980 to 2010. 1 = middle class; 2 = lower/aging; 3 = Black/poor; 4 = upper income; 5 = immigrant/renter.

stable. In particular, no upper income neighbourhoods have been superseded by Black/poor neighbourhoods in the Piedmont region.

(3) *Lower/aging* neighbourhoods: Most *lower/aging* sectors in suburban and some in exurban settings have been upgraded into *middle class* neighbourhoods. Several *lower/aging* neighbourhoods in the central city of Charlotte have changed into *Black/poor*. A few have been upgraded into *upper income* neighbourhoods or changed into immigrant/renter neighbourhoods both in central cities and suburbs.

(4) *Middle class* neighbourhoods: Elsewhere within the background of *middle class* suburbia it is possible to discern two broad trajectories of change: deteriorating *middle class* tracts and strong and rising *middle class* tracts. Generally speaking, *middle class* ecologies formed rings in the suburbs of the region; the interior perimeter of these *middle class* rings was adjacent to neighbourhoods that were previously *middle class*, but have shifted to *immigrant/renter* and *Black/poor* neighbourhoods; the outer suburban exterior perimeter of the *middle class* rings, in particular in CGR MSA, tended to consist of neighbourhoods that were previously *middle class* but have emerged as *lower/aging* tracts. These outcomes confirm the invasion-succession model to a certain extent, where neighbourhoods will change as higher income residents in outer rings are invaded and finally replaced by lower income residents from inner rings. Finally, certain sectors of the *middle class* rings have changed into *upper income* tracts, a phenomenon mainly found in tracts of 1980s suburbs.

Changing Spatial Patterns: Neighbourhood Growth

In addition to these patterns of ecological succession, it was possible to identify another aspect of metropolitan change by looking at patterns of ecological transformation: the growth of a neighbourhood type through occupation of a territory that was formerly dominated by other types.

Table 5.5 shows neighbourhood growth by type from 1980 to 2010. Patterns of growth process can be categorized into four patterns: clustering, sectoral growth, border accretion, and greenfield expansion. The ecological changes of each typology of tracts in the Piedmont region follow one or a mix of the four patterns.

(1) Clustering: The *immigrant/renter* tracts that are a distinctive feature of the 2010 maps generally emerged from neighbourhood social ecologies that were previously dominated by White populations or, in a few small patches, by *Black/poor* populations. By 2010, these newly emerged *immigrant/renter* tracts had generally developed a cluster pattern within the central cities as well as in the suburbs.
(2) Border accretion: Most new *Black/poor* areas emerged from *middle class* tracts on the border of stable *Black/poor* districts. Many new *upper income* areas have evolved from formerly *middle class* tracts on the suburban border of stable *upper income* tracts.

Spatial Transformations in the Piedmont Region   101

Table 5.5 Neighbourhood growth from 1980 to 2010

	Types of neighbourhood change[a]	Total tracts	Suburbs	Cities	Suburbs (%)	Cities (%)
Middle class growth	11	122	120	2	98.36	1.64
	21	35	35	0	100.00	0.00
	31	5	5	0	100.00	0.00
	41	38	33	5	86.84	13.16
	51	10	10	0	100.00	0.00
Lower/aging growth	12	18	17	1	94.44	5.56
	22	69	66	3	95.65	4.35
	32	2	2	0	100.00	0.00
	42	3	1	2	33.33	66.67
	52	3	3	0	100.00	0.00
Black/poor growth	13	20	12	8	60.00	40.00
	23	6	6	0	100.00	0.00
	33	47	18	29	38.30	61.70
	43	0	0	0	-	-
	53	5	3	2	60.00	40.00
Upper income growth	14	30	30	0	100.00	0.00
	24	2	1	1	50.00	50.00
	34	4	4	0	100.00	0.00
	44	57	45	12	78.95	21.05
	54	16	14	2	87.50	12.50
Immigrant/ renter growth	15	23	18	5	78.26	21.74
	25	7	4	3	57.14	42.86
	35	10	2	8	20.00	80.00
	45	29	28	1	96.55	3.45
	55	68	30	38	44.12	55.88

[a] Each two-digit number represents a neighbourhood type in each census year from 1980 to 2010. 1 = middle class; 2 = lower/aging; 3 = Black/poor; 4 = upper income; 5 = immigrant/renter.

(3) Sectoral growth: The expansion of *upper income* ecologies has tended to occur along the outer border of the original *upper income* tracts, but many of them are due to the transformation of outlying *middle class* and *lower/aging* districts through upscale new developments in sectoral patterns.

(4) Greenfield expansion: Most new *middle class* districts have evolved from formerly *lower/aging* tracts, or, to a lesser extent, *upper income* tracts in underdeveloped suburban areas. Several *middle class* tracts have evolved from *immigrant/renter* and *Black/poor* suburban

tracts. In addition, tracts that became *lower/aging* also exhibited a pattern of greenfield expansion. Most of these areas transitioned from former *middle class* tracts in outer suburbs, and several from the other types of tracts.

Conclusions

In the Piedmont region, as in other US metropolitan regions, external forces such as structural economic change, secular changes in social organization and demographic structure, as well as immigration, have clearly had a significant influence in shaping trajectories of ecological change. The Piedmont region has morphed rapidly into polycentric metropolitan regions with traditional city centres and suburban employment centres, and has become an exemplar of contemporary sunbelt suburbanization.

Given the joint action of postwar external forces and historical concentration of poverty and racial minorities, the Piedmont has evolved rapidly into segmented, diversified, and polarized socioecological settings and a more pronounced polycentric metropolitan form. Across the Piedmont metro region, relatively stable spatial patterns of historical trends at the macro scale have shown to be juxtaposed with growing fragmentation and diversification at a finer grained level. Segregation as reflected in long-standing structural inequalities may be the reason that the Piedmont region stays largely stable over time; while new dynamics of socioeconomic change and the emerging new-economy industries are likely the reason that this region is growing into more segmented and fragmented patterns.

Investigation of neighbourhood distribution reveal something similar to the spatial patterns of the Chicago School and the typical patterns of Murdie's factorial ecology models. Framed by core cities whose social ecology still bears some resemblance to the textbook factorial ecology model of North American cities, the social ecology of Piedmont has demonstrated that, in certain cases, urban phenomena do match those of classic models. It reminds us to reevaluate the contemporary importance of traditional theories. However, it may also question the Los Angeles School's fundamental claim that Los Angeles should be considered the paradigmatic postmodern American city. Urban development in the Piedmont region suggests that similar processes might shape spatial patterns in the way that they have in other parts of the United States, such as Los Angeles. However, just as Hanchett (1998) has pointed out, the particular historical context and racial background in the Piedmont will more likely give this process a distinctive Southern flavour.

Further, examination of the changing patterns of metropolitan transformation revealed factors that have generally been ignored by any school. By investigating the structural dimensions of metropolitan change – succession and growth patterns over the past three decades – our study shows some interesting patterns. Despite the variations in patterns of ecological succession among the five types of tracts, a commonality exists. Overall, there are several trajectories of neighbourhood ascent within central cities and inner-ring suburbs; however, urban-side tracts (near city centres or inner-ring suburbs) are more likely to have experienced a downward socioeconomic transition, while suburb-side tracts tended to experience an upward socioeconomic transformation. It is consistent to the invasion-succession model that neighbourhoods will change as higher income residents in outer rings are invaded and finally replaced by lower income residents from inner rings. This may also reflect certain aspects of central city and inner ring suburban decline and the prevailing process of suburbanization. Meanwhile, this study also identified four types of patterns resulting from ecological growth: clustering, border accretion, sectoral growth, and greenfield expansion. This implies the selective operation of a variety of sociospatial processes, including segregation/congregation, filtering, invasion/succession, redevelopment and gentrification, and greenfield development.

Without comparing the changing spatial patterns of neighbourhood distribution and transformation in other metropolitan areas, it is difficult to generalize these patterns to other regions in the United States. This study only uses the North Carolina Piedmont region as the basis for empirical analysis. The literature on urban spatial form shows differences between the East and West in the United States (Lang, Popper, & Popper, 1995, 1997) and a clear split in the eastern and western halves of the sunbelt (Lang, 2003; The Brookings Institution, 2000). The North Carolina Piedmont urban complex is a clear exemplar of the southeastern form, which shares a rough spatial equivalent in regions such as Nashville and Atlanta. Yet this form is different from that in the Southwest, and even from, say, relatively nearby South Florida. Due to the limited generalizable nature of the findings within this study, future research is needed to compare the patterns we found across regions, such as western (or Midwest and Northeast) US metropolitan complex of comparable urban scale and population. By discovering what similarities or differences may exist, we can discover whether and to what extent that the Chicago School theories about the social ecologies of the early 20th century explain today's metropolitan structure.

NOTE

1 We are grateful to David Bieri, Virginia Tech, for assistance with data compilation and analysis.

REFERENCES

Alonso, W. (1964). *Location and land use: Toward a general theory of land rent.* Cambridge, MA: Harvard University Press.
Berube, A., Katz, B., & Lang, R.E. (2006). *Redefining urban and suburban America: Evidence from census 2000.* Washington, DC: The Brookings Institution.
The Brookings Institution. (2000). *Adding it up: Growth trends and policies in North Carolina.* The Brookings Institution.
Burgess, E.W. (1925). The growth of the city: An introduction to a research project. In R. LeGates & F. Stouts (Eds.), *The inner city reader.* New York, NY: Routledge.
Clark, W.A.V. (2009). Changing residential preferences across income, education, and age: Findings from the multi-city study of urban inequality. *Urban Affairs Review, 44*(3), 334–55.
Frey, W.H. (2005). *Metro America in the new century: Metropolitan and central city demographic shifts since 2000.* The Brookings Institution.
Frey, W.H. (2010). *A demographic lull at census time.* The Brookings Institution.
Hanchett, T.W. (1998). *Sorting out the new south city-race, class and urban development in Charlotte, 1875–1975.* Chapel Hill: The University of North Carolina Press.
Hanlon, B. (2009). A typology of inner-ring suburbs: Class, race, and ethnicity in U.S. suburbia. *City & Community, 8*(3), 221–46.
Hayden, D. (2003). *Building suburbia: Green fields and urban growth, 1820–2000.* New York, NY: Pantheon Books.
Hoyt, H. (1939). *The structure and growth of residential neighborhoods in American cities.* Washington, DC: Federal Housing Administration.
Hughes, J. (1990). City review: North Carolina. *National Real Estate Investor, 32,* 131–4.
Katz, B. (2010). *State of metropolitan America: On the front lines of demographic transformation.* Brookings Institution Metropolitan Policy Program.
Kitchen, P., & Williams, A. (2009). Measuring neighborhood social change in Saskatoon, Canada: A geographic analysis. *Urban Geography, 30*(3), 261–88.

Kneebon, E. (2009). *Job sprawl revisited: The changing geography of metropolitan employment.* Metropolitan Policy Program at Brookings.

Knox, P.L. (2008). *Metroburbia USA.* New Brunswick, NJ: Rutgers University Press.

Koo, J. (2005). How to analyze the regional economy with occupation data. *Economic Development Quarterly, 19*(4), 356–72.

Landis, J. (2009). The changing shape of metropolitan America. *Annals of the American Academy of Political and Social Science, 626*(1), 154–91.

Lang, R.E. (2003). *Edgeless cities: Exploring the elusive metropolis.* Washington DC: Brookings Institution Press.

Lang, R.E., Popper, D.E., & Popper, F.J. (1995). Progress of the nation: The settlement history of the enduring American frontier. *The Western Historical Quarterly, 26*(3), 289–307.

Lang, R.E., Popper, D.E., & Popper, F.J. (1997). Is there still a frontier? The 1890 US census and the modern American West. *Journal of Rural Studies, 13*(4), 377–86.

Marcińczak, S., & Sagan, I. (2011). The socio-spatial restructuring of Łódź, Poland. *Urban Studies, 48*(9), 1789–809.

Meade, M. (2008). The population of the Carolinas. In D.G. Bennett & J.C. Patton (Eds.), *A geography of the Carolinas* (pp. 161-86). Boone: Parkway Publishers.

Mikelbank, B.A. (2004). A typology of U.S. suburban places. *Housing Policy Debate, 15*(4), 935–64.

Mikelbank, B.A. (2011). Neighborhood déjà vu: Classification in metropolitan Cleveland, 1970–2000. *Urban Geography, 32*(3), 317–33.

Morenoff, J.D., & Tienda, M. (1997). Underclass neighborhoods in temporal and ecological perspective. *Annals of the American Academy of Political and Social Science, 551*, 59–72.

Murdie, R.A. (1969). *Factorial ecology of metropolitan Toronto, 1951–1961: An essay on the social geography of the city.* Chicago, IL: University of Chicago.

Phelps, N.A., Wood, A.M., & Valler, D.C. (2010). A postsuburban world? An outline of a research agenda. *Environment and Planning A, 42*(2), 366–83.

Singer, A. (2004). *The rise of new immigrant gateways.* The Brookings Institution.

Smith, H.A., & Furuseth, O.J. (2004). Housing, hispanics and transitioning geographies in Charlotte, North Carolina. *Southeastern Geographer, 44*(2), 216–35.

Stoll, M.A. (2005). *Job sprawl and the spatial mismatch between blacks and jobs.* Washington, DC: The Brookings Institution.

Turner, W.R. (2002). Development of streetcar systems in North Carolina. Retrieved from https://www.nctrans.org/About-Us/History/Streetcar-Systems.aspx

Weitz, J., & Crawford, T. (2012). Where the jobs are going: Job sprawl in U.S. metropolitan regions, 2001–2006. *Journal of the American Planning Association, 78*(1), 53–69.

PART 2

Changing Political Economies
of Suburbanization

6 The Strange Case of the Bay Area

RICHARD WALKER AND ALEX SCHAFRAN

Introduction

The San Francisco Bay Area is frequently underestimated in size, misdefined by its odd spatial footprint, and falsely praised for its liberal social order. We intend to correct some common errors about the metropolis and emphasize its distinctive features. At the same time, we hope to cast light on North America metropolitan growth and change, including suburbanization, through a closer look at one of the continent's largest and most dynamic city regions. Using the new 12-county US Census definition, the Bay Area is the fourth largest urban area in the United States, with 8.5 million people in 2015. It is a hybrid whose iconic centre, San Francisco, is an East Coast city at the heart of a sunbelt metropolis. Despite the relative density of the core, it spreads out across a low-density metropolitan area of more than 50 by 100 miles.

Our purpose here is to take a critical look at the Bay Area through the lens of its geography. There is considerable strangeness to it, so we organize the discussion in terms of 10 peculiarities: unclear boundaries, strange and upended centrality, affluent and developer suburbs, an expansive greenbelt, speculative and inflated suburbs, and continuing but inverted inequality. Regional differences still matter in the supposedly homogenous territory of North America.

Nevertheless, a case study is a step toward comparison, generalization, and explanation. The peculiarities of the Bay Area are chiefly exaggerations (or advanced forms) of fundamental forces at work in North American cities: extensive sprawl, employment clustering, expansion by absorption, low-density housing, large-scale property developers, financial booms and bubbles, environmental politics, race and class segregation, and more. Moreover, the rapid growth of population and prosperity belie some troubling developments in the

Bay Area. Contradictions abound: great prosperity yet an epicentre of foreclosures, urban limit lines but endless sprawl, a dynamic central city eclipsed by one of its satellites. It is a region known for its tech economy, innovation, and political liberalism, but it is also a leader in things like the production of social inequality, generation of housing bubbles, and exurban sprawl.

Elusive Metropolis

One strange thing about the Bay Area is that no one agrees on what it is, even the US Census Bureau: the 5-county San Francisco-Oakland metropolitan statistical area (MSA), 9-county San Francisco-Oakland-San Jose consolidated metropolitan area of the 2000 census, or enlarged 12-county combined statistical area (CSA) of 2015. Wikipedia cites a 13-county region that includes Stockton in the Central Valley, and San Francisco Bay Area Planning and Urban Research (SPUR) speaks of a megaregion of 17 or more counties (map 6.1).

Locals have little notion of a unified Bay Area. San Franciscans barely see realms beyond the borders of their small city (also a county). Silicon Valley enjoys worldwide fame as an entity apart. Oakland and the inner East Bay enjoy fierce loyalty from their inhabitants, who bristle against the arrogance of the West Bay, and the outer East Bay feels worlds apart the other way. The North Bay thinks it is still a rural paradise. These local differences are inscribed in a political landscape ruled over by a dozen or more counties, around 150 municipalities and nearly 500 special districts.

Defining metropolitan regions is not simple. The Census Bureau defines MSAs around central cities, yet metropolitan areas have been multinodal for more than a century (Muller, 2001). CSAs are based on unifying threads of commuting and commerce, but urban regions grow up around distinct centres or satellites that never fully merge. Urbanized areas are defined by density of buildings and people, but there is no clear line between the exurban fringe and rural areas (Berube, Singer, Wilson, & Frey, 2006).

Nonetheless, it is easy to fault the Census Bureau over its treatment of the Bay Area. It joins San Francisco and Oakland in one MSA, but splits off Silicon Valley even though San Jose has been no more independent than Oakland. The bureau inscribes three mini-MSAs in the North Bay, which is just as much a hinterland of San Francisco as Los Angeles County is to L.A. city (conjoined as one MSA). In demarcating the CSAs, the bureau brings together the north, south and central Bay Area as one, then throws in the long-separate MSA of Stockton.

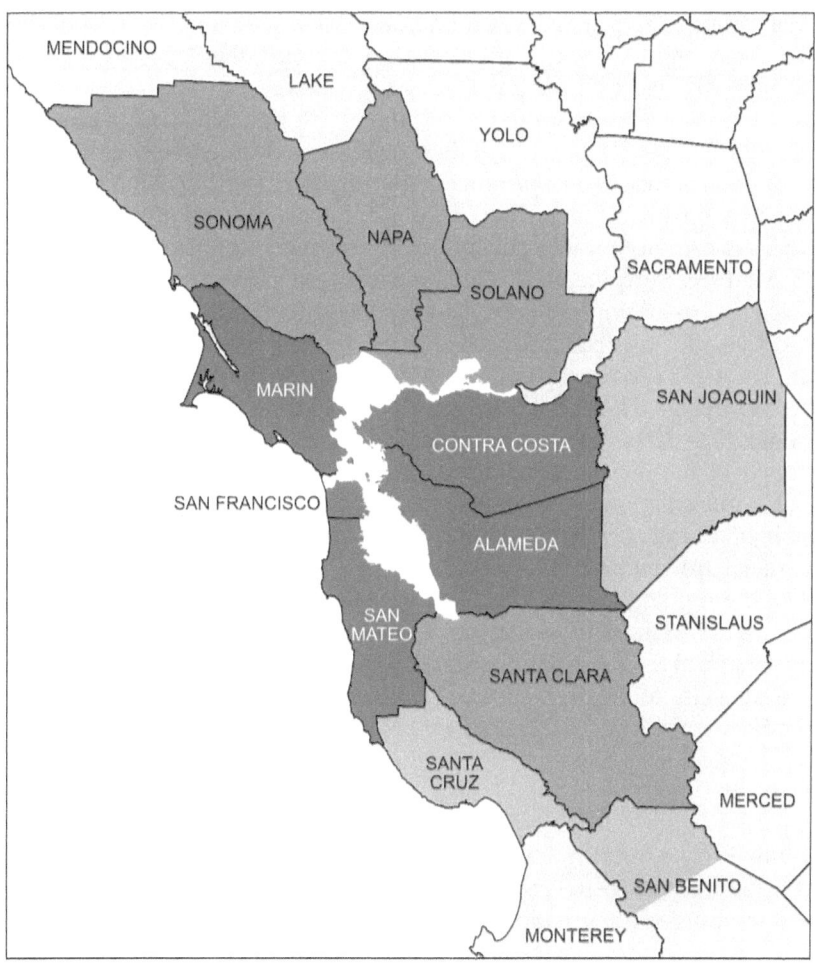

Map 6.1 The many Bay Areas. Source: Schafran (2012b).

Cerberus City

One thing that makes the Bay Area strangely disconnected is that it is the only North American metropolis with three big cities: San Francisco, Oakland, and San Jose. Some have two, such as Minneapolis-St. Paul, and some small metros are quad cities, such as Davenport-Bettendorf-Rock Island-Moline. The Bay Area is a three-headed monster, a Cerberus city. While New York City is unchallenged within its CSA with around 33% of the populace, and Los Angeles dominates its city region with more than 20%, San Francisco and San Jose are at roughly 10% of the Bay Area, and Oakland is at half that. As a result, the Bay Area is traditionally divided in three parts, roughly balanced in population – West Bay, East Bay, and South Bay – and the North Bay is not much smaller. No wonder people cannot get their heads around the whole region.

The Cerberus city is a product of growth by clustering. All metropolitan areas grow in this fashion, through the multiplication of industrial districts or employment clusters, even in diversified urban economies (Scott, 1988; Storper, 2013). The three principal cities are based on massive clusters of employment: San Francisco for financial and business services, plus tourism and conventions; the East Bay for metalworking, oil, and food products; and Silicon Valley for electronics (Walker, 2004). To complicate things the 680 Corridor (Outer East Bay) is replete with huge office parks for corporate managers, and the North Bay is a major pole of economic activity around agriculture and wine.

The suburbanization of industry has always been crucial to the expansion of North American metropolitan areas (Lewis, 2004). Moreover, regional growth has been a process of accretion by means of "edge cities," which got going long before the term was popularized (Garreau, 1991). Garreau used the 680 Corridor as one of his examples, but what he missed was that Oakland was San Francisco's edge city of 1900 and Silicon Valley the edge city of the postwar era. Two more have sprouted up since 1975: the 680 Corridor and the Santa Rosa/Napa complex in the North Bay.

Flip City

While known worldwide as the San Francisco Bay Area, the old core city has been exceeded by 20% by San Jose. San Francisco has priority in the popular imagination because it dominated the region, the state, and the entire Pacific Coast for almost a century. An early start means it has greater density, walkable dimensions, historic sites, and cultural institutions that attract millions of tourists. It is located more centrally to the region than San Jose, but less so than Oakland. Nonetheless, San Francisco no longer dominates the economic life of the region. The old city lost most of its manufacturing base, port, banks, and corporations over the second half of the 20th century (Walker, 2006). Silicon Valley has long been the global heart of electronics, information, and biotech and the Bay Area's principle node of employment, corporate headquarters, and value added (Storper, 2013, Walker & Lodha, 2013).

The reversal of fortunes goes farther. For more than a century, people have commuted from the East Bay or the Peninsula into San Francisco's downtown, even as Oakland grew its own metro region and the South Bay became Silicon Valley. Today, commuting numbers are roughly equal between the South and Central Bay Areas, epitomized by the scores of Google, Facebook, and Apple buses leaving each morning from San Francisco. The city's main engine of growth for the last 20 years has been a fast-rising tech cluster of social media, the sharing economy, and business aids provided by companies such as Twitter, Uber, and Yelp.

In short, the electronics age has inverted the metropolitan area, making San Francisco and the East Bay subordinate to Silicon Valley. No other metropolis across North America has flipped urban centrality on its head. The Census Bureau has renamed the Bay Area as the San Jose-San Francisco-Oakland CSA, but San Jose is not the centre of Silicon Valley, let alone the whole region. The electronics industry grew up in the northwestern corner of Santa Clara County, and San Jose is mostly a residential suburb of the cluster. This illustrates a key fact of metropolitan geography: urban centrality is not simply a matter of maximum access across the city region but is tied to employment clustering and industrial growth or shrinkage.

Suburban City

The Bay Area is one of the most suburbanized cities in North America, notwithstanding San Francisco and the East Bay's more compact form and relatively high density. The vast majority of the built-up area is low-rise, rarely more than four stories, even in San Francisco and Oakland. Sprawl has been the name of the game. As regards the two main factors in suburbanization – cheap transport and single-family homes – the Bay Area was an innovator in both decentralizing technologies. In addition, the region's singular affluence has enabled its addiction to suburbia.

Nineteenth-century transport development aided the physical expansion of the region, as well as the centrality of San Francisco. Bay Area suburbanites commuted by ferry and rail from such idyllic enclaves as Alameda and San Rafael. Much of the region's long-term framework was established along the rail lines and bayside ports of the time, a chain of depots, industrial satellites, and farm towns connected by transport networks owned by San Francisco capitalists (Vance, 1964). Onto this foundation were mapped some of the continent's most extensive trolley systems. San Francisco gained a streetcar network that open up its western suburbs and strengthened downtown. The East Bay's Key System was second in length only to Los Angeles' network. The result was a sprawling but unified region even before cars and highways came along. The latter developed quickly, aided by high rates of car ownership, an expansive state road system, and some of the earliest freeways in the world. In the postwar era, interstate freeways and Bay Area Rapid Transit provided a regional framework for expansion (Scott, 1985).

The bay region is a long-time mecca of the single-family home. The Victorian city of San Francisco not only featured miles of single-family row houses, but also large sections of workers' cottages. The Western Addition and Mission District are today part of the central city but were suburban homes in the 19th century, as were the row houses of the western and southern flanks of San Francisco that filled in from 1910 to 1950. Oakland was laid out as a city of freestanding homes on spacious lots on wide streets; then came streetcar suburbs fanning out along radial trolley lines. The rest of the region, whether built before or after the Second World War, is awash in detached suburban dwellings – more than 60% of the total housing stock (Walker, 1995). Only San Francisco has a housing stock that is majority multiple units (Schafran, 2012b).

The Bay Area has had another prime condition for suburbanization: high incomes (Storper, 2013; Walker & Lodha, 2013). This has created a large class of people who could afford single-family homes and cars. While Boston's working class lived in three-deckers (Warner, 1962), Oakland's lived in streetcar cottages that they owned (Walker, 1995). Virtually every corner of the Bay Area had elevated homeownership rates until recently, when California began to fall behind the United States because of inflated land values (figure 6.1). Extremely affluent homeowners, of which the Bay Area has plenty, have a further effect on sprawl, because they tend to gather in spacious jurisdictions with minimum lot sizes, like Los Altos Hills and Atherton.

Developer's Suburbia

While suburbanization is often treated as a phenomenon of cultural preferences in housing and automobility, capital has its own logic for building cities and suburbs (Walker, 1981). This is marked in the Bay Area, which has been capital rich since the Gold Rush and has had a penchant for investing in land, especially at the urban fringe (Brechin, 1999). Supply-driven suburbanization made an early shift here from a "weave of small patterns" (Warner, 1962) to planned developments to "community builders" (Weiss, 1987).

In the usual story of North American suburbanization, large unified housing schemes began at the turn of the last century, as with Kansas City's Country Club District, and big builder-developers like Levitt became dominant after the Second World War. In the Bay Area, big capital in the property sector was a step ahead. San Francisco already had mass homebuilding in the 19th century, when the Realty Associates were building thousands of Victorian row houses (Moudon, 1986). Large planned residential developments emerged in the early 1890s, with San Mateo's Hillsborough and Mill Valley in Marin County (Brechin, 1999).

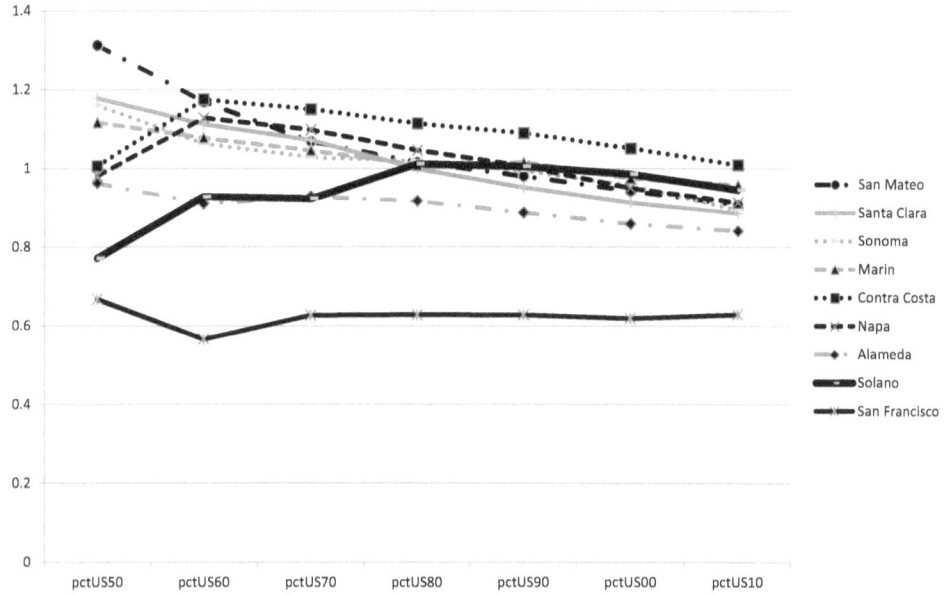

Figure 6.1 Bay Area homeownership rates, 1950–2010, by county, as a percentage of the national average. Source: US Census 2012, Minnesota Population Center (2011).

After 1900, such developments were prominent, built by companies such as Baldwin & Howell in San Francisco and Mason-McDuffie in Berkeley (Brechin, 1989; Loeb, 2002). The gigantic Realty Syndicate in Oakland, fuelled by "Borax" Smith's enormous mining fortune, was the nation's largest single developer at the time (Hildebrand, 1981).

By the 1940s, California had the biggest array of new community developers in the country, such as Doelger, Bohannon, and Eichler in the Bay Area (Brechin, 1990; Walker, 1995). Only Florida could compare with California in the realm of giant housing developers, and several of those moved into California by the 1960s, such as T. Jack Foster and Lennar Corporation. California generated some of the world's biggest realty companies, as well, such as Coldwell Banker (today's CBRE) and Cornish & Carey. Today, the biggest suburban developers in the Bay Area are Shea, Shappell, and Seeno/Discovery, all local firms that grew fat over several housing booms (Schafran, 2012b).

Capitalist developers do more than give consumers what they want. First, they stretch suburbia as far as possible, leaping outward in search of better profits (land rents) from purchasing land cheaply and converting it to urban uses. Ancillary to that, they leverage governments and private providers to approve land use changes and push infrastructure outward as far as possible (Clawson,1971; Walker, 1981). Sprawl pays, and it has done so since San Francisco built its first Victorian suburbs in the 1870s.

Second, property capitalists carve out gigantic swaths of land to develop. Large developers have kept down costs through simplicity of construction and repetition of forms, while increasing revenue through higher density and better design (Loeb, 2002). While this has led to many innovations in floor plans, fixtures, and construction methods, a numbing redundancy is evident in suburbs from Fairfield to Livermore (Owens, 1973). Giantism in suburban development can be found all around the Bay Area, from Doelger's postwar boxes in San Francisco's Sunset District to the endless mini-mansions of the outer East Bay's Dougherty Valley.

Third, developers popularize styles of housing. A Bay Area style has been distinctive, even if mass-produced. Bay Victorian was an exuberant twist on the prevailing architecture of the late 19th century (Moudon, 1986). Craftsman and shingle-style flourished in the early 20th century, making a different kind of show of woody abundance. Even through the Modernist era of the mid-20th century, local builders added redwood siding, shingles, and decks (Woodbridge, 1976). Between the wars, Mediterranean styles flourished, thanks to fear of fire, covering wood framing with stucco to get a bright, sunny look (Gebhard & Von Breton, 1968).

The ability to mass-produce detached suburban homes was made into an architectural virtue in California, and local styles spread around the country. The California bungalow became an icon of American suburbia in the 1910s and '20s, the ranch house and split-level in the 1930s to '50s, and the redwood look in the 1960s and '70s (King, 1984; Woodbridge, 1976). Since the 1980s, the California neo-Tuscan style has been all the rage, with tile roofs, stucco siding, minimal eaves, and bloated footprints (Harris & Dostrovsky, 2008).

Ecoburbia

The Bay Area is the greenest metropolis in the United States – or, rather, a patchwork of green, brown, gold, and blue. San Francisco Bay is the centrepiece of the regional greensward and larger than the built-up area; the amount of parkland (reserved space) is almost double that and is larger than Yosemite National Park; and that is again exceeded by the remaining open space in agriculture and forests – much of it

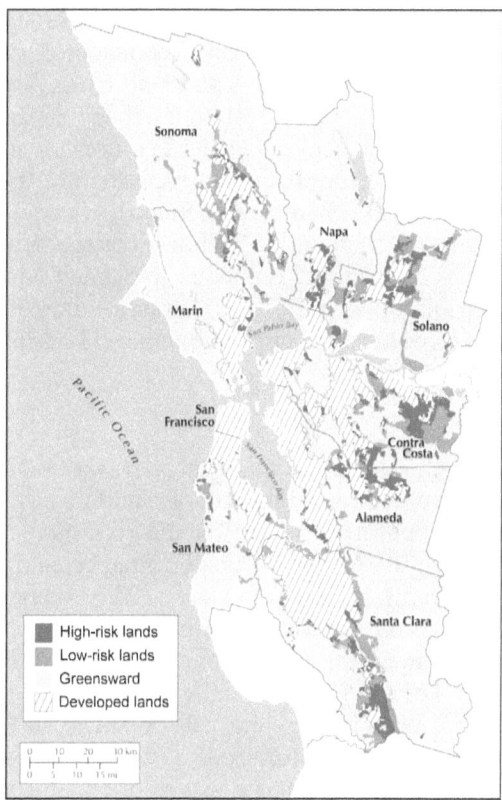

Map 6.2 Protected areas on the San Francisco Peninsula. Source: Created by Molly Roy for Richard Walker, *Pictures of a Gone City*, published by PM Press and reproduced with permission.

under development covenants (map 6.2). To the west is the Pacific Ocean with wide-open vistas, marine reserves, and a protected coastline (Walker, 2007).

The Bay Area's greensward partly depends on contours of the natural landscape (Christensen, Denning, & Mcdonald, 2010), but the vast amount of protected open space is a product of a century of political battles. US cities have a robust history of parks and conservation, but what is unique to the Bay Area is the success of mass environmentalism, from John Muir and the Sierra Club to Save the Bay and the Greenbelt Alliance. This movement has combined a passion for local landscapes with a regional planning vision of how to channel urban development into less destructive paths (Walker, 2007).

As a consequence, the Bay Area has become more compact behind urban-limit lines, better able to plan around endangered areas, and better protected by an array of park agencies and land trusts. The highest land values in the country have done the rest. Today, the urbanized region is the third densest in the nation, behind Greater Los Angeles and New York.[1] Yet neither environmental controls nor densification of the urban heartland has meant an end to urban sprawl. Suburban nodes may be more compact than before and surrounded by more open space, but they extend over a vast territory. This sprawl is often been blamed on supply restrictions for the greenbelt (Frieden, 1979) but it is much more the product of a booming economy, in-rushing finance, and frenzied housing bubbles.

Boomburbs

Metropolitan growth and suburban expansion takes place in waves as housing cycles propel new building outward and upward (Whitehand, 1987). Those cycles are, in turn, fed by the rise and fall of credit in the financial system; urbanization runs on finance, and urban booms and busts follow credit cycles quite closely. This financial dimension of growth has not been sufficiently appreciated in the literature on American suburbs, though the great boom and bust of the 2000s woke up a lot of people (Lang & LeFurgy, 2006; Shiller, 2008).

The Bay Area has been in the vanguard of American (sub)urbanization because it has been at the forefront of the credit machine of capitalism, which has spurred immense waves of expansion (figure 6.2). San Francisco has been a centre of capital accumulation since the Gold Rush and one of the top banking centres in the United States. Much of that capital has gone into financing urban development, through infrastructure expansion, land acquisition, building and home buying (mortgages) (Walker, 1995). In the postwar era, when California's suburbs became the talk of the nation, the state's banks, such as Bank of America, Crocker, and Security Pacific, and savings and loan institutions, such as Amundsen and Golden West Savings, were among the largest and most prolific lenders in the nation (Abrahamson, 2013).

By the 1970s, California home prices began to outpace the country dramatically, and a big reason for this, besides the state's general growth and prosperity, was easy credit. In the 1980s, the housing market became red hot, led by Southern California-based Drexel Burnham, American Savings, and Lincoln Savings (Pizzo, Fricker, & Muolo, 1989; Stein, 1992). When the bubble burst in 1989, the federal government had to rescue the financial system by means of the Restitution Finance Corporation (RFC), which injected $300 billion into bailouts and sold

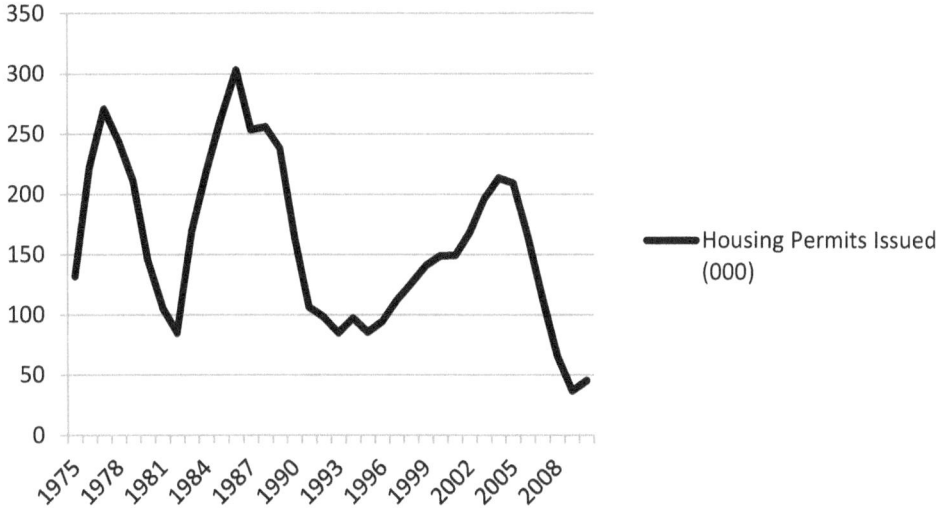

Figure 6.2 California house construction, 1975–2010. Source: Construction Industry Research Board & California Department of Finance.

off failing savings and loans to emerging interstate banking empires, such as Nationwide Bank in the South and Bank of America in the West. The RFC's focus was on California, Texas, and Florida (Mayer, 1990). In the mid-1990s, the market revived smartly, California housing prices doubled, and commercial real estate went through the roof. San Francisco and Silicon Valley office space briefly passed New York in dollar value per square foot, so this time it was the Bay Area's turn to blow up the most spectacularly, after which the region fell into a deep recession from 2000 to 2003 (Walker, 2006).

In the 2000s, the greatest of all housing bubbles began its run-up, with California again in the vanguard. At the peak in 2006, median home prices in the state were double the national average and the Bay Area's were double that. California is well off, and the Bay Area is rich, and much of its fabulous new wealth was plowed back into real estate; nevertheless, it is not that disproportionately wealthy. Inflated land prices also derive from the enormous flow of finance in California, which was the single biggest playground for the funny-money being generated on Wall Street (Bardhan & Walker, 2011). The result of this gargantuan credit boom was an exceptionally well-oiled process of suburban building and outward expansion throughout the state, with tens of thousands of new single-family homes added (map 6.3).

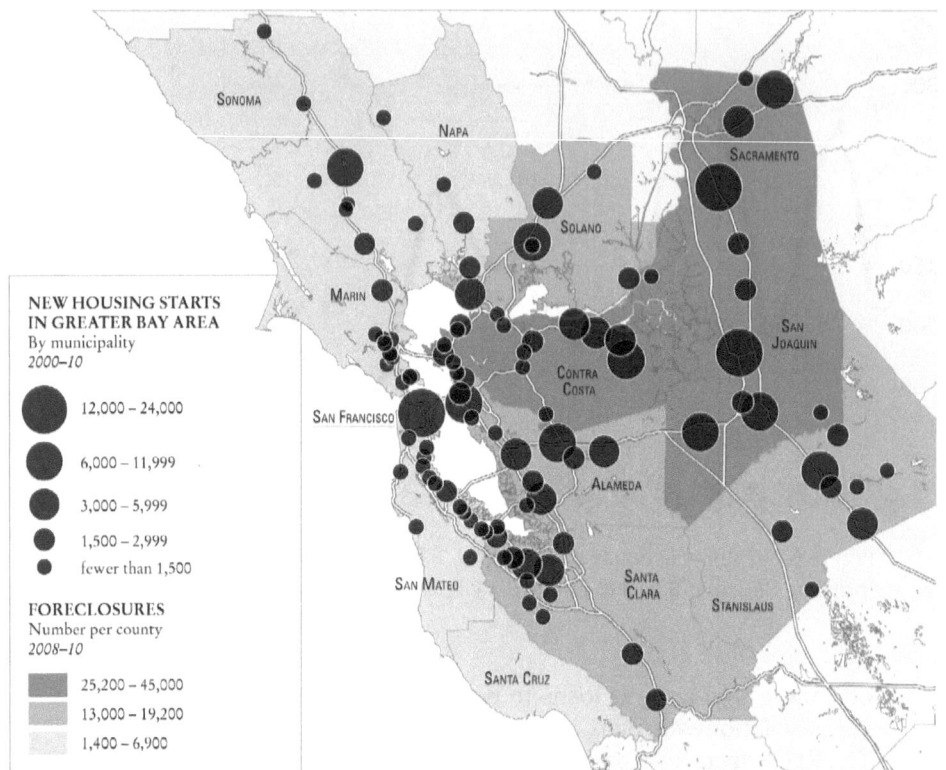

Map 6.3 New housing starts in the Bay Area, by city, 2000–2010. Reproduced with permission from *The Atlas of California*, by Richard A. Walker and Suresh K. Lodha, © University of California Press, 2013. www.myriadeditions.com. Map courtesy of Darin Jensen and Mike Jones, University of California, Berkeley.

Inland Empire

The housing boom of the 1990s and 2000s blew the bay metropolis up like a balloon, bursting across the landscape of Northern California. Everyone knows about the postwar "exploding metropolis" (Editors of Fortune, 1958), but present-day urban regions like the Bay Area exceed the cities of that era by a degree of magnitude (Morrill, 2012). Today, the Bay Area has its own inland empire, just like Greater Los Angeles (map 6.3). A new generation of suburbia has spilled into the topographic zone of the Central Valley, traditionally a completely different region. The Bay Area commuting field now reaches east into San Joaquin,

Stanislaus, and Merced counties, northeast into Yolo and Sacramento counties, south to Santa Cruz and San Benito counties, and north to Lake and Mendocino counties (see also map 6.1). This has created eye-popping commutes of up to two hours back to the old Bay Area.

The inland empire of Northern California is little understood and makes defining the Bay Area even harder. The bay region is fusing with two large MSAs in the Central Valley, Sacramento (~2 million) and Stockton (~1 million). The result is an urban maze on a gargantuan scale, leading to speculation that there is a new "megopolis" of 15 to 20 counties, more than 10 million people, and more than 10,000 square miles of territory – almost the size of North America's largest city region, Greater Los Angeles (Morrill, 2012; Metcalf & Terplan, 2007). Such a Northern California megapolis is no more far-fetched than the Census Bureau's union of Washington and Baltimore into a single CSA (Knox, in this volume). This is more than a definitional dispute. Recognizing the new inland empire affects how we analyse crucial social issues in the metropolis, such as the impact of housing foreclosures, the suburbanization of poverty, and the new geographies of race and class, to which we now turn.

Busted Burbs

The most important thing about the exploding metropolis of our time is not simply that it covers so much territory, but that it conceals a dramatically altered social geography, which, in turn, reveals some essential facts about the social order of the region, state, and country.

The Bay Area has been a place of great opportunity, given its repeated waves of innovation and growth. It has the second-highest percentage of degree-carrying workers of any US large metro area and is the pin-up city for partisans of the creative class (Florida, 2002). It is North America's richest major metro area, with the most millionaires and billionaires per capita (Walker & Lodha, 2013). Yet, the bay metropolis is a trendsetter for the vast gulf of inequality opening up in the advanced economies (California Budget Project, 2011). Even though average incomes are the highest in the country (United States Census Bureau, 2012), one third of the region's workers earn minimum wage, wages for production and service workers have been flat for years, and tech companies have led the way in contingent employment and offshoring of jobs (Bardhan, 2008; Benner, 2002).

This vast wealth divide has a decided effect on the landscape of city and suburbs. Not only is the Bay Area the least affordable metropolitan area in the United States but as high-income earners have crowded into San Francisco it has overtaken Manhattan as the highest rent city centre in the country (Brahinsky, 2012). The West Bay is replete with elite suburbs, Marin is the richest county in the United States, and the Napa Valley has become a playground for the nation's rich. The wealthy districts are slightly less prominent in the East Bay, but still form a wedge from Piedmont to Danville. The gentrification of San Francisco, Oakland, and Berkeley is stunning and has evoked repeated protests against evictions, tony high rises and Google buses – and the fiercest Occupy movement outside of New York (Mahler, 2012; cf. Harvey, 2012).

Equally striking is the new social geography of the working class. Astronomical housing prices in the central Bay Area have driven working people – especially couples with growing families – to far-out suburbs, where they hoped to afford their middle class dream house. But the Great Recession laid waste to fringes of the Bay Area and the dreams of tens of thousands of families. California was the chief locus of the housing bust, with one million foreclosed homes, six million underwater mortgages, and a $3 trillion loss in home values; and the inland metros were tops nationally in foreclosure rates (30% to 50%,) and house-price declines (40% to 50%) (Bardhan & Walker, 2011; Schafran, 2012a). The recession left the Golden State with more than two million unemployed (12.4% in 2010), 10 million in poverty (25% in 2012), and the largest state budget deficit. The most afflicted counties were in the inland areas, with unemployment rates of 20% to 30%, widespread hunger, and the spectacular bankruptcies of Stockton and San Bernardino (Schafran & Wegmann, 2012). Yet the suffering of the inland empires is generally seen as something apart from the prosperity of the coastal cities.

Babylon Burbs

The pattern of racial settlement, segregation, and inequality has been inverted in the Bay Area, as in many other American metros (Niedt, 2013). The same forces that have made the region so dynamic have flipped its racial geography upside down in the recent years. But for all the success of the Bay Area and so many of its favoured firms and residents, it is no closer to racial justice than before. We call this "the road from Babylon to Brentwood" (Schafran, 2012b).

In the postwar era, when a "white noose suburbs" surrounded Oakland and San Francisco, African Americans were trapped in the ghettoes of what they called an "American Babylon" (Self, 2003). The Bay Area's liberal reputation belies the degree to which Blacks lived in segregated neighbourhoods. Yet, since the 1980s the region's core African American neighbourhoods have witnessed an exodus, with San Francisco showing the fastest decline of Black population of any major US city (Brahinsky, 2012). While better-off African Americans have moved to integrated neighbourhoods such as the Oakland Hills, the primary destination for the working class majority has been far-flung suburbs and satellites, such as Antioch, Brentwood, and Modesto (Schafran, 2009).

They were joined by hundreds of thousands of Latinos and Asians, who make up the Bay Area's working class. People of colour constitute the vast majority of new residents in the outer rings of the metropolis, and many towns such as Concord that now have non-White majorities were known in the past for their explicit racism. But the Great Recession revealed the fragility of the new open-housing suburbs, and high foreclosure zones were the most likely to have witnessed an influx of Latinos and African Americans during the previous two decades (Schafran & Wegmann, 2012).

The Bay Area is notable for its large Asian origin population, many of whom have done well as business owners, tech workers, and professionals (Saxenian, 2006). One of the striking characteristics of the region is suburban clusters of Chinese and Indians in Silicon Valley suburbs such as Cupertino, Saratoga, and Fremont, or East Bay communities such as San Ramon and Hercules. But the majority of Asian Americans are working class, such as the Filipinos of Daly City and Vietnamese of East San Jose. The latter were heavily represented in the high foreclosure cities like Stockton, Vallejo, and Fairfield.

The restructuring of the racial geography of the metropolis is, therefore, deeply entwined with the foreclosure crisis (Schafran, 2012a), which brought "the greatest loss of wealth for people of colour in recent U.S. history" (Rivera, 2008; Wyly, Moos, Hammel, & Kabahizi, 2009). The hopes of the suburban good life dashed on the shoals of the Great Recession were dreams of colour-blind opportunity. Whereas millions in the White working class saw themselves levitated into the great American middle class through suburbanization during the postwar era (Beauregard, 2006), communities of colour got there too late. In age of neoliberalism, the suburban American dream was based on predatory debt rather than federally backed mortgages, and proved to be out of reach for the new working class of colour.

Conclusion

The portrait of the Bay Area offered here has been organized around the distinctive geography of the metropolis, whether as a three-headed beast, a region inverted by the tech boom, or a greenbelt jostling with unchecked sprawl. This is an exercise that all big city regions need to undertake: an accounting of how they exhibit unique features based on their history and political economy. Such observations are not intended to validate parochial interests but instead to point to the difficulty of generalizing from a few "classic" cities like Chicago or New York. Moreover, they sometimes enhance our ability to grasp the causal forces at work in metropolitan suburbanization.

This has, therefore, been simultaneously a study in the determinants of urban growth and suburban sprawl in the large cities of North America. Such forces consist of more than the evolution of transportation, consumer choice in housing, or the smooth operation of the rent gradient. Metropolitan expansion rests on an economy, but one built on distinct sectors that unfolds through the build-up of employment clusters. Suburbanization rides the rails and highways but depends on consumer income and abundant mortgage credit, which surges periodically and deeply to propel waves (even tsunamis) of expansion. Developers drive the process of growth from the supply side, in search of land rents and scale economies.

Finally, we hope to expose some of the contradictions evident in the Bay Area's metropolitan geography: between a region growing denser at its core and continuing to expand into new suburban frontiers; between a region generating great riches, yet reshuffling the geography of inequality rather than reducing it; and between a region with abundant capital that nevertheless blew up on a surfeit of fictitious credit. These parallel deeply disturbing elements of North American society: growing class inequality, racial injustice, a bloated financial system, and a failure of environmental controls. The modern metropolis is not just a fascinating geographical study but an object lesson in what has gone wrong across the continent, making the Bay Area less strange than at first it seems. In fact, it is all too typical.[2]

NOTES

1 See http://www.census.gov/geo/reference/ua/uafacts.html.
2 For much fuller studies of the character and geography of the Bay Area, see Alex Schafran (2018), *The Road to Resegregation: Northern California and the Failure of Politics* (Berkeley: University of California Press); Richard Walker (2018), *Pictures of a Gone City: Tech and the Dark Side of Prosperity in the San Francisco Bay Area* (Oakland: PM Press).

REFERENCES

Abrahamson, E. (2013). *Building home: Howard F. Ahmanson and the politics of the American dream*. Berkeley: University of California Press.
Bardhan, A. (2008). Globalization, job creation and inequality: Challenges and opportunities on both sides of the offshoring divide. In K. Hall (Ed.), *Studies in inequality and social justice: Essays in honor of Ved Prakash Vatuk* (pp. 111–30). Meerut, India: Archana Publications.
Bardhan, A., & Walker, R. (2011). California shrugged: The fountainhead of the Great Recession. *Cambridge Journal of Regions, Economy and Society, 4*(3), 303–22.
Beauregard, R. (2006). *When American became suburban*. Minneapolis: University of Minnesota Press.
Benner, C. (2002). *Work in the new economy: Flexible labor markets in Silicon Valley*. Oxford: Blackwell.
Berube, A., Singer, A., Wilson, J., & Frey, W. (2006). *Finding exurbia: America's fast-growing communities at the metropolitan fringe*. Washington, DC: Brookings Institution.
Brahinsky, R. (2012). *The making and unmaking of southeast San Francisco* (Doctoral dissertation). Department of Geography, University of California, Berkeley.
Brechin, G. (1989, September). St. Francis Wood: A misty haven for San Francisco haves. *San Francisco Focus*, 20–5.
Brechin, G. (1990, June). Mr. Levitt of the sunset. *San Francisco Focus*, 23–6.
Brechin, G. (1999). *Imperial San Francisco: Urban power, earthly ruin*. Berkeley: University of California Press.
California Budget Project. (2011). A generation of widening inequality. Retrieved from https://calbudgetcenter.org/wp-content/uploads/111101 _A_Generation_of_Widening_Inequality.pdf
Christensen, J., Denning, C., & Mcdonald, R. (2010). Did land protection in Silicon Valley reduce the housing stock? *Biological Conservation, 143*, 1087–93.
Clawson, M. (1971). *Suburban land conversion in the United States: An economic and governmental process*. Baltimore, MD: Johns Hopkins University Press for Resources for the Future.
Editors of Fortune. (1958). *The exploding metropolis*. New York, NY: Doubleday.
Florida, R. (2002). *The rise of the creative class: And how it's transforming work, leisure, community and everyday life*. New York, NY: Basic Books.
Frieden, B. (1979). *The environmental protection hustle*. Cambridge, MA: MIT Press.
Garreau, J. (1991). *Edge city: Life on the new frontier*. New York, NY: Doubleday.
Gebhard, D., & Von Breton, H. (1968). *Architecture in California, 1868–1968*. University of California, Santa Barbara Art Museum.

Harris, R., & Dostrovsky, N. 2008. Style for the zeitgeist: The stealthy revival of historicist housing since the late 1960s. *The Professional Geographer, 60*(3), 314–32.
Harvey, D. (2012). The urban roots of financial crises: Reclaiming the city for anti-capitalist struggle. *Socialist Register, 48*, 1–34.
Hildebrand, G. (1981). *Borax pioneer: Francis Marion Smith*. San Diego, CA: Howell-North Books.
King, A. (1984). *The bungalow: The production of a global culture*. London: Routledge and Kegan Paul.
Lang, R., & LeFurgy, J. (2006). *Boomburbs: The rise of America's accidental cities*. Washington, DC: Brookings Institution Press.
Lewis, R. (Ed.). (2004). *Manufacturing suburbs: Building, work and home on the metropolitan fringe*. Philadelphia, PA: Temple University Press.
Loeb, C. (2002). *Entrepreneurial vernacular: Developers' subdivision in the 1920s*. Baltimore, MD: Johns Hopkins University Press.
Mahler, J. (2012, August 1). Oakland: The last refuge of radical America. *New York Times Magazine*, https://www.nytimes.com/2012/08/05/magazine/oakland-occupy-movement.html
Mayer, M. (1990). *The greatest-ever bank robbery: The collapse of the savings and loan industry*. New York, NY: Charles Scribner's.
Metcalf, G., & Terplan, E. (2007, November/December). The northern California megaregion. *The Urbanist*. Retrieved from http://www.spur.org/sites/default/files/publications_pdfs/SPUR_The_Northern_California_Megaregion.pdf
Minnesota Population Center. (2011). National historical geographic information system: Version 2.0. Minneapolis: University of Minnesota.
Morrill, R. (2012). Megalopolis and its rivals. *Newgeography.com*. Retrieved 21 March 2012 from http://www.newgeography.com/content/002788-megalopolis-and-its-rivals
Moudon, A. (1986). *Built for change: Neighborhood architecture in San Francisco*. Cambridge, MA: MIT Press.
Muller, E. (2001). Industrial suburbs and the growth of metropolitan Pittsburgh, 1870–1920. *Journal of Historical Geography, 27*(1), 58–73.
Niedt, C. (Ed.). (2013). *Social justice in diverse suburbs*. Philadelphia, PA: Temple University Press.
Owens, B. (1973). *Suburbia*. San Francisco, CA: Straight Arrow Books.
Pizzo, S., Fricker, M., & Muolo, P. (1989). *Inside job: The looting of America's savings and loans*. San Francisco, CA: McGraw-Hill.
Rivera, A. (2008). *Foreclosed: State of the dream 2008*. Boston, MA: United for a Fair Economy.
Saxenian, A. (2006). *The new argonauts: Regional advantage in a global economy*. Cambridge, MA: Harvard University Press.
Schafran, A. (2009, Summer). Outside endopolis: Notes from Contra Costa County. *Critical Planning*, 11–33.

Schafran, A. (2012a). Origins of an urban crisis: The restructuring of the San Francisco Bay area and the geography of foreclosure. *International Journal of Urban and Regional Research, 37*(2), 663–88.

Schafran, A. (2012b). *The long road from Babylon to Brentwood* (Doctoral dissertation). Department of City and Regional Planning, University of California, Berkeley.

Schafran, A., & Wegmann, J. (2012). Restructuring, race and real estate: Changing home values and the California metropolis, 1989–2010. *Urban Geography, 33*(5), 630–54.

Scott, A. (1988). *Metropolis: From the division of labor to urban form*. Berkeley and Los Angeles: University of California Press.

Scott, M. (1985 [1959]). *The San Francisco Bay Area: A metropolis in perspective*. Berkeley: University of California Press.

Self, R. (2003). *American Babylon: Race and the struggle for postwar Oakland*. Princeton, NJ: Princeton University Press.

Shiller, R. (2008). *The subprime solution: How today's global financial crisis happened, and what to do about it*. New York, NY: Penguin.

Stein, B. (1992). *A license to steal: The untold story of Michael Milken and the conspiracy to bilk the nation*. New York, NY: Simon & Schuster.

Storper, M. (2013). *The keys to the city*. Princeton, NJ: Princeton University Press.

United States Census Bureau. (2012). Statistical abstract/National data book. Table 683. Personal Income by Selected Large Metropolitan Area: 2005 to 2009. Retrieved from http://www.census.gov

Vance, J. (1964). *Geography and urban evolution in the San Francisco Bay Area*. Berkeley: University of California Institute of Governmental Studies.

Walker, R. (1981). A theory of suburbanization: capitalism and the construction of urban space in the United States. In M. Dear & A. Scott (Eds.), *Urbanization and urban planning in capitalist societies* (pp. 383–430). New York, NY: Methuen.

Walker, R. (1995). Landscape and city life: Four ecologies of residence in the San Francisco Bay Area. *Ecumene, 2*(1), 33–64.

Walker, R. (2004). Industry builds out the city: Industrial decentralization in the San Francisco Bay Area, 1850–1950. In R. Lewis (Ed.), *Manufacturing suburbs: Building work and home on the metropolitan fringe* (pp. 92–123). Philadelphia, PA: Temple University Press.

Walker, R. (2006). The boom and the bombshell: The new economy bubble and the San Francisco Bay Area. In G. Vertova (Ed.), *The changing economic geography of globalization* (pp. 121–47). London: Routledge.

Walker, R. (2007). *The country in the city: The greening of the San Francisco Bay Area*. Seattle: University of Washington Press.

Walker, R., & Lodha, S. (2013). *The atlas of California: Mapping the challenge of a new era*. Berkeley: University of California Press & London: Myriad Editions.

Warner, S. B. (1962). *Streetcar suburbs: The process of growth in Boston, 1870–1900*. Cambridge, MA: Harvard University Press.

Weiss, M. (1987). *The rise of the community builders: The American real estate industry and urban land planning*. New York, NY: Columbia University Press.

Whitehand, J.W.R. (1987). *The changing face of cities: A study of development cycles and urban form*. New York, NY: Oxford University Press.

Woodbridge, S. (Ed.). (1976). *Bay area houses*. New York, NY: Oxford University Press.

Wyly, E., Moos, M., Hammel, D., & Kabahizi, M. (2009). Cartographies of race and class: Mapping the class-monopoly rents of American subprime mortgage capital. *International Journal of Urban and Regional Research, 33*(2), 332–54.

7 Vancouverism as Suburbanism

ELLIOT SIEMIATYCKI, JAMIE PECK,
AND ELVIN WYLY

Introduction: Vancouver Looks Up[1]

Some cities are known for schools of thought (Frankfurt, Chicago, Los Angeles), or even shapes and colours (Vienna), but rare is the city with its own "ism." As one of the truth spots for the "new urbanism" in the 21st century, Vancouver's mantra of "Living First" has found validation across the global-city discourse-complex for its winning combination of density, livability, and sustainability – all rendered seductively real in the forest of glass-walled condominium towers that has colonized the downtown core since the late 1980s (figure 7.1). Vancouverism – the term used to describe Vancouver's urban political-economic model of dense, amenity-enriched residential development in the downtown core – has been artfully crafted by a self-conscious network of "city builders" who have gone on to promote the livability paradigm around the world, in the company of a growing ensemble of "globe-trotting public officials, journalists, and urbanists at large" (Barnes, Hutton, Ley, & Moos, 2011, p. 322). Vancouverism has become deeply influential among decisive circuits of policy elites as well as investors (McCann, 2011). This is a city that likes looking at itself in the mirror, and with but a few notable exceptions, it continues to be impressed with what it sees.

Today, Vancouverism names an aspiration to the vanguard class of cities from which a world of emergent connections can be understood. To be sure, Vancouver "exemplifies the city as a space of flows and recurrent restructuring rather than [some] durable construct of stable industries, labour, social class, and communities" (Barnes et al., 2011, p. 291). In social and cultural terms, Vancouver's transnationalism is a pronounced East/West encounter between a Pacific-North American site and the dynamic diasporas of Asia; but in terms of the spatial imaginaries of planning discourse, the city's built form is often defined in

Figure 7.1 "Vancouverism" from above. Photo courtesy of E. Wyly.

stark binary opposition to the worst excesses of North American (sub) urbanism, as conventionally expressed in Los Angeles. If Los Angeles is remembered as "72 suburbs in search of a city," as Dorothy Parker famously had it, then Vancouver believes that it has found the secret of downtown *living*. While Los Angeles sprawls to its smog-bound horizon, so the preferred narrative goes, Vancouver's green-glassed condo towers reach for the pristine skies, with snow-capped mountains as their picture-perfect backdrop. This, in essence, is the "Vancouver achievement" (Punter, 2003), the triumph of "City Making in Paradise" (Harcourt, Rossiter, & Cameron, 2007). Now living its dream of urban exceptionalism, "Vancouver and [its trademarked planning paradigm of] Living First are turning the traditional idea of a downtown on its head" (Hern, 2010, p. 45), the city having been widely credited with "reversing North America's post-Second World War romance with the suburbs" (Berelowitz, 2005, p. 220).

Yet there are troubles in Canada's Lotus Land, for all the new-urbanist hubris, and many of them revolve around the city's millennial pact with an innovative mode of development to which we apply the tongue-in-cheek neologism "suburbocentrism." This is based on an audacious residential reclamation and recapitalization of the urban core that has engendered a systemic affordability crisis. Vancouver's space of flows may look cosmopolitan and sustainable in the central core of downtown condo towers of this storied "city of glass," but its development model is deeply intertwined with the structural imperatives that have long defined capitalist suburbanization (Barnes et al., 2011; Walker, 1981). Here, crucial elements of the contemporary North American *suburban* "polycentric habitus that has increasingly replaced hierarchical oppositions of center to periphery" (Kolb, 2011, p. 155) have congealed into an iconic built form that is marketed as distinctively *urban*, even as the city's liminal position at the US/Canada border zone and within transnational networks complicates conventional modes of comparative analysis or the application of "imported" urban theories (see Dear, Burridge, Marolt, Peters, & Seymour, 2008, p. 104). Much like Los Angeles, Vancouver cannot be understood from surface appearances but must instead be "unskinned," if its underlying rationality is to be understood.

Like many cities, Vancouver has long displayed an ambivalent attitude towards its suburbs. As the city grew to maturity after the Second World War, "suburbophobic" impulses intensified, leading to an official campaign against sprawl in the 1950s and grassroots mobilizations against freeway development in the 1960s. It is this crucial era – culminating in a pivotal series of political and policy changes in the late 1960s into the mid-1970s – that institutionalized the pursuit of a new kind of urbanity as an overriding metropolitan objective (Ley, 1980). Vancouver can certainly claim some real achievements, at least in comparison to the crass hostility to both "urbanism" and "planning" that might be said to characterize mainstream American *metropolitik*. But Vancouver's moment of municipal experimentation, back in Canada's liberalist heyday, seems to have degraded into a kind of leveraged buyout, negotiated at the nexus of transnationalizing real estate imperatives and the distinctive rhythms of the country's federal and provincial neoliberalization. Constituted through a powerful self-image as the antidote to (sub)urban sprawl, the Vancouver model would not be a US-style antiurban "Sin City" (Lees & Demeritt, 1998), but neither, in truth, has it proved to be the antisuburban utopia that is celebrated in promotional readings of Vancouverism, and subsequently parlayed into "one of Canada's leading exports" (Saunders, 2013). We argue here that Vancouver's fix is grounded in a comprehensive *reabsorption* of suburban logics, values, and aesthetics into the production and functioning of the built environment – even if some first appearances can be deceiving.

Tracing this paradoxical form of recombinant suburbanism necessitates a critical-historical reconstruction of Vancouver's (sub)urbanization – this chapter's somewhat sprawling task. Its two sections trace the roots and consequences of the city's love/hate relationship with the suburbs. This begins with a discussion of the city's early development, from almost unregulated growth to the simultaneous emergence, in the 1960s, of sprawl and "suburbophobia." Next we turn to the genesis of "livability" politics in 1970s Vancouver and the ensuing recentralization of suburban values, or suburbocentrism. Here, we reflect on the intersecting aesthetic, cultural, and political-economic bases of Vancouverism as a stealth form of devolved, polycentric suburbanization and the slowly metabolizing crises of metropolitan governance and affordability that have followed in its wake. In the conclusion, Vancouver's recombinant sub/urbanism is positioned in relation to moving modalities of North American suburbanization.

From Unregulated Growth to Suburbophobia

Vancouver has always considered itself to be a city in direct, unmediated dialogue with nature, a pervasive narrative that has been persistently disrupted by the uncooperative realities of exurban sprawl. Indeed, what might be considered the city's founding myth is that this pioneering place was cut from "virgin forest untroubled by civilisation" (Mattocks, 1932, p. 49; see also Mawson, 1913), when in fact the European settlement of the Lower Mainland during the second half of the 19th century entailed a protracted process of colonial displacement and dispossession, affecting the Musqueam, the Squamish, the Tsawwassen, and the Kwantlen, among other long-established inhabitants of the region (Harris, 1997). The persistent (self-)image of Vancouver as a youthful, almost innocent city, striving for an ever-more harmonious relationship with its bucolic surroundings, was consequently premised on a brazen act of colonial erasure that is repeated with each generation's modernist amnesia (Blomley, 2004).

In its first 50 years as a European settlement, Vancouver went from a sawmill camp of a few hundred people (dominated by a White European male elite and a labouring class of Chinese, Japanese, and Aboriginal workers) to a city of a quarter million by the early 1930s. These early decades brought a thousand-fold expansion, the "momentum of [which quickly] spilled well beyond its legal bounds," into Richmond, Surrey, Coquitlam, and Delta, "relative growth rates in these outer districts far exceed[ing] those in Vancouver proper" right from the start (Wynn, 1992, p. 69). The city dabbled with but never truly committed to the notion of urban planning during these formative years. A city plan commissioned in the late 1920s from Harland Bartholomew, a planning consultant based in St. Louis, Missouri, was never formally adopted by the newly established Town Planning Commission, but many of its underlying principles – such as the overlaying of an arterial transport network on the inherited gridiron street pattern, and the measured distribution of parks, retail districts, and community facilities – was effectively subsumed into the city's "spatial vocabulary" (Berelowitz, 2005, p. 62). Following the conventional pattern of downtown zoning for commerce and industry, complemented by residential subdivisions, the Bartholomew plan drew praise from the planning profession: "It is satisfactory to note that by far the largest part of the residential area of Greater Vancouver is zoned for single family houses" (Mattocks, 1932, p. 54). Overriding these aspirations to suburban order, however, the governing principle in Vancouver's early development was "slapdash recklessness [since] the city's growth was subject neither to control nor to anticipation" (Morley, 1961, p. 220). Practically unregulated development remained the norm, echoing Mattocks's earlier complaint that "Vancouver has *just grown* ... producing the usual crop of difficulties" (1932, p. 49, emphasis added).

A belated response came in 1949, when the Lower Mainland Regional Planning Board of British Columbia was established, with a remit to develop a shared planning framework across the various municipalities and unorganized areas of the Lower Fraser Valley, stretching from Vancouver to Hope. The spectre of urban sprawl dominated the deliberations of the Regional Planning Board during its first decade. A widely circulated pamphlet on the subject, summarizing the "most important and urgent study ever made by this Board" (LMRPB, 1956a, p. 1), presented an impassioned case against the proliferation of "sprawl areas," the effect of which was likened to a termite infestation (map 7.1). The case was made principally on the grounds of fiscal equity: areas beyond the metropolitan pale were found to be in "consistent deficit," offloading tax burdens onto "unsuspecting neighbors" (LMRPB, 1956b, p. 2, 1956a, p. 9). The infestation that was suburban sprawl, it was now determined, must be tackled:

> *There is little to be said for sprawl. It is socially inconvenient and economically stupid. It is born of ignorance and short sight. It promises "country living" and low costs, but destroys the country and hides the costs. It is unfair both to the farmer and the true urbanite ... Nobody wins when sprawl takes over* [...] The citizens and governments of the Lower Mainland appear to be blissfully unaware of the existence of sprawl as a major problem. Or if it is recognized at all, it is accepted without concern as a natural or inevitable thing, or one which will remedy itself in time as the area "fills in." [...] These are dangerous fallacies. Sprawl will never remedy itself. (LMRPB, 1956a, pp. 15–16, original emphasis)

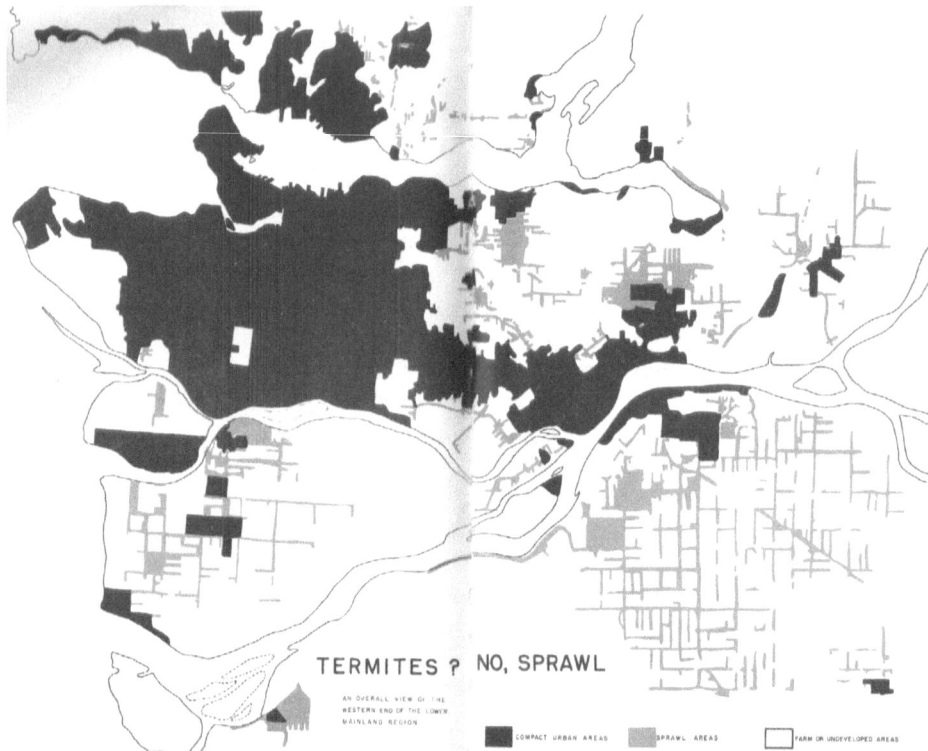

Map 7.1 Sprawl as infestation. A metaphoric/cartographic representation of "suburbophobia" in Vancouver's regional planning during the 1940s and 1950s. Source: Lower Mainland Regional Planning Board.

The board made a strong case for residential zoning (necessitating the appointment of some of the first planning officers in suburban and exurban jurisdictions), in order to secure minimum densities for service delivery, stressing that *"unless this is done there can be no sanity and no economy in municipal affairs"* (LMRPB, 1956a, p. 16, original emphasis).

The pursuit of suburban ideals, enabled by rising real incomes, car ownership, and the decentralization of employment and services, continued to animate a centrifugal culture among homeowning Vancouverites. No matter that this often proved to be a forlorn pursuit, as the Regional Planning Board disapprovingly observed, since suburbanization was "built on illusions" – of fresh air, open spaces, and low taxes – as if those populations "umbilically attached to [the] great metropolis [could realistically] escape urban cost and condition by moving one step from it" (LMRPB, 1963a, p. 20). The board's investigation of life in the sprawl belt had discovered that "the new 'pioneer' on the urban frontier, while he (sic) may be content to dispense with some urban conveniences, remains essentially city-oriented and dependent on city-style services" (LMRPB, 1963b, p. 41). Suburbia was the "habitat of the young married couple with a large family of young children, a commuting husband, and one or two cars;" *homo sprawlus* was employed in the skilled trades or in the submanagerial white-collar sector, and while one eighth of "frontier wives" were found to be working, they were to remain largely invisible, beyond proxy discussions of schooling and shopping; families were typically "forced to the fringe for financial reasons" (LMRPB, 1963b, pp. 7, 35).

If the 1960s was Vancouver's decade of "high suburbanism," echoing the classic North American pattern, it also represented something of an inflection point, setting the stage for a bold performance that would finally validate the city's culture of exceptionalism. Bill Rathie won the 1962 mayoral election with the exhortation "Let's get Vancouver moving!" This he proposed to do with an extensive program of downtown office development, urban renewal, and freeway construction. A plan to superimpose a freeway network had been fomenting for decades in Vancouver, deep in what UBC professor Walter Hardwick referred to as the "technostructure" of engineers and transportation planners, with the connivance of a hired band of American consultants. But once it became properly public, the freeway proposal triggered concerted opposition, first from the immediately threatened Strathcona neighbourhood and soon from a gathering network of west-side professionals (including Hardwick himself), which together eventually defeated the proposal. The "spectre of American-style urban decay" and the energy generated by the freeway controversy led to the formation of a centrist, reform-oriented political party – TEAM, The Electors' Action Movement – "whose members articulated a growing sense that Vancouver was different from American cities, that it was developing along its own trajectory, and that it needed made-in-Vancouver plans" (North & Hardwick, 1992, p. 208). Walter Hardwick was one of the first TEAM councilors, elected to the city council in 1968, along with Art Philips, who would become the city's mayor in 1973. TEAM dominated municipal politics in the 1970s, institutionalizing community involvement and turning planning processes away from the "urban renewal and freeways" model that was "all the rage" in the immediate postwar period (Harcourt et al., 2007, p. 34). The ostensibly American path towards an overengineered, modernist-rationalist pattern of development was effectively blocked in Vancouver, in favour of a "more humane and aesthetic city," focused on a new vision of the downtown core consistent with the values and interests of TEAM's liberal, middle-class electoral base (Ley, 1980, p. 251). Vancouver's livable city had been born, both as a model and a movement.

Between Livable City and Suburbocentrism

It was under the sway of TEAM's new vision of liberal urbanism that "Vancouver began softening the hard distinction between downtown and bedroom suburbs, a process still underway today [in this] the largest city in North America without a freeway clogging its arteries" (Harcourt et al., 2007, p. 54). TEAM articulated and sharpened the anticar, antipollution ethos, similar to the one that had fuelled earlier concerns over a metaphorical termite infestation. A raft of provincial and metropolitan-scale responses (the Official Regional Plan of 1966, the Agricultural Land Reserve of 1973, the first "Livable Region" strategic plan of 1976), began to lock in the new consensus, while inside the city limits, TEAM mobilized to prevent what Hardwick called "bulldozer renewal" and "urbanicide."

This was the time, too, when a cultural reevaluation of higher density lifestyles began in earnest. While there was "widespread condemnation of the West End in the city at large" (the site of a handful of residential high rises), Hardwick contended that this was not an opinion "shared by those who live there," finding it necessary to add that "density in itself is not bad" (1974, p. 203). Hardwick liked to take his own children, plus assorted students from his geography classes, to the "industrial sewer" that was False Creek South in the early 1970s, on reconnaissance visits to what he hoped would ultimately become a model microneighborhood for 30,000 urban pioneers – a residential vision eventually realized by the TEAM coalition in the face of opposition from the city's planning establishment (Harcourt et al., 2007, pp. 97–100 Punter, 2003). Completed in 1977, False Creek South stands as the site of germinal authenticity for a socially conscious variant of Vancouverism comprising a mixed-income, primarily residential high-density development, pedestrian and bicycle-friendly but anticar, and oriented to the waterfront. In David Ley's reading, False Creek South represented

> a landscape metaphor of liberal ideology, of the land use implications of the transition from industrial to post-industrial society, from an ethic of growth and production of goods to an ethic of amenity and the consumption of services. (1980, p. 252)

An especially deep recession in the early 1980s punctuated these trends, marking the onset of a protracted phase of restructuring in the region's staples economy and the subsequent exodus of most of the city's headquarters functions (Hutton, 2004). Expo '86, in Vancouver's centennial year, helped pull the city out of the economic doldrums, but perhaps more importantly cleared the space – literally and metaphorically – for a yet more ambitious experiment in "city building." Expo itself resulted in a loss, but the received account is that the city won: following the event, the publicly owned site was sold off by its provincial owners (in a single, controversial transaction) to Hong Kong billionaire Li Ka-Shing, whose real estate investment firm, Concord Pacific, duly launched one of the largest urban-redevelopment projects in North America (Harris, 2011; Olds, 2001). The residential condo towers of False Creek North – incorporating not only capital and expertise imported from Asia, but also aesthetic, cultural, and design sensibilities – would transform Vancouver's skyline, crystalizing the development model that lay claim to the status of Vancouver*ism*. As Ley, Hiebert, and Pratt (1992, p. 239) observed, the dramatic conversion of degraded industrial land to premium residential space encapsulated the dominant ideology of the period, based on "privatization, polarization, and internationalization," noting that "only strenuous intervention from the city [council had] secured sites for social housing," for which sustainable sources of funding nevertheless proved difficult to find. A future that Walter Hardwick had feared, that of an exclusive "executive city," seemed to be taking shape.

If the condo complex of False Creek North represented a concrete manifestation of Vancouverism, the City's "Living First" planning policy statement of 1991 was its Magna Carta (Boddy, 2006). According to an architect of the Living First policy, city planner Larry Beasley, this bold experiment in "downtown living" was facilitated by a confluence of factors: rezoning in favour of residential over commercial and industrial uses; highly conducive market conditions, including strong demand from wealthy immigrants and retirees; a design ethos oriented to walkability, housing intensity, and mixed use; and innovations in planning practice, enabling a share of development premiums to be captured for the purpose of "public" reinvestment in neighbourhood facilities such as parks and community centres (Beasley, 2000). All of this had been facilitated by the establishment in the late 1960s of the condominium as a legal form of subdivided private ownership, which in "detach[ing] the legal category of land from the surface of the earth," pressed up against what Arthur Erickson (1964) had called "the weight of heaven" to enable a skyward surge in speculative development

Table 7.1 Geographies of population change, Metropolitan Vancouver, 1971–2011

	Population, 1971 ('000s)	Population, 2011 ('000s)	Change (%)	Share of regional growth (%)
City	426.3	603.5	42	11.4
Vancouver (C)	426.3	603.5	42	
Inner suburbs	*357.0*	*654.0*	*83*	*22.4*
Burnaby (C)	126.0	223.0	77	
West Vancouver	36.0	43.0	19	
Richmond (C)	62.0	190.0	206	
North Vancouver (C)	32.0	48.0	50	
North Vancouver (DM)	58.0	84.0	45	
New Westminster	43.0	66.0	53	
Outer suburbs	*285.0*	*982.0*	*245*	*66.2*
Coquitlam	53.0	126.0	138	
Delta	46.0	100.0	117	
Port Coquitlam	20.0	56.0	180	
Port Moody	11.0	33.0	200	
Surrey	99.0	468.0	373	
White Rock	10.0	19.0	90	
Langley (DM)	22.0	104.0	373	
Maple Ridge	24.4	76.0	211	
Total	1,068.0	2,239.5	110	100.0

Source: Authors' calculations from Statistics Canada, Census of Canada, 1971–2011.

(Harris, 2011, p. 696). A rarity as recently as the mid-1970s, condos accounted for almost 40% of owner-occupied households in Vancouver by 2006. There are now more than 100,000 condo units in the city, a growth rate that has merely slowed in the last decade, as zoned areas have become almost completely built out.

But while all the attention was focused on the downtown condo towers, garden-variety suburbanization continued apace. The metropolitan population more than doubled from 1971 to 2011, but this growth was almost entirely driven by the suburbs proper – especially the outer belt, where the population more than tripled (table 7.1). Nine tenths of all net metropolitan population growth between 1971 and 2011 took place in the suburbs, while more than two thirds of international migrants settled in suburbia (Metro Vancouver, 2008). The central city has fared well by North American standards, its population having increased by 57% since the early 1960s, but while the City of Vancouver was home to almost half of metro-area residents in 1961, this had fallen to little more than a quarter by 2011 (table 7.2). The downtown core has

Vancouverism as Suburbanism 139

Table 7.2 Population shares, Metropolitan Vancouver, 1961–2011

	Population in '000s (metro share)		
	City of Vancouver	Inner suburbs[a]	Outer suburbs[b]
1961	384.5 (47.2%)	265.2 (32.6%)	165.2 (20.2%)
1971	426.3 (39.9%)	356.8 (33.4%)	284.6 (25.7%)
1981	414.0 (33.5%)	405.8 (32.8%)	415.7 (33.7%)
1991	472.0 (30.3%)	481.4 (30.9%)	603.4 (38.8%)
2001	545.7 (28.4%)	580.8 (30.1%)	803.0 (41.5%)
2006	578.0 (28.2%)	605.7 (29.5%)	867.9 (42.3%)
2011	603.5 (27.0%)	654.1 (29.2%)	981.8 (43.8%)
Change (growth), 1961–2011 (in '000s)	*219.0 (57.0%)*	*388.9 (147.0%)*	*816.6 (494.3%)*

Source: Authors' calculations from Statistics Canada, Census of Canada, 1971–2011.
[a] Burnaby (C), West Vancouver, Richmond (C), North Vancouver (C), North Vancouver (DM), New Westminster; [b] Coquitlam, Delta Port, Coquitlam, Port Moody, Surrey, White Rock, Langley (DM), Maple Ridge

certainly densified in residential terms, underlining the sense in which the Living First strategy has achieved its goals: Vancouver's downtown population, which had sat at about 6,000 through the 1970s and 1980s, surged by 22,000 during the 1990s to top 43,000 by 2006. This coming of the "new middle class" filled the green-glass condo towers, accentuating what some have characterized as a vertical variant of gentrification (Blomley, 2004; Harris, 2011; Ley, 1996).

Since the early 1990s, the downtown employment base has both expanded and professionalized, as the share of blue-collar jobs collapsed (Hutton, 2004). The City of Vancouver's 47% employment growth over the period 1971–2006 might be considered respectable by comparison to many of the city's peers. Vancouver's central business district employment grew by 15.8% between 1981 and 1996, compared with 8.0% in Montreal and 0.3% in Toronto (Coffey & Shearmur, 2006). In 2006, 29% of the Vancouver metropolitan area's jobs were within 5 km of the city centre, compared with 28% in Montreal and 20% in Toronto (Shearmur & Hutton, 2011). But this growth is modest relative to explosive rates in the suburbs, especially outer suburbs like Surrey, Port Coquitlam, Delta, and Langley, where employment levels (albeit from a low base) increased fivefold (table 7.3). Altogether, the suburbs accounted for 93% of metro-regional job growth between 1971 and 2006. During this period, the City of Vancouver's share of metro jobs fell from 57% to 30% (table 7.4).

For every 100 new residents arriving in the city since 1971, only 70 new jobs have been created. In the suburbs, extremely high rates of

Table 7.3 Geographies of employment change, Metropolitan Vancouver, 1971–2006

	Employment, 1971 ('000s)	Employment, 2006 ('000s)	Change (%)	Share of regional growth (%)
City	225.1	331.3	47	7.0
Vancouver (C)	225.1	331.3	47	
Inner Suburbs	116.0	309.0	166	25.0
Burnaby (C)	43.0	115.0	167	
West Vancouver	7.0	16.0	129	
Richmond (C)	24.0	108.0	350	
North Vancouver (C)	10.0	25.0	150	
North Vancouver (DM)	11.0	22.0	100	
New Westminster	21.0	23.0	10	
Outer Suburbs	54.0	296.0	448	68.0
Coquitlam	9.0	39.0	333	
Delta	8.0	47.0	488	
Port Coquitlam	3.0	18.0	500	
Port Moody	2.0	6.0	200	
Surrey	16.0	119.0	644	
White Rock	3.0	6.0	100	
Langley (DM)	7.0	41.0	486	
Maple Ridge	6.0	19.0	217	
Total	395.1	936.3	137	100.0

Source: Authors' calculations from Statistics Canada, Census of Canada, 1971–2006.

Table 7.4 Employment shares, Metropolitan Vancouver, 1971–2006

	Employment in '000s (metro share)			
	City of Vancouver	Inner suburbs	Outer suburbs	No fixed workplace
1971	225.1 (57.1%)	115.5 (29.3%)	53.1 (13.6%)	n/a
1981	301.0 (47.6%)	197.6 (31.2%)	133.2 (21.2%)	n/a
1991	333.3 (41.3%)	264.6 (32.8%)	207.6 (25.9%)	n/a
1996	308.4 (33.6%)	270.1 (31.5%)	232.3 (25.3%)	89.0 (9.6%)
2001	310.6 (31.8%)	295.9 (30.4%)	266.4 (27.3%)	102.6 (10.5%)
2006	331.3 (30.4%)	309.3 (28.5%)	311.0 (28.7%)	135.8 (12.4%)
Change (growth), 1961–2011 (in '000s)	106.2 (47.2%)	193.8 (167.8%)	257.9 (485.7%)	n/a

Source: Authors' calculations from Statistics Canada, Census of Canada, 1971–2006.

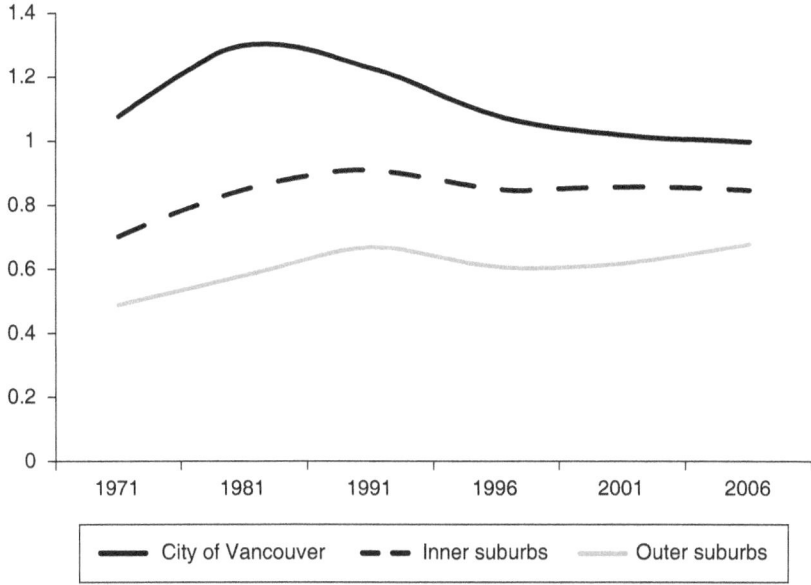

Figure 7.2 Ratio of jobs to labour force, Vancouver and suburbs, 1971–2006. Source: Authors' calculations from Statistics Canada, Census of Canada, 1971–2006.

population growth have also been outstripping employment growth (for every 100 new residents arriving in the suburbs since 1971, 54 new jobs were created), but these rates are gradually converging. As figure 7.2 reveals, employment: population ratios are also converging across the metro area, as the urban core becomes more residential and as the employment base of the suburbs expands. In other words, downtown is beginning to look more like a suburban bedroom community, while the bustling employment centres of suburbia look more like downtowns.

This raises questions about the economic foundations of the Vancouver model, which in the absence of propulsive industries and higher order corporate functions remains something of a puzzle, even to the locals (see Bula, 2011). A persistent shortfall in high-paying jobs, value-adding functions, leading industries, and prominent corporate players is precariously matched with a dependency on real estate, speculative capital, and personal-services jobs. Yet Vancouverism undeniably *looks like* a successful formula, and as a globally circulating urban aesthetic, the city's winning looks have also been instrumental in its rise to prominence as an urban-development imaginary and planning paradigm

(Berelowitz, 2005). The implicit claim is that success was effectively "designed in" to the Vancouver model in both policy and architectural terms. Rather less is said about the fortuitous and inherently nonreplicable location of Vancouver at the confluence of accelerating flows of wealthy migrants and speculative capital – a buoyant market that allowed planners to guide development capital towards favoured sites and schemes in the urban core.

Conclusion: Suburbocentrism Under Stress

In class and cultural terms, indeed on most registers other than the literal count of population density itself, downtown Vancouver has effectively been suburbanized, a phenomenon sometimes styled as "residentialization" or "de-downtownification." Some simply call it "the city that acts like a suburb" (McMartin, 2012). Trevor Boddy (2006, p. 22) likewise complains that "too much housing may be killing peninsular downtown Vancouver," the Living First planning hegemony having promoted the unimaginative replication of "mono-form, mono-class, crank-the-handle towers." The condo towers of the downtown area may seem like the antithesis of the low-density suburban "Vancouver special," but there is arguably a parallel in their shared modes of mass production and indeed in the very logic of repetitive development (Bogdanowicz, 2006). In the core itself – and the necklace of subcentres along the SkyTrain transit lines – Vancouverism has become a medium for repetition rather than innovation. The tower-and-podium condo format has been pushed to the point of parody, disparaged by prominent local architects as a "single building typology … qualified only by its incessant sameness," the (mass) product of an excessively "formulaic" living-first planning regime (Kluckner, 2012, pp. 41–2).

This is not simply a matter of repetitive condo development; it also extends to the multidimensional (re)production of suburban spaces. Here the polarity of Wirth's (1938, p. 1) maxim – "While the city is the characteristic locus of urbanism, the urban mode of life is not confined to cities" – has been inverted, and Vancouver has come to represent the distinctively Canadian vanguard of "suburbanism as a way of life" (cf. Walks, 2013a). Market imperatives, mass-produced novelty in urban design and marketing, pervasive privatism, and friendly looking half-measures to soften the sharp edges of deepening inequalities have come to constitute Vancouver's centripetal suburbanism. If the condos of Vancouverism were once seen as edgy and innovative by virtue of their challenge to North America's picket-fence hegemony, their subsequent mass commodification – materially and culturally – solidifies the harsh instrumentalism of capital-gains accumulation and reductive marketing cliché. Reassuringly standardized – right down to the granite countertops, top-of-the-range stainless-steel appliances, and hardwood floors – Vancouver condos have become highly fungible and slickly marketed investment commodities (Harris, 2011).

Governance processes and rhythms of social life in Vancouver's suburbanized core offer a cautionary tale of the long-term consequences of etching the tower-and-podium paradigm into the (sub)urban fabric. Hern (2010, p. 46) observes that downtown is increasingly populated by "people with little attachment and few civic bonds to the city" – close to a Mumfordesque definition of suburbanites, which has since found expression in an ongoing public debate about social isolation and anomie in the "cold city" of Vancouver (Vancouver Sun, 2012). Nevertheless, Larry Beasley, now retired from the city and a consulting urbanist with an international practice, continues explicitly to defend "the suburban values of predictability, cleanliness and lack of architectural variety in the high-rise zones shaped by his 'Living First' strategy, arguing that a key motor of Vancouver's downtown success is making itself attractive to those who grew up in the urban fringes" (Boddy, 2006, p. 21).

Vancouver's conspicuous wealth is mostly generated elsewhere, and then cycled through the highly charged real estate market. The frequently voiced concern is that downtown Vancouver will evolve into something like a bedroom community or vertical Levittown – albeit a high-class, transnationally cosmopolitan one – sustaining its own ecosystem of personal services, but not much more, as the metro area's commercial, industrial, and employment base continues to be hollowed out. With downtown becoming "a place where huge numbers of people live but not many work," it has been estimated that Vancouver's net commuting balance could be a centrifugal one as soon as 2020 (Ehrenhalt, 2006). Some have suggested that Vancouver possesses a Monte Carlo–style "resort economy," a dynamic recognized by some of those behind the city's development model:

> People used to say this was a setting in search of a city. You don't hear that anymore. This city is now a city people come to *for the city*…. It's the ambience of the place. But tens of thousands of people are coming here … bringing their wealth here and they are leaving it here and spending it here. Someone recently said, "No one knows what drives Vancouver." Well what drives Vancouver is that people make wealth in unpleasant places and they come here and spend their wealth in a pleasant place – that's it! (retired city-council official, Vancouver, interview August 2010)

What looks to many like a housing bubble, especially in the oversupplied condo market, has hovered for several years at the bursting point, sustaining an ongoing discursive industry of predictions and projections. And so the bubble continues, sustaining Canada's highest household debt ratios (Walks, 2013b), while carving "a new geography of rising income inequality and polarization" into the metropolitan landscape (Ley & Lynch, 2012, iv). "Tens of thousands of middle-class households are getting a hard lesson in diminished expectations," reported the usually boosterish *Vancouver* magazine; unless they "want to raise their children in a one-bedroom condo, their salaries will qualify them to be no more than permanent renters – members of what "real-estate-wise, might be called Vancouver's Generation Fucked" (Bridge, 2011, n.p.).

Until the closing decades of the last century, Vancouver seemed to be moving predictably along the grooves of North American (sub) urbanism, its urban core comprising the usual array of commercial and industrial functions, as the monoculture of low-density, single-family subdivision spread laterally from the inner suburbs out to what Walter Hardwick disapprovingly called the "sprawl belt," constrained only, it seemed, by the limits of the city's "super-natural" setting – water to the west, mountains to the north, the US border to the south. With barely a few tweaks, this did not appear to be wildly different from the Chicago-school model of urban modernization (Hutton, 2004; Ley, Hiebert, & Pratt, 1992). Yet in the past two decades, in particular, Vancouver's development model has departed from what is sometimes portrayed as an inevitable or "mature" stage of edge-city postindustrialism. More than interurban reorganization, more even than postmodern juxtaposition, Vancouver's development pattern reveals an "overdriving," complexification, and recombination of extant sub/urban forms.

Classically American forms of suburbanization always reflected much more than the decentralized outcome of lifestyle choices; they represented an active rejection of both urban cultures and metropolitan governance (Kotkin, 2005), what might be seen as a victory of one Chicago school over another – a victory of the economic strain of free-market neoliberalism over the sociological strain of urban-political liberalism (Peck, 2011). Vancouver's homegrown form of suburbocentrism is sharply interventionist at the neighbourhood scale, while reworking the reigning model of weakly regulated "under-governance" at the metropolitan-region scale. Yet the city's suburbocentric hegemony, like all hegemonies, remains incomplete, contradictory, and contestable. Vancouver's long-gestating affordability crisis has been intensifying the contradictions of "livability" for generations. Four decades after Ley (1980, p. 258) suggested that postindustrial urbanism might be "too naive, not recognizing that its humane philosophy might be coopted by the calculus of the marketplace," real estate accumulation has become the de facto economic base, the unapologetic calculative ideology, and the governing mode of spatial organization for the metropolis. In this sense, Vancouver embodies a complex yet relentless centralized involution of the "suburban solution" that restarted American capitalism after the Great Depression and the Second World War (Walker, 1981).

For more than a decade now, Vancouver has consistently ranked as the world's second most expensive city in terms of the ratio of house prices to local incomes (behind only Hong Kong), and by early 2016 the median price for an existing single-family home surpassed the $2 million mark (Sherlock, 2016). Using standard underwriting scenarios, only 5.3% of the city's working households currently qualify to purchase the median single-family home, and so once again many continue to be "forced to the fringe," or beyond, "for financial reasons" (LMRPB, 1963b). One of Canada's leading banks now rates Vancouver as "dangerously unaffordable" – RBC's (2016) affordability index (which measures the proportion of median pretax household income required to cover housing-related costs at the median market price) surging in late 2015 to a record 109% for single-family detached houses (compared to 51% nationally and 71% in the country's second most-expensive market, Toronto) and 44% for condos (compared to 35% nationally and 37% for Toronto).

The affordability crisis that has for many years immiserated submedian households increasingly threatens the viability of a resort economy reliant on low-wage service workers, while also clawing steadily upward into the professional classes. Reading about a doctor who left the city because he could not afford to raise a family, a young Vancouver professional took to Twitter with a simple complaint; Evelyn Xia's #donthaveamillion quickly went viral, catalyzing growing discontent over the realization that local incomes will never be a match for the transnational circuits of planetary sub/urbanization. The flows of surplus residential capital that have sustained downtown speculation for a quarter century remain distinctly vulnerable to disruption, even as for the time being they continue to turbocharge a real estate growth machine buffeted by populist backlashes against rezoning, speculative property flipping, and tax evasion and money laundering. A distant echo of the campus-based activism that propelled the TEAM mobilizations of the late 1960s, 10 professors from the University of British Columbia and Simon Fraser University recently published a study calling for a modest real estate surcharge on absentee foreign owners in order to address the affordability crisis. "[T]he market is a horror show [and] a disaster for those who don't own," explained one of the study's authors, the combination of a devalued Canadian dollar and ever-more integrated speculative investment circuits meaning that "the local real estate market 'has a target on its back' for international capital" (Lee-Young & Shaw, 2016, p. A12). As Vancouver becomes what the local urban planner Andy Yan has dubbed a "hedge city," it is both dependent on and vulnerable to fickle flows of extralocal investment, flows that always threaten to gravitate towards yet more favourable hotspots of capitalizable "livability." This suburbocentric mode of development may therefore be many things, but stable it is not.

NOTE

1 This is an abridged version of: Peck, J., Siemiatycki, E. & Wyly, E. (2014). Vancouver's suburban involution. *City, 18*(4–5): 386–415, reproduced by permission of *City* and Taylor & Francis.

REFERENCES

Barnes, T., Hutton, T., Ley, D., & Moos, M. (2011). Vancouver: Restructuring narratives in the transnational metropolis. In L. Bourne, T. Hutton, R. G. Shearmur, & J. Simmons (Eds.), *Canadian urban regions: Trajectories of growth and change* (pp. 291–327). Don Mills, ON: Oxford University Press.
Beasley, L. (2000, April). "Living first" in downtown Vancouver. *American Planning Association, Zoning News.*
Berelowitz, L. (2005). *Dream city.* Vancouver: Douglas & McIntyre.
Blomley, N.K. (2004). *Unsettling the city.* New York, NY: Routledge.
Boddy, T. (2006). Downtown's last resort. *Canadian Architect, 51*(9), 20–2.
Bogdanowicz, J. (2006). Vancouverism. *Canadian Architect, 51*(8), 23–4.
Bridge, T. (2011, November). Going, going, gone. *Vancouver Magazine.*
Bula, F. (2011, October). The future of Vancouver's economy. *Vancouver Magazine,* 35–9.
Coffey, W.J., & Shearmur, R.G. (2006). Employment in Canadian cities. In T. Bunting & P. Filion (Eds.), *Canadian cities in transition* (pp. 249–71). Oxford: Oxford University Press.
Dear, M.J., Burridge, A., Marolt, P., Peters, J., & Seymour, M. (2008). Critical responses to the Los Angeles school of urbanism. *Urban Geography, 29*(2), 101–12.
Ehrenhalt, A. (2006, July). Extreme makeover. *Governing,* 104–8.
Erickson, A. (1964). The weight of heaven. *Canadian Architect, 9*(3), 48–63.
Harcourt, M., Rossiter, S., & Cameron, K.D. (2007). *City making in paradise.* Vancouver: Douglas & McIntyre.

Hardwick, W.G. (1974). *Vancouver*. Don Mills, ON: Collier-Macmillan.
Harris, C. (1997). *The resettlement of British Columbia*. Vancouver: UBC Press.
Harris, D.C. (2011). Condominium and the city. *Law and Social Inquiry, 36*(3), 694–726.
Hern, M. (2010). *Common ground in a liquid city*. Edinburgh: AK Press.
Hutton, T.A. (2004). Post-industrialism, post-modernism and the reproduction of Vancouver's central area: Re-theorising the 21st-century city. *Urban Studies, 41*(10), 1953–82.
Kluckner, M. (2012). *Vanishing Vancouver*. North Vancouver, BC: Whitecap.
Kolb, D. (2011). Many centers: Suburban habitus. *City, 15*(2), 155–66.
Kotkin, J. (2005). *The new suburbanism*. Costa Mesa, CA: Planning Center.
Lees, L., & Demeritt, D. (1998). Envisioning the livable city: The interplay of "sin city" and "sim city" in Vancouver's planning discourse. *Urban Geography, 19*(4), 332–59.
Lee-Young, J., & Shaw, R. (2016, February 18). Experts say push to collect buyer data has pitfalls. *Vancouver Sun*, A1, A12.
Ley, D. (1980). Liberal ideology and the postindustrial city. *Annals of the Association of American Geographers, 70*(2), 238–58.
Ley, D. (1996). *The new middle class and the remaking of the central city*. Oxford: Oxford University Press.
Ley, D., Hiebert, D., & Pratt, G. (1992). Time to grow up? From urban village to world city, 1966–91. In G. Wynn & T. Oke (Eds.), *Vancouver and its Region* (pp. 234–66). Vancouver: UBC Press.
Ley, D., & Lynch, N. (2012). *Divisions and disparities in lotus-land: Socio-spatial income polarization in greater Vancouver, 1970–2005*. Toronto: Cities Centre, University of Toronto.
LMRPB [Lower Mainland Regional Planning Board of British Columbia]. (1956a). *Economic aspects of urban sprawl*. New Westminster, BC: LMRPB.
LMRPB. (1956b). *Urban sprawl in the lower mainland*. New Westminster, BC: LMRPB.
LMRPB. (1963a). *The urban frontier, Part 1*. New Westminster, BC: LMRPB.
LMRPB. (1963b). *The urban frontier, Part 2*. New Westminster, BC: LMRPB.
Mattocks, R.H. (1932). A plan for the city of Vancouver, British Columbia. *Town Planning Review, 15*(1), 49–55.
Mawson, T.H. (1913). Vancouver: A city of optimists. *Town Planning Review, 4*(1), 6–12.
McCann, E. (2011). Urban policy mobilities and global circuits of knowledge: Toward a research agenda. *Annals of the Association of American Geographers, 101*(1), 107–30.
McMartin, P. (2012, October 9). Vancouver: The city that acts like a suburb. *Vancouver Sun*, A1.
Metro Vancouver. (2008). *2006 Census Bulletin 6*. Burnaby, BC: Metro Vancouver.

Morley, A. (1961). *Vancouver*. Vancouver: Mitchell Press.

North, R.N., & Hardwick, W.G. (1992). Vancouver since the Second World War: An economic geography. In G. Wynn & T. Oke (Eds.), *Vancouver and its Region* (pp. 200–33). Vancouver: UBC Press.

Olds, K. (2001). *Globalization and urban change*. Oxford: Oxford University Press

Peck, J. (2011). Neoliberal suburbanism: Frontier space. *Urban Geography, 32*(6), 884–919.

Punter, J. (2003). *The Vancouver achievement*. Vancouver: UBC Press.

RBC. (2016). *Housing trends and affordability*. Toronto, ON: RBC.

Saunders, D. (2013, February 23). The world wants Vancouverism. Shouldn't Canada? *Globe and Mail*, F2.

Shearmur, R., & Hutton, T. (2011). Canada's changing city-regions: The expanding metropolis. In L.S. Bourne, T. Hutton, R.G. Shearmur, & J. Simmons (Eds.), *Canadian urban regions* (pp. 99–124). Oxford: Oxford University Press.

Sherlock, T. (2016, February 2). Statistics show homes getting even further out of reach for most. *Vancouver Sun*, A9.

Vancouver Sun. (2012, June 28). More to healthy life than diet, exercise. *Vancouver Sun*, A16.

Walker, R. (1981). A theory of suburbanization: Capitalism and the construction of urban space in the United States. In M. Dear & A.J. Scott (Eds.), *Urbanization and urban planning in capitalist society* (pp. 383–429). New York, NY: Methuen.

Walks, A. (2013a). Suburbanism as a way of life, slight return. *Urban Studies, 50*(8), 1471–88.

Walks, A. (2013b). Mapping the urban debtscape: The geography of household debt in Canadian cities. *Urban Geography, 34*(2), 153–87.

Wirth, L. (1938). Urbanism as a way of life. *American Journal of Sociology, 44*(1), 1–24.

Wynn, G. (1992). The rise of Vancouver. In G. Wynn & T. Oke (Eds.), *Vancouver and its region* (pp. 69–145). Vancouver: UBC Press.

8 Montreal: An Ordinary North American Metropolis?

CLAIRE POITRAS AND PIERRE HAMEL

Introduction

Several characteristics distinguish the Montreal region from other similarly sized North American urban areas: the predominance of French, a relatively high level of urban density, the proportion of commuters using public transit (23.5% in 2016, third among North American cities, behind New York and Toronto), a sociospatial cleavage or division based on language rather than race, a rate of homeownership below the North American average, a high number of single-person households in the central city,[1] a lower degree of racial and economic segregation, and the economic importance of manufacturing. This final characteristic helped establish Montreal as the metropolis of Canada into the 1930s and beyond (Higgins, 1986).

During its most dynamic period of industrialization, which lasted from 1880 to 1930, the Montreal region served as Canada's key port and rail hub (Gournay & Vanlaethem, 1998). Even today, manufacturing remains an important sector of the Montreal economy, although its relative importance has diminished as that of business services sector has increased (Polèse, 2012). Since the 1960s, however, the Montreal region has lost its status of the country's transportation hub; its port, railroads, and airport no longer play the role they once did. As a result, the city has gone from being the metropolis of Canada to the metropolis of Quebec (Linteau, 2000). This repositioning has had several significant sociopolitical consequences for the city and the surrounding region. Not only has the central city's sphere of influence within North America become smaller, but both its economic and sociopolitical influence within the larger metropolitan region have been diminished.

There are numerous factors in play. Looking back to the 1960s, decisions made by the Quebec government have certainly contributed to making the central city less dynamic (Hamel & Poitras, 2004). The introduction of various administrative reforms, including the creation of new administrative divisions, has reduced the city's political influence within its metropolitan area. While increasing their demographic weight, suburban municipalities have also gained more economic, political, and cultural importance. A growing number of suburban municipalities now boast large concert venues and other cultural amenities. The retail sector has also become significantly less centralized in recent decades, as households have embraced the new retail formats offered by redesigned shopping districts, including "lifestyle centres" and "power centres." As a result, residents no longer need to travel to the Island of Montreal or downtown to find exclusive goods and specialized services. This decentralization of retail activity has allowed suburban municipalities, including outer suburbs, to become increasingly autonomous in relation to downtown Montreal.

In the last two decades, has suburban Montreal become more diversified in terms of urban form, as well as social and ethnic composition? What are the main characteristics of these suburban areas? In terms of employment opportunities and geography, does the metropolitan area now have multiple poles of economic activity, including ones centred in suburbs? What characterizes present-day relations between Montreal and its suburbs? What are the key elements of Montreal's suburban development? Has a certain form of suburban culture led to a reimagining of 21st-century urban metropolitan life, and has this new vision given rise to new values? To begin answering these questions, we examined the evolution of the metropolitan structure and trends associated with the process of suburbanization, as well as its demographic, socioeconomic, and sociocultural impacts.

Key Geographic and Social Characteristics of the Montreal Metropolitan Area

In 2014, the Montreal metropolitan area had a population of more than four million living in 90 different municipalities. This represented about 47% of the population of Quebec. It should also be noted that these numbers reflect the boundaries established by Statistics Canada, the federal government agency responsible for defining census metropolitan areas (CMAs) based on daily commutes between residences and places of work.

Quebec's demographic vitality is largely dependent on the five administrative regions that are partially or entirely located within the Montreal region.[2] The Montreal Archipelago is dominated by two

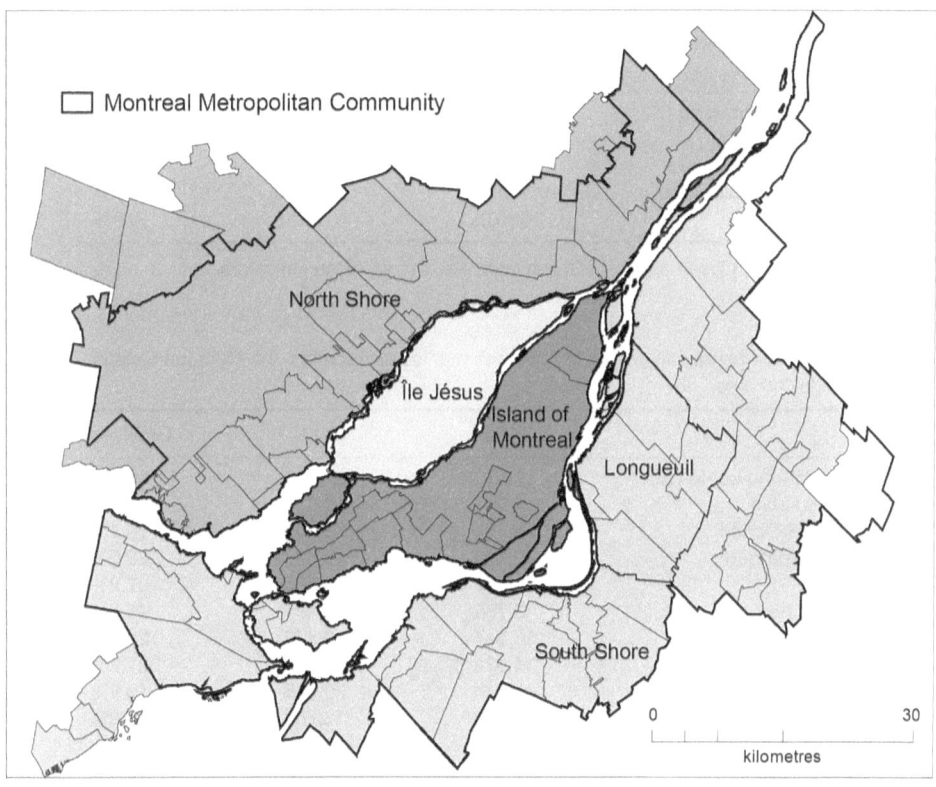

Map 8.1 The Montreal metropolitan area and its main divisions. Source: INRS-Urbanisation Culture Société

large islands: the Island of Montreal (home to Greater Montreal) and Île Jésus (home to the City of Laval). Along with the City of Longueuil, located on Montreal's South Shore, they are surrounded by northern and southern outer suburbs (see map 8.1 and table 8.1). It is important to note that the population of the Island of Montreal itself increased in recent years (by more than 20,000 inhabitants in 2013–14).

However, the boundaries established by Statistics Canada tend to mask some of the internal diversity of the Montreal metropolitan area (Filion et al., 2010). For example, some communities in the northern and southern outer suburbs that have substantially increased their populations are located outside the boundaries of the Montreal Metropolitan Community (MMC). The latter is a regional political and administrative authority created in 2000 by the Quebec government to plan the

Table 8.1 Population distribution in the Montreal region by area

	2006	2011	Growth (%)
Longueuil urban area	385,533	399,095	3.4%
Island of Montreal	1,854,440	1,886,480	1.7%
Northern outer suburbs	493,971	550,065	11.3%
Southern outer suburbs	429,971	471,850	9.7%
Laval	368,709	401,555	8.9%

Source: MMC, Observatoire du Grand Montréal, metropolitan indicators.

Table 8.2 Demographic growth in suburban municipalities within the Montreal census metropolitan area

	2006	2011	Growth (%)
Sainte-Marthe-sur-le-Lac	11,311	15,689	38.8%
Saint-Amable	8,398	10,870	29.4%
Vaudreuil-Dorion	25,789	33,305	29.1%
Saint-Colomban	10,136	13,080	29.0%
Pincourt	11,197	14,305	27.8%
Mascouche	33,764	42,491	25.8%
Candiac	15,947	19,876	24.6%
Mirabel	34,626	41,957	21.2%
L'Assomption	16,738	20,065	19.9%
Mont-Saint-Hilaire	15,720	18,200	15.8%
Mercier	10,121	11,584	14.5%
Saint-Lazare	17,016	19,295	13.4%
Chambly	22,608	25,571	13.1%
Blainville	46,493	53,510	15.1%
Bois-des-Filion	8,383	9,485	13.1%
Terrebonne	94,703	106,322	12.3%
Sainte-Anne-des-Plaines	13,001	14,535	11.8%
Brossard	71,154	79,273	11.4%

Source: Statistics Canada, Focus on Geography Series, 2011 Census, Montreal census metropolitan area.

development of the metropolitan area and manage shared resources. And while the central city still accounts for the largest portion of the region's population, some suburban areas located on its periphery are gaining significant ground. In fact, municipalities experiencing the highest rates of growth are located in the northern and southern outer suburbs (see table 8.2). In short, within the larger metropolitan context, certain suburban municipalities have experienced significant growth in recent years.

Like elsewhere in North America, suburban growth and the negative effects of urban sprawl on both the environment and the vitality of the central city have become important issues in Montreal. However, because Montreal has experienced a slower rate of urban development since the 1960s, in comparison to many other North American metropolitan regions, the direct repercussions of peripheral growth have been less of a threat to the vitality of the central city.

For decades, provincial authorities have considered measures to contain urban sprawl or keep households and businesses within the urbanized parts of the metropolitan area. Numerous studies, policies, and programs seeking to counter urban sprawl have been launched since the 1960s (Sénécal & Vachon, 2012). And the Montreal agglomeration – the designation used by political and administrative authorities to refer to the 15 municipalities and the 14 boroughs located on the Island of Montreal – has seen its demographic weight increase slightly in recent years, after several decades of decline. However, this trend is not the result of government policies seeking to limit development outside of already-urbanized areas. After more than 50 years of analysis and discussion, no progress has been made in developing effective systems of governance capable of managing urban growth and urban sprawl.

The most recent metropolitan planning exercise was undertaken by the MMC. It involved the development of a metropolitan land use and development plan, which was adopted in December 2011. The focus was on transit-oriented development.

Specifically, the plan looked at increasing residential density at the points of origin of daily commuters, rather than at their destination (figure 8.1). This would allow for somewhat denser urban development, while protecting agricultural and environmentally sensitive areas. Local political representatives from the Montreal region – in particular, mayors of suburban municipalities – managed to agree on ways of developing the metropolitan area that would encourage the emergence of a more polycentric urban form.

What urban form the Montreal metropolitan area should take has long been a subject of debate. To some extent, it relates to the issue of periurban agriculture. Indeed, the Montreal region encompasses the most fertile farmlands in Quebec. As a result, more than 57% (221,000 ha) of the territory of the MMC has been designated as agricultural land (Chacine, 2011). Various businesses are involved in fruit and vegetable production, and periurban producers provide fresh produce to the region's households. However, the passage of the provincial *Act to Preserve Agricultural Land* in 1978 failed to slow the erosion of protected farmlands (figure 8.2).

154   Claire Poitras and Pierre Hamel

Figure 8.1 Commuter train station in the town of Deux-Montagnes located on Montreal's North Shore. Photo by Denise Caron, 2016.

Figure 8.2 Saint-Joseph-du-Lac in the Lower Laurentians where hills, orchards, fields, and housing meet. Photo by Denise Caron, 2016.

As a result, the protection of agricultural lands in the metropolitan area has remained a public issue in a context of strong real estate pressures. In Quebec, property taxes represent almost 65% of municipal revenues, and local municipalities can broaden their property tax base by encouraging residential, commercial, and industrial development. Local elected officials are therefore faced with a dilemma that is difficult to resolve: they must choose between encouraging new construction by granting construction permits in nonurbanized areas, on the one hand; and maintaining agricultural lands and natural environments, on the other hand.

Despite decades of plans and strategies developed by public officials to limit urban sprawl and dispersal, the Montreal region has not experienced reconcentration or densification. Filion et al. (2010) show this clearly in their article on the subject. Their analysis questions the common view that Montreal is a compact urban region. Other major Canadian cities such as Vancouver and Toronto launched reconcentration or densification initiatives beginning in the 1970s and 1980s, efforts that included the construction of numerous residential high rises. By contrast, the Montreal metropolitan area has progressively become less dense.

This is a surprising assessment, given the fact that the Montreal metropolitan area, when compared to other major Canadian urban centres, seems to perform better in terms of the compactness of its urban fabric and usage rates for public transit. In 2011, 22.9% of the metropolitan area's population used public transit to commute to work (Statistique Canada, 2011). For the Island of Montreal, the rate was 34.1%.

However, density decreases as you move away from the historic centre of the metropolitan area. Granted, a certain form of densification is underway. But compared to Toronto, residential high rises are largely absent from the metropolitan landscape. Multifamily housing units recently built in the outer suburbs of Montreal tend to be small duplexes or buildings with four storeys or less. High-density residential construction (buildings with five storeys or more) is mainly concentrated in municipalities located on the Island of Montreal and close to downtown, including Westmount and Côte-Saint-Luc, as well as in Saint-Lambert, located across the St. Lawrence River from downtown (Marchand, 2015, p. 10). These communities have aging populations and offer a more diverse choice of housing options than newer suburban municipalities. The latter tend to focus on attracting young families who are more sensitive to the cost of housing and face more complex financial challenges than households without children. The further they move from downtown, the less they have to pay for housing.

In recent years, all of Canada's major cities have experienced a decrease in demographic density alongside a process of economic realignment. However, the Montreal region has been particularly affected by the departure of households, especially young families, for the suburbs. In other words, even as the process of suburbanization has evolved, especially through the diversification of types of households and economic activities, the model of the residential suburb that took shape in the United States during the years following the Second World War remains an everyday reality for households in the Montreal region.

From this perspective, the Montreal metropolitan area has developed differently from Toronto and Vancouver, with a lower degree of centralization. The Montreal region could be said to have followed a model of "spatial decentralization," "a density trajectory ... marked by a sharp decline in its traditional compact built environment and the development of low-density outer suburbs" (Filion et al., 2010, p. 555). That being said, when compared to North American benchmarks, the Montreal region as a whole retains a relatively high degree of density. To understand variations in urban density across different metropolitan regions, the researchers cited above have identified some key geographic and social characteristics that influence urban density: "topography, inherited built environment, urban culture, demographic and market trends, political institutions, land-use policies and patterns, and transportation" (Filion et al., 2010, p. 555). In the case of Montreal, their analysis has highlighted several factors that have influenced both the preservation of older neighbourhoods and the development of new suburban communities. Compared to other major Canadian cities, the Montreal region is also distinguished by its island location and numerous bridges, which increase traffic congestion problems.

Within a territory of 4,360 square kilometres, where industrial and agricultural activities coexist and intermingle with vast stretches of detached housing and protected ecological areas, what criteria can be used to define a boundary between suburbs and the central city? Leaving aside controversies over where to draw the outer boundaries of the metropolitan area and over how well administrative boundaries reflect the socioeconomic activities that take place within them, a study of the daily practices of households in the Montreal region since the end of the 1970s has shown that the central city no longer holds the same meaning for the aspirations, practices, and representations of suburbanites (Fortin & Bédard, 2003). A more recent study of spatial representation in Quebec cinema by sociologist Andrée Fortin (2015) has confirmed this trend. What is more, before the 1960s, Montreal's suburbs displayed a greater diversity in terms of social composition and economic activities than they do today (Sénécal & Vachon, 2012). More recently, the process of suburbanization has nevertheless given rise to new expressions of sociospatial organization.

Montreal's Ethnoburbs

The City of Laval covers a vast territory that includes all of Île Jésus, as well as a few smaller islands. Altogether, it measures 247.09 square kilometres, giving the city a population density of 1,625 individuals per square kilometre (Statistique Canada, 2015). The city's population is slightly more than 400,000 inhabitants, making it the third largest city in Quebec, after Montreal and Quebec City. According to the 2011 census, 24.6% of Laval residents are immigrants, of whom at least 21.2% have been established for two generations or more. Although the presence of immigrants in Laval is not a new phenomenon, it is significant that the proportion of immigrants in the city's population increased significantly between 2001 and 2011, going from 15.5% to 24.6%. With the exception of some suburban municipalities on the Island of Montreal, households whose members were born outside the country – most often in France, Morocco, Lebanon, and Romania – are fairly common in the Montreal region. Laval's sociodemographic characteristics are therefore fairly similar to those of the central city. Nevertheless, there is one characteristic that distinguishes Laval from Montreal: its population is younger than that of the rest of Quebec. In other words, the suburban city continues to attract the sorts of households that are associated with the suburban ideal of family life, as it emerged in the years following the Second World War.

Until the mid-1990s, with some rare exceptions, most immigrant households settled in the urban neighbourhoods of Montreal. The relatively recent settlement of immigrants in the suburbs upon their arrival in the country has changed sociodemographic dynamics in Laval (figure 8.3).

When faced with the arrival of significant numbers of immigrants in the 1990s, the municipal government introduced "a municipal policy on diversity management very different from Montreal's, one that could be described as republican" (Germain & Alain, 2005, as cited in Jean & Germain, 2014, p. 11). In the City of Laval, the visible minority population is very diverse, with representatives from all of the following groups: Chinese, South Asian, Black, Arab, West Asian, Filipino, Southeast Asian, Latin American, Japanese, and Korean.

From Bedroom Community to Ethnoburb[3]

Founded in 1958, the City of Brossard was developed on former agricultural land. At the outset, the city's urban planning commission hired a private firm to prepare a development plan and oversee implementation. A few years later, in 1972, municipal authorities created an urban planning department charged with overseeing development and applying zoning by-laws. In the 1960s and 1970s, real estate developers

Figure 8.3 Laurentian Boulevard in the City of Laval. Photo by Denise Caron, 2016.

working in Brossard were also active in Montreal's other suburban communities. In addition to the main features of the houses themselves, arguments used to attract prospective home buyers highlighted the proximity of major transportation arteries and shopping centres. Newspaper advertisements from the 1970s featured a variety of housing options, including ranch-style houses, semidetached houses, condominium townhouses, and split-level houses. Advertisements therefore sought to encourage a fairly diverse range of social groups to settle in Brossard. Starting in the 1980s, condominiums in larger apartment buildings were added to the mix of available housing.

Originally conceived as a residential area, Brossard has experienced considerable social and urban change over the years. Nevertheless, the original planning concept developed to regulate the municipality's development remains in place. The comprehensive plan that governs the city's development designates neighbourhoods with different letters of the alphabet (for example: B, S, T) that determine the first letter of all street names within a neighbourhood. The varied housing stock (single-family detached homes in the traditional "Canadian" style, semidetached homes, small apartment buildings, residential high rises) is grouped by type, ensuring a certain degree of coherence within the urban landscape. Since its founding, Brossard has applied a relatively strict set of urban planning regulations, and real estate developers have built the city neighbourhood by neighbourhood. It is therefore easy to identify the different periods of construction within the built environment and the urban fabric.

Diversity also characterizes the city's commercial properties, built to accommodate a car-driving clientele. Several different types of retail areas – all typically suburban – have been developed in Brossard over the decades. The 1950s were the era of the "strip," exemplified by Taschereau Boulevard. Then came the classic enclosed malls, such as the Champlain Mall, opened in 1975 and subsequently expanded. Finally, the first decade of the 21st century saw the opening of Quartier DIX30, one of the new breed of shopping centre – the lifestyle centre – at the intersection of two major expressways. Designed around an emulation of an urban shopping district, Quartier DIX30 is an imposing commercial development, onto which office buildings and residential complexes have been grafted to serve the needs of different types of households (single-person households, families, seniors, etc.). By encouraging mixed land use in the area – though not at the building, street, or block level – municipal authorities have sought, on the one hand, to get away from the "bedroom community" image long associated with Brossard; and, on the other hand, to diversify property tax revenues, a leading preoccupation for municipal officials. With these objectives in mind, local authorities gave the green light to the development of new neighbourhoods in the late 1990s. To make the new residential and commercial subdivisions accessible, they had to build new boulevards or extend existing ones. By prioritizing shopping and office spaces, the city "wishe[d] … to distance itself from its image as a bedroom community … " (Bureau d'audiences publiques sur l'environnement, 1999, p. 3). At the same time, creating infrastructure capable of supporting 20,000 new jobs directly addressed the fact that a large proportion of residents did work outside the city's boundaries. The municipality also supported the development of public transportation services to improve access to Montreal's central business district, which remains an important destination for commuters.

Originally built as a bedroom community, Brossard has successfully diversified its economic base. It has become a suburban city with a broad-based economy and a booming new commercial area onto which high-density housing has been grafted. In more recent years, municipal authorities have also authorized the development of new, "premium" residential subdivisions, where houses and lots are larger than in older sections of the city. This has been part of a strategy to stay competitive with newer suburbs, while preserving Brossard's appeal as a planned community.

The demographic changes that have occurred in Brossard in recent years have led local authorities to reexamine the residential fabric of the city, with an eye to attracting more affluent households. At the turn of the 21st century, one real estate developer in the area began promoting a new residential subdivision with a spatial organization inspired by gated communities. The subdivision therefore features a prominent entrance gate, landscaping that clearly marks its perimeter, winding streets, and strict controls on architectural styles. Working from a set of technical specifications approved by the City of Brossard, the developer ensured that the appearance of the houses and their surroundings met the municipality's criteria. Ultimately, however, these criteria were intended to maintain property values, as made clear in the following passage from a promotional document:

> "Domaines de la Rive-Sud" is a residential project comprised solely of detached single-family houses in a distinctive, country-like setting. The planning concept values the natural elements highlighted by sinuous streets and magnificent landscaping. The plan favours large lots so as to ensure the desired flexibility for architectural design of houses. The project's brand image is based on the application of norms and criteria that guide the execution of each residential project, for greater harmonization and quality control. When applied to the development of the site, to the architecture and to the choice of materials, these quality standards will ensure the integrity of the overall environment as well as its long-term value. (Domaine de la Rive-Sud, Technical Specifications, January 2005)

Designed to attract potential new home buyers, the promotional language crafted by 21st-century real estate developers is not unlike that used by those who developed the original planned suburbs of the 1950s and 1960s. The primary goal remains to convince future residents that residential planning norms will help maintain property values and, by extension, preserve a certain homogeneity in the area's social and functional composition.

According to a study by Anne-Marie Séguin (2011), this type of subdivision, somewhat inspired by gated communities, is not particularly common in suburban Montreal. By analysing the advertising discourse used by residential real estate developers in daily newspapers between 2004 and 2009, Séguin shows how promoters only rarely appealed to the private residential enclave model. To begin with, the term "gated community" and its French equivalent, as well as the notions of closure and controlled access, were absent from the advertising discourse. Instead, the advertisements highlighted amenities (pools, exercise rooms, other common services) to convince potential home buyers to choose a particular development. At the same time, the availability of such amenities was rather commonplace in both condominium-style developments and low-rise residential developments. Meanwhile, security issues were rarely brought up.

Above all, when promoting residential subdivisions developed in suburban municipalities, developers focused on the advantages offered by their geographic locations. In this way, the arguments put forward by real estate promoters were not that different from the ones used by suburban municipalities themselves to convince households to settle in a particular area.

French Flight Instead of White Flight

Within a predominantly English-speaking continent, the Montreal region stands out from the rest of Quebec with regard to the use French as a first language. In 2011, more than 92% of individuals living elsewhere in Quebec had French as a first language. By contrast, in Greater Montreal, the percentage was 62.3% (MMC, Observatoire métropolitain, 2011). In particular, this difference can be explained by the historic presence of an English-speaking community and by the arrival of new immigrants whose first language was neither English nor French. To ensure the survival of French, public authorities in Quebec have been seeking for many years to ensure that new immigrants become integrated into the French-speaking community.

Compared to other North American metropolitan areas, the Montreal region stands out for another reason. Contrary to what was observed in the United States during the 1950s and 1960s, the phenomenon of White flight – the exodus of the White middle and upper classes from the central city to the suburbs – did not occur in Montreal. However, if White flight did not occur (Marois & Bélanger, 2014), there was nevertheless a form of sociospatial segregation among immigrants, based on the areas where they chose to settle. Marois & Bélanger (2014) have examined how various factors helped determine immigrants' choice of residence.

One thing that sets the Montreal region apart from other metropolitan areas in North America is the relationship between English- and French-speaking groups, which has produced a sort of dual city (Levine, 1991). However, language dynamics in the Montreal region have changed in recent years with the arrival of many immigrant households who speak neither French nor English as a first language. The analysis undertaken by Marois & Bélanger (2014) to explain why immigrants choose a particular municipality shows that beyond the availability of services and housing that meet a household's needs, the linguistic composition of a community was a more important factor than the presence of visible

Table 8.3 Proportion of the population in Quebec having knowledge only of French, 2011

Greater Montreal	36.2%
Island of Montreal	28.0%
Laval	36.6%
Longueuil urban area	40.0%
Northern outer suburbs	54.9%
Southern outer suburbs	43.5%
Rest of Quebec	65.6%

Source: MMC, Observatoire du Grand Montréal and Statistics Canada, Census of Population, 2011.

minorities. In contrast to what occurred in American cities, in Montreal "it was not so much [W]hites who tended to leave the central city for the suburbs as visible minorities" [our translation] (Marois & Bélanger, 2014, p. 442). Ultimately, language seems to have proven a bigger consideration than race: "Our results show ... that spatial segregation was strongest in terms of language. For French-speakers, English-speakers, and other language groups, the attractiveness of a municipality depended on the proportion of individuals belonging to the same group: the greater the proportion, the more those moving within the metropolitan area were attracted to a particular municipality" (Marois & Bélanger, 2014, p. 461). This was particularly striking among English- and French-speaking households: "What is most remarkable about the spatial distribution of households in the Montreal region, especially in the suburban municipalities located outside of the Island of Montreal, is the omnipresence of a population having knowledge only of French according to the Canadian census of 2011" (Marois & Bélanger, 2014, p. 461). For example, municipalities in the northern outer suburbs such as Terrebonne, Mascouche, and Mirabel, which experienced strong demographic growth in recent years, have a large proportion of residents who speak only French (table 8.3 and map 8.2).

Also for example, during the last two decades, Mirabel, which remains home to significant agricultural activity, has been transformed into a sort of boomburb with growth rates among the highest in the Montreal region (figure 8.4). The population rose from 34,626 inhabitants in 2006 to 41,957 inhabitants in 2011, which corresponds to a growth rate of 21.2% (Statistics Canada, Focus on Geography Series, 2011 Census).

Naturally, the central part of Greater Montreal is highly urbanized, but substantial outlying areas are protected under the *Act to Preserve Agricultural Land*. In 2011, protected areas recognized for the quality of their soil covered 58% of all lands in the metropolitan area (MMC, 2011). Debates on the protection of agricultural lands in the Montreal region are not new. Since the 1960s, when demographic growth was expected to remain very strong, various plans and blueprints have been developed by municipal and regional stakeholders to protect agricultural interests. However, since the adoption of the *Act to Preserve Agricultural Land*, multiple amendments have been made to zoning regulations, allowing suburban development to encroach on agricultural areas.

Since the adoption of the MMC's metropolitan land use and development plan in 2011, local officials and real estate developers have had to adjust their own plans and bring them in line with newly established regulations designed to protect natural areas and agricultural lands in Greater Montreal.

Alongside the significant growth of various suburban municipalities, real estate promoters have actively promoted new residential developments. In collaboration with the municipalities themselves, developers appeal to classic suburban ideals to attract new residents. Thus, local officials in Mirabel and real estate promoters responsible for the construction of new subdivisions use traditional sales arguments to attract potential buyers, especially young families: "For many citizens, Mirabel represents a strategic location between work and leisure activities. For others, greenery and open spaces are the main attraction" [our translation] (City of Mirabel, 2015). Among other arguments used to attract new residents were low taxes, proximity to key roads and expressways, local employment opportunities, and the presence of numerous parks. In short, the suburban ideal to which many young families aspire is very much alive. However, it should be noted that a significant number of Mirabel households are led by couples who are not legally married. This reflects another particularity of Quebec as a whole, which also characterizes the social landscape of the Montreal region (Pelletier, 2012).

Does Montreal have Boomburbs?

Are there boomburbs in the Montreal metropolitan area? The authors responsible for this neologism (Lang & LeFurgy, 2007) have defined boomburbs as suburban municipalities or communities that have a significant population (more than 100,000 inhabitants), that are not the largest municipality in their metropolitan area, and that have experienced a growth rate of more than 10% in recent years. None of the suburban municipalities in the Montreal metropolitan area meet all of these criteria, especially the one regarding population size. However, the Canadian censuses of 2006 and 2011 show that 18 of the

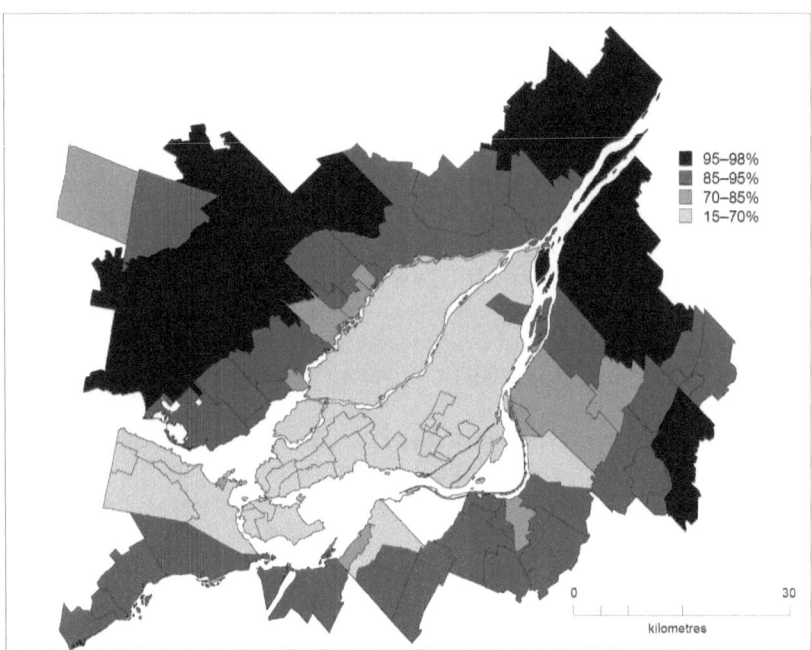

Map 8.2 French reported as mother tongue in the Montreal Census Metropolitan Area, 2011. Source: INRS-Urbanisation Culture Société.

82 municipalities that make up the Montreal metropolitan area have at least some features of a boomburb. Furthermore, the communities that most resemble boomburbs are almost all located in the northern and southern outer suburbs (see table 8.2, above).

Suburbs located outside the Island of Montreal also include areas of poverty, which often include concentrations of households led by single mothers. Thus, certain areas developed following the Second World War, especially some parts of the inner suburbs, including Longueuil on the South Shore and Laval to the north of the Island of Montreal, are home to low-income households. This helps explain why, in the Montreal metropolitan area, the proportion of renters is relatively high. In 2011, the proportion of households in the metropolitan area living in rented accommodations was 45.3%, compared to 32.9% for the rest of Quebec. However, the metropolitan area is home to a rather diverse range of situations and, as always, it is important to remain wary of oversimplifications. In particular, a high percentage of renters is not necessarily a sign of poverty. For example, in the well-to-do inner suburb of Westmount, located on one of the slopes of Mount Royal in the centre of the Island of Montreal,

Figure 8.4 A new residential area in Sainte-Marthe-sur-le-Lac where the population grew by 38.8% between 2006 and 2011. Photo by Denise Caron, 2016.

46.5% of households rent their accommodations. In other words, inner suburbs also include a large proportion of renters. In comparison, the more recent suburban communities accommodate a much larger proportion of homeowners. This is yet another example of how newer suburban municipalities display certain characteristics of the bedroom communities that first appeared in the years following the end of the Second World War.

Suburbia as Cultural Space

Over the last two decades, Montreal's suburbs have experienced several major changes, including the construction of major new cultural amenities. For example, concert venues and new libraries that offer experiences and services beyond those provided by their traditional counterparts, where silence and reading reigned, have provided suburban municipalities with important new spaces for social interaction. These new libraries, designed as vast spaces bathed in natural light and with the capacity and mandate to host a wide diversity of activities, now play the role of socio-community centres offering a range of services to communities that are increasingly socially diverse. In this way, their new role reflects a new suburban reality. It also shows how municipal decision makers have committed financial resources to attracting and retaining families by offering better cultural and recreational infrastructure.

The centre city no longer holds a monopoly on culture. Granted, a study undertaken a few years ago on the location of artists' studios within the City of Montreal (Bellavance & Latouche, 2008) has shown how these spaces are concentrated in former industrial areas of the centre city, using repurposed buildings. But the suburbs have also become centres for both cultural expression and affirmation. In the Montreal region, Isabelle Hayeur's work involving site-specific installations and public art projects is particularly noteworthy. Hayeur uses video to illustrate the effects of suburban development – especially the construction of McMansions – on the environment and landscape of metropolitan Montreal in areas where traces of older agricultural settlement are still visible. Likewise, contemporary films by André Forcier (*Coteau Rouge*, 2011) and Xavier Dolan (*Mommy*, 2014) showcase the suburbs as an offbeat landscape, home to tumultuous human relationships, especially in the working-class neighbourhoods of Montreal's South Shore. This particular area has also been acknowledged as a source of inspiration by the Canadian indie rock group Arcade Fire. From this perspective, research undertaken by Andrée Fortin has clearly shown how, beginning in the first decade of the 21st century, suburban life emerged as the everyday reality of characters populating Quebec films that also showcase the suburbs of Montreal (Fortin, 2015). Furthermore, the suburbs' increasingly large role in Canadian cultural production is clearly demonstrated in the work of Alison Bain (2013). Along with producing works in and about the suburbs, artists have also been active in penurbia – the rural areas and towns on the outer periphery of major cities. These spaces also make up part of the metropolitan area and contribute to its reconfiguration.

Conclusions

Going back to the 1960s, we can see that the destiny of North American rustbelt city regions has also been met in the Montreal metropolitan area. Yet Montreal differs in that in recent decades its central core has proven to be comparatively more dynamic and accessible. This is related both to economic and linguistic issues. Due to anemic economic development, extensive suburbanization outside the urbanized historical core did not occur before the 1980s. Since then, sociospatial change and peripheral expansion has been marked by French flight, instead of White flight. The dual city – as expressed by cultural markers produced on the one hand by French speakers and, on the other, by English speakers – has been replicated in the expansion of suburban municipalities located outside the Island of Montreal.

In retrospect, the historical linguistic division characterizing Montreal since the introduction of the British Regime in the 1760s continues today at the metropolitan scale, bringing with it new challenges, as the general cultural context within which English and French speakers are evolving is considerably different. With this mind, while Montreal may be revealing itself to fit within the canon of North American metropoles, its linguistic duality continues to highlight its distinct identity.

NOTES

1 Between 2001 and 2011, in Greater Montreal (the territory of the MMC), the proportion of households composed of people living alone increased from 30.9% to 32.7%. This category of household had the strongest growth during the period (CMM, 2013, p. 2).
2 For the purposes of planning and coordination of public services, the Quebec government has divided the province into 20 administrative regions. The Montreal metropolitan area covers, in whole or in part, five of these regions.
3 Some elements of this section were previously published in French in a book chapter (Poitras, 2012).

REFERENCES

Bain, A. (2013). *Creative margins: Cultural production in Canadian suburbs*. Toronto, ON: University of Toronto Press.
Bellavance, G., & Latouche, D. (2008). Les ateliers des artistes dans l'écosystème montréalais. Une étude de localisation. *Recherches sociographiques, 49*(2), 231–60.
Bureau d'audiences publiques sur environnement. (1999). *Élargissement et construction de boulevards à Brossard*. Montréal, QC: BAPE.
Chacine, G. (2011). L'autre agriculture urbaine en zone métropolitaine: une recherche-action sur les opportunités de mise en valeur et développement de l'agriculture périurbaine montréalaise. *VertigO. La revue électronique en sciences de l'environnement, 11*(1).
City of Mirabel. (2015). A propos de Mirabel. Retrieved 6 September 2019 from https://ville.mirabel.qc.ca/ville/a-propos-de-mirabel
Communauté métropolitaine de Montréal. (2011). *Perspective du Grand Montréal*. Montréal, QC: Communauté métropolitaine de Montréal (8p).
Communauté métropolitaine de Montréal. (2013). Évolution récente de la composition des ménages et du logement du Grand Montréal.

Perspective du Grand Montréal, Montréal: Communauté métropolitaine de Montréal (8 p).

Filion, P., Bunting, T., Pavlic, D., & Langois, P. (2010). Intensification and sprawl: Residential density trajectories in Canada's largest metropolitan regions. *Urban Geography, 31*(4), 541–69.

Fortin, A. (2015). *Imaginaire de l'espace dans le cinéma québécois*. Québec: Presses de l'Université Laval.

Fortin, A. (2013). Un nouveau récit collectif dans le cinéma québécois : la centralité de la banlieue. *Sociologie et sociétés, 45*(2), 129–50.

Fortin, A., & Bédard, M. (2003). Citadins et banlieusard. Représentations, pratiques et identités. *Canadian Journal of Urban Research/Revue canadienne de recherche urbaine, 12*(1), 124–42.

Gournay, I., & Vanlaethem, F. (Eds.). (1998). *Montréal metropolis, 1880–1930*. Montreal, QC: Canadian Centre for Architecture.

Hamel, P., & Poitras, C. (2004). Déclin et relance économique d'une agglomération métropolitaine. Le discours et les représentations des élites économiques à Montréal. *Recherches sociographiques, 45*(3), 457–92.

Higgins, B. (1986). *The rise and fall of Montreal*. Moncton, NB: Institut canadien de recherche sur le développement régional.

Jean, S., & Germain, A. (2014). La diversité croissance des quartiers de classe moyenne dans la métropole montréalaise. Des jeunes familles perplexes. *Canadian Ethnic Studies/Études ethniques du Canada, 46*(2), 5–25.

Lang, R.E., & LeFurgy, J.B. (2007). *Boomburbs: The rise of America's accidental cities*. Washington, DC: Brookings Institution Press.

Levine, M. (1991). *The reconquest of Montreal: Language policy and social change in a bilingual city*. Philadelphia, PA: Temple University Press.

Linteau, P.-A. (2000). *Histoire de Montréal depuis la Confédération*, deuxième édition augmentée. Montréal: Boréal.

Marchand, J.-F. (2015). Y a-t-il une densification résidentielle dans la Communauté métropolitaine de Montréal (CMM)? Regard sur les données de recensements de 2001 à 2011. Veille Stratégique métropolitaine. Montréal: Secrétariat à la région métropolitaine (20 mars), 12 p.

Marois, G., & Bélanger, A. (2014). Déterminants de la migration résidentielle de la ville centre vers la banlieue dans la région métropolitaine de Montréal: Clivage linguistique et fuite des francophones. *The Canadian Geographer/Le géographe canadien, 58*(2), 141–59.

Pelletier, D. (2012). Accéder au meilleur quartier possible : type de famille et ségrégation résidentielle croisée à Montréal. *Cahiers québécois de démographie, 41*(2), 257–98.

Poitras, C. (2012). Les banlieues résidentielles planifiées dans la région de Montréal après la Seconde Guerre mondiale. Un modèle en redéfinition? In

D. Fougères (Ed.), *Histoire de la Montréal et de sa région, tome II, De 1930 à nos jours* (pp. 899–924). Québec: Presses de l'Université Laval and INRS.

Polèse, M. (2012). Montréal économique: de 1930 à nos jours. Récit d'une transition inachevée. In D. Fougères (Ed.), *Histoire de Montréal et de sa région, tome II, De 1930 à nos jours* (pp. 959–1004). Québec: Presses de l'Université Laval and INRS.

Séguin, A.-M. (2011). L'attrait pour le modèle de l'enclave résidentielle privée ... que nous révèle le discours publicitaire sur le logement neuf? *Cahiers de géographie du Québec, 55*(4), 109–29.

Sénécal, G., & Vachon, N. (2012). L'expansion métropolitaine: vers une polycentralité assumée. In D. Fougères (Ed.), *Histoire de Montréal et de sa région, tome II, De 1930 à nos jours* (pp. 867–98). Québec: Presses de l'Université Laval and INRS.

Statistique Canada. (2011). *Commuting to work: Results of the 2010 general social survey*, by Martin Turcotte. Ottawa, ON: Statistique Canada.

Statistique Canada. (2015). *Profils statistiques par région et MRC géographiques.*

9 New York's Suburbs in a Globalized Metropolitan Region

JAMES DEFILIPPIS AND CHRISTOPHER NIEDT

Introduction

The New York metropolitan area both mirrors the larger stories of American suburbanization and is fundamentally different from them. It is not simply New York's size and scale that make it unique, although that certainly matters. With more than 22 million people, stretched over 32 counties in four different states,[1] it is an enormous region that includes almost 7% of the American population. Beyond its size, there are several features that distinguish the metropolitan region within the North American urban system. This chapter describes the historical development of the region and its suburbs, and analyses a set of dynamics that has reshaped the region's suburbs in the last 20 years.

There are several stories told in this chapter. First, the historical development of region's suburbs is one of the imperfectly realized efforts of the region's elites to control and shape the development of the region in ways that both served themselves and mitigated the tensions inherent in capitalist urbanization (and thereby eased the political conflicts that arise from such tensions). Second, while the 1898 consolidation of New York City's five boroughs swallowed up a set of towns that were suburbs of New York, including Brooklyn – at that time limited to today's Brooklyn Heights, the first commuter suburb in North America (Jackson, 1987) – the political fragmentation that has marked mass suburbanization unmistakably marks this region as well. Such fragmentation has enabled exclusion and segregation, which have been contested by local and regional movements for civil rights. Third, the suburbs of New York have always existed in a global context. This was always a global metropolitan region, and contemporary globalization has remade the region's labour markets and demographic makeup. Fourth, this demographic transformation has brought with it contentious politics, as immigrants face anti-immigrant policies and political sentiments in the region's suburbs. Fifth, as the older suburban downtowns struggle with economic decline, there have been pronounced efforts to rebuild them. Finally, all of these issues have become focal points for social movements in the suburbs that have challenged elite dominance within them.

Taming the City and Suburbs: The Elite Project of Regional Planning

New York's prominence, gigantic size, and complexity have invited competing visions for the rationalization and reordering of metropolitan space. Formal metropolitan or regional planning – grand plans for the transformation of infrastructure, landscape, and the governance of both – have typically been elite projects, while seldom attracting the elite's unanimous support. Rather, struggles around regional planning have reflected the incoherence of elite priorities, with their varied ideological commitments, economic interests, and attitudes towards a race- and class-stratified public. With the eventual development of local, submetropolitan elites and growth machines, geographic divisions constituted an additional challenge to the coalition-building required to realize elite regional aspirations.

These tensions were apparent well before contiguous development had neared the present-day boundaries of New York City. Scobey (2002) describes how the development of real estate markets enabled a coalition of Victorian reformers, property speculators, and political bosses to promote a particular form of bourgeois urbanism in mid-19th-century New York – one that stressed moral environmentalism as a way to overcome the ill effects of crowded residential districts. Facing resistance from landlords and the political machine in downtown Manhattan, this coalition looked outward – to Central Park and Upper Manhattan, Brooklyn, Staten Island, and eventually the Bronx – seeking greenfield sites for a bourgeois utopia that would "counteract the confinement and moral corrosiveness of the older city with a topography of domestic and civic virtue" (217). Of course, this topography was often populated by farmers, who did not at all fit into visions of the regional planners. Thus Kings County (Brooklyn) had an Olmsted park and "parkways" planned and built into what was still agricultural land in the mid-19th century (see Linder & Zacharias, 1999), as they were being swept away by what we would now call "suburban sprawl." The 1898 creation of "Greater New York City" more fully incorporated these suburban towns into the city, despite the existence of many neighbourhoods in the four "outer boroughs" that remain quite "suburban" in their demographics, commuting patterns, densities, and built environments.

Many of these proto-planners, supported by elite organs like the *New York Times*, envisioned a place for the working class – Charles Loring Brace suggested developing "cheap workingmen's trains [to] suburbs laid out in New Jersey or on Long Island expressly for working people" (Scobey, 2002, p. 248) – yet segregated workers into "a landscape of civilizing hierarchy into which they were welcomed as subordinates" (249). These efforts enjoyed limited success after the downfall of local machine leader Boss Tweed and the economic turmoil of the Panic of 1873, which compromised the political and fiscal basis for the coalition. But bourgeois visions were realized far outside the city limits in early "bourgeois utopias" like Llewellyn Park (Fishman, 1987).

These early plans, like Olmsted's design for the West Bronx, incorporated some minimal infrastructural integration between city and suburb (Scobey, 2002), yet actual metropolitan planning as such – that foregrounded regional planning rather than suburban community building – did not begin in earnest until the Progressive Era (1900–1920). Catalyzed by the work of Patrick Geddes, first-wave regionalists included "decentralists" such as Lewis Mumford who favoured opening up and dispersing crowded urban centres, while "metropolitanists" favoured revitalizing the central city as the core of an expanding urban region (Fishman, 2000). Both schools of thought were steeped in cultural representations of pathogenic immigrant ghettos and industrial pollution, and both agreed that a conscious segregation of land uses and a controlled extension of the metropolis would dissipate unhealthy urban environments.[2]

Regional planning continued to be an elite project through the 20th century. For the most part it was dominated by the Regional Plan Association (RPA), a business and efficiency-centred organization. In many ways, their work played a far greater role in shaping the trajectory of the suburbs in the region than is generally acknowledged. The RPA's first regional plan was released in 1929, shortly before the stock market crashed. It was, as Robert Fishman (2000) argued, one of "recentralization" – at least of certain activities and people. Manhattan, and the expansion of its central business district, was the central concern of the plan. Thus, the plan called for the removal of manufacturing and port activities from Manhattan – as both had "a chance for unlimited growth in the New Jersey part of the district, where industrial sub-centers would naturally occur, rather than Westchester or in Long Island" (quoted in Fitch, 1993, p. 69). Long Island would be further sheltered from the incursion of manufacturing and

New York's Suburbs in a Globalized Metropolitan Region 173

Figure 9.1 New housing near South Farmingdale by the Southern State Parkway. Courtesy of the Hofstra University Special Collections, Long Island Parkways Collection.

low-cost housing for such workers by the use of parkways and parks that would act as buffers to prevent the expansion of such activities into Nassau County. Manufacturing came anyway, but perhaps not to the extent it would have otherwise. And, of course, there would have to be highways and parkways to open up the newly protected suburban towns of Westchester and Long Island (figure 9.1). These transformations are strongly associated with Robert Moses, the public official who is central to many histories of New York's midcentury urban and suburban development. Yet despite Moses's tremendous institutional power – as well as his capacity for individual grievance and practical compromise – he was ultimately the technocratic executor for a broadly shared vision of modernist planning.

By the 1960s, regional planning had changed direction. It was now increasingly concerned with the losses of (wealthy and White) people in the centre, and planners realized that the emphasis on roads had facilitated the White flight of the postwar period. Accordingly, it shifted its infrastructural emphasis from roads to rail. But the rapid growth of the suburbs posed a challenge to any and all efforts to plan the region from the centre. This is particularly true because the powers of local home rule were sacred to so many suburbanites. Planning continued, however, to emphasize deindustrialization at the core in order to accommodate an expanded CBD for the region. By then, however, the suburbs had developed a set of regional centres with their own local elites, which made coherent regional planning from the centre increasingly difficult. On Long Island, Nassau and Suffolk's political machine repeatedly resisted regional planning proposals – such as those initiated by bodies such as the RPA and the Tri-State Regional Council – and often sought to restrict the powers of regional bodies or extricate Long Island from their purview altogether. The ability to tame the suburbs via a regional planning framework that emphasized the centre wasn't just hampered by the presence of suburban elites with their own agendas, but also by the sheer political and administrative fragmentation of the suburbs.

Segregation and Fragmentation

The postwar expansion of the region's suburbs followed the pattern that is common in so many metropolitan areas. The failure to construct a meaningfully coherent regional government meant that much of the postwar boom in the suburban towns and counties was one of increased political fragmentation and segregation. By the early 1960s, the region had more than 1,400 different governments in three different states (Wood, 1961). As of 2015, the number of jurisdictions had grown even more and now includes more than 2,000 different governmental units, including counties, towns, cities, special government units, authorities, and other entities (Levine, 2015, p. 238).

While Wood (1961) famously concluded that governments didn't actually matter that much in shaping the development of the region, that argument is specious at best. There were, and are, pronounced and significant differences between the myriad of different political jurisdictions in the region, and these have often been the result of government action. Westchester County towns, for instance, were largely very successful in their efforts to keep industries – and their working class employees – out of the county. Overlaid with strong racial anxieties, this fear of "Bronxification" has been central to the public policies of the towns in that county. This was largely accomplished via the well-known mechanism of large-lot zoning (Duncan & Duncan, 2004), and this practice of exclusionary zoning has continued: Westchester County reached a desegregation agreement with the US Department of Housing and Urban Development in 2009 (Roberts, 2009), and the issues surrounding the implementation of that agreement continue as of this writing. And the most working class of the Westchester suburbs, Yonkers, had for so long implemented racially discriminatory housing and education policies that the NAACP and the federal Justice Department brought a lawsuit against the town in 1980. In 1985 the courts agreed with the plaintiffs that Yonkers had been guilty of 40 years of discriminatory policies and instituted a federally appointed monitor to implement desegregation of the town's public housing projects and its public schools (Belkin, 1999). That federal monitor governed housing and education integration in the town until 2007, when the town and the NAACP finally reached a settlement (Santos, 2007).

The New Jersey suburbs had comparable forms of discrimination, which were also challenged in court. In 1975, the State Supreme Court of New Jersey found in favour of the plaintiff (the South Burlington NAACP) against Mount Laurel Township (see Massey, Albright, Casciano, Derickson, & Kinsey, 2013). While this case emerged from Philadelphia suburbs of south Jersey, the court's decision, and its subsequent one in 1983 (colloquially called *Mount Laurel II*), transformed the landscape of affordable housing policy statewide. While there is not the space here to fully explain the history and developments associated with the Mount Laurel cases, the judicial and legislative remedies adopted after this pair of cases did press some suburban jurisdictions to accommodate the development of new affordable housing (though legislative loopholes allowed others to maintain exclusionary zoning).

The rapidly growing suburbs of Long Island were certainly enacting their own forms of segregation and discrimination. Prewar suburbs such as Garden City regulated occupancy through restrictive covenants, while Levittown – built with African American labour (figure 9.2) – was closed to African American residents by Levitt and his sales agents (Kushner, 2009). Even as the era of direct "de jure" segregation ended, discrimination continued through the familiar panoply of real estate agent steering, individual landlord refusals, local exclusionary zoning, weak government enforcement of fair lending law, NIMBY neighbourhood opposition, and extralegal terrorism. Long Island's civil rights groups – including active chapters of the NAACP and CORE, as well as testing groups like Long Island Housing Services – fought to open Long Island's homogenous communities from the 1960s forward. More recently, groups like ERASE Racism have successfully pursued lawsuits against municipalities that have adopted exclusionary zoning ordinances and area landlords that continue to discriminate against applicants.

On Economic Transformations, Globalization, and the "Great Inversion"

The New York suburbs have long included a set of cities that were centres in their own right, even if their existence as such greatly depended

Figure 9.2 African American construction workers during the construction of Levittown, NY – homes that African Americans were unable to purchase due to discrimination. Courtesy of the Hofstra University Special Collections, Levittown Construction Photographs Collection.

upon their proximity to the metropolitan centre, and their place within a much larger urban agglomeration. For a long time these second-order centres were primarily manufacturing based. Paterson, NJ, for instance, was founded by Alexander Hamilton in 1792 for the express purpose of building a manufacturing economy independent of the British, and would play a central role in the development of American manufacturing, before deindustrialization in the postwar period devastated the city. Newark, NJ, has a similar history. It has long been the largest city in the state and an important transportation and manufacturing hub for the region. The region's elite played a role in the making of Newark as a transport hub by their decision to move the port to the western side of the Hudson River. Despite the location of the port (and airport), deindustrialization and White flight into other parts of Essex County in the postwar period, coupled with the large uprisings/riots in 1967, left Newark one of the poorest cities in the country. The deindustrialization of these two regional centres was not anomalous and instead represented the larger deindustrialization that the region underwent from the 1960s onward. To some extent, the expulsion of manufacturing from the five boroughs of New York City is one that the city's elite – including those engaged in regional planning – have long been quite comfortable with, and in many ways actively encouraged. Still, the loss of manufacturing in the New Jersey parts of the region was largely unplanned and cost the region significantly in economic diversity and vitality.

More recent regional centres have emerged, however; but these regional centres are very much of the post-industrial variety. The oldest of the newer, professional-services-centred regional centres is Stamford, CT. A newer and different kind of regional centre has emerged in Hudson County, New Jersey, just across the Hudson from lower Manhattan. Stamford began its transformation from disinvested regional city to a hub of office development when the General Telephone & Electronics Corporation (GTE) moved its headquarters there from New York City in 1973. In the more than 40 years since then, Stamford has become one of the major office hubs in the region, and includes the headquarters for Thompson Corporation (Thompson Reuters) and NBC Sports among its downtown occupants. The story of Hudson County (Jersey City, mostly) is a bit different in that its development occurred later – from the 1990s onwards – and was driven less by the relocation of corporate headquarters than by the relocation of back offices of larger firms headquartered just across the river in Manhattan. In the last 20 years, though, Jersey City has developed more than 18 million square feet of Class A office space, which is the 12th largest amount for cities in the United States (Healy, n.d.).

But it is the loss of manufacturing employment and the concurrent rise in what used to be called the FIRE (Finance, Insurance, Real Estate) sectors that dominates the region's labour-market and land-development policies and trajectories. These are, of course, central to the dynamics involved in the myriad of processes that constitute economic globalization. New York's role as the quintessential "global city" (Sassen, 1991) is obviously the story here. And while the region's economy has always been "globalized," the empirical content of "globalized city-region" has certainly changed as different regimes of accumulation have been dominant. The current period is one in which a set of dynamics has yielded a significant centripetal force within region. These dynamics include the centralization of capital and command and control functions that has taken place in global cities, and the deindustrialization and the residentialization/gentrification of formerly industrial areas within the centre.

Suburbanites and suburban elites, meanwhile, have become increasingly concerned about the loss of young people to the stronger, more vibrant centre. During the past two decades on Long Island, for example, policymakers, foundations, housing advocates, and the local media have developed a consensus that the region must provide affordable housing or face an exodus of its young adults. This threat of a "youth drain" has served as a pillar of regional affordable housing advocacy, advocacy that seeks to make common cause between regional liberal and conservative elites in the service of redevelopment and expanded Transit-Oriented Development (TOD) rental housing opportunities.

To some extent the trend is real, and suburbs like Nassau and Suffolk counties witnessed a sharp decline in the absolute number of young adults age 25 to 44 between 2000 and 2010: 17.0% and 14.3% (respectively) in just 10 years; the same statistics in the exurban suburban counties of New Jersey were even higher. But set in context, the change is less dramatic than it appears. First, the declines partly reflected the aging of the baby boomers and the smaller size of the younger cohort nationwide: the absolute number in the 25 to 44 age group also fell across the nation, state, and most of New York City. This generational shift was particularly important for explaining the decline on Long Island, which had small child and adolescent cohorts in 1990. The declines also took place amid the build out and levelling off of the overall population. When we examine younger adults as a share of the total population (table 9.1), the share fell 5.0 and 5.8 percentage points in Nassau and Suffolk counties, compared to 1.7 points in New York City, 3.6 points in the United States, and 3.5 points in New York State. This shift does not seem as dramatic as the headline figures of absolute decline would suggest.

It is true that the absolute numbers of young adults rose in Hudson and Kings Counties and remained stable as a share of the total population, seemingly giving support to the argument of housing advocates that young adults are fleeing the suburbs for the city. Yet this generalization obscures that "youth" of different racial/ethnic groups were "draining" from opposite sides of the city line. In the suburbs, the declines in the number of non-Hispanic White youth in Nassau and Suffolk counties were considerably more pronounced than the declines in young adults generally. Although non-Hispanic White young adults increased in Manhattan and Brooklyn, and fell only modestly in Hudson County, the percentage of Latino and Asian youth increased substantially across much of the New York suburbs – particularly on Long Island during the same decade. African American young adults declined modestly in both absolute terms and as a share of the total population in many of the same areas. But the number of

Table 9.1 Change (pct. pts.) in young adults as share of total population, New York, 2000–2010 (lowest five in light grey, highest in dark grey)

	Total	NH White	Latino	NH Black	NH Asian	NH Other	All PoC
Southern and western (New Jersey) suburbs							
Bergen County, New Jersey	-4.7%	-6.6%	1.4%	-0.2%	0.8%	-0.1%	1.9%
Essex County, New Jersey	-2.6%	-2.7%	1.5%	-1.4%	0.1%	-0.2%	0.1%
Hudson County, New Jersey	0.4%	-0.7%	-0.4%	-0.6%	2.6%	-0.5%	1.1%
Hunterdon County, New Jersey	-9.1%	-9.9%	0.7%	-0.2%	0.4%	0.0%	0.9%
Mercer County, New Jersey	-3.7%	-6.2%	1.9%	-0.6%	1.3%	-0.1%	2.5%
Middlesex County, New Jersey	-4.5%	-7.2%	1.2%	-0.5%	2.1%	-0.1%	2.7%
Monmouth County, New Jersey	-6.4%	-6.9%	1.1%	-0.5%	0.0%	0.0%	0.5%
Morris County, New Jersey	-6.7%	-8.2%	0.9%	-0.1%	0.7%	0.0%	1.5%
Ocean County, New Jersey	-3.9%	-4.8%	1.0%	-0.1%	0.1%	0.0%	0.9%
Passaic County, New Jersey	-4.2%	-4.4%	1.3%	-0.9%	0.3%	-0.4%	0.2%
Somerset County, New Jersey	-7.5%	-10.0%	1.1%	-0.1%	1.5%	0.0%	2.6%
Sussex County, New Jersey	-7.6%	-8.6%	0.7%	0.1%	0.1%	0.0%	1.0%
Union County, New Jersey	-3.8%	-5.0%	2.1%	-0.8%	0.1%	-0.2%	1.2%
Warren County, New Jersey	-7.2%	-8.9%	1.0%	0.4%	0.4%	0.0%	1.7%
Northern (New York/Connecticut) suburbs							
Fairfield County, Connecticut	-5.0%	-6.9%	1.6%	-0.2%	0.5%	-0.1%	1.9%
Hartford County, Connecticut	-4.2%	-5.8%	1.0%	-0.1%	0.8%	-0.1%	1.6%
Litchfield County, Connecticut	-7.1%	-7.8%	0.7%	0.0%	0.1%	0.0%	0.7%
New Haven County, Connecticut	-4.2%	-6.0%	1.5%	-0.1%	0.4%	0.0%	1.7%
Dutchess County, New York	-6.3%	-7.0%	1.0%	-0.5%	0.2%	0.0%	0.7%
Orange County, New York	-4.8%	-6.9%	1.7%	0.1%	0.2%	0.1%	2.1%
Putnam County, New York	-7.9%	-9.5%	1.6%	0.0%	0.1%	0.0%	1.6%
Rockland County, New York	-4.4%	-5.7%	1.9%	-0.4%	0.0%	-0.2%	1.3%
Sullivan County, New York	-4.4%	-4.7%	0.8%	-0.7%	0.1%	0.1%	0.4%
Ulster County, New York	-5.6%	-6.0%	0.5%	-0.2%	0.1%	0.0%	0.4%
Westchester County, New York	-5.0%	-5.8%	1.7%	-0.7%	0.1%	-0.2%	0.8%

(Continued)

Table 9.1 (Continued)

	Total	NH White	Latino	NH Black	NH Asian	NH Other	All PoC
Long Island							
Nassau County, New York	-5.0%	-6.6%	1.3%	-0.2%	0.7%	-0.1%	1.6%
Suffolk County, New York	-5.8%	-7.7%	2.0%	-0.2%	0.2%	-0.1%	1.9%
New York City							
Kings County, New York	-0.1%	1.4%	-0.1%	-1.6%	0.6%	-0.5%	-1.5%
New York County, New York	-2.0%	-0.6%	-1.1%	-1.1%	0.9%	-0.1%	-1.4%
Bronx County, New York	-2.5%	-1.1%	0.5%	-1.4%	0.0%	-0.4%	-1.3%
Queens County, New York	-2.3%	-1.6%	0.3%	-1.0%	0.9%	-0.8%	-0.6%
Richmond County, New York	-4.1%	-5.3%	1.3%	-0.3%	0.3%	-0.2%	1.1%
New York City	-1.7%	-0.6%	0.0%	-1.3%	0.6%	-0.5%	-1.1%
Metro Area	-3.7%	-4.1%	0.8%	-0.7%	0.6%	-0.2%	0.5%
NY	-3.5%	-3.6%	0.5%	-0.7%	0.4%	-0.2%	0.0%
US	-3.6%	-4.5%	1.0%	-0.4%	0.3%	0.0%	0.8%

Source: Author's calculations based on Summary File 1, U.S. Decennial Census, 2000 and 2010.
NH = Non-Hispanic; PoC = people of colour

Latino and Black young adults fell across New York City and Hudson County. This pattern of racial succession – mostly affluent Whites have entered the cities as gentrifiers, displacing people of colour, who settle in the suburbs – is similar to the "great inversion" hypothesis popularized by Alan Ehrenhalt (2012).

This distinction is not merely academic but is also significant for the politics of affordable housing in the suburbs. The concern commonly articulated in the pages of local media and in regional Smart Growth meetings – that Long Island must build affordable housing to keep "our" children from leaving the area – betrays a subtle racial politics. This politics overlaps with anxieties that a loss of college graduates will jeopardize economic development: the number of college-educated 25- to 44-year-olds grew 5.1% and 10% in Nassau and Suffolk County, paling in comparison to a 64.3% and 70.6% increase in Kings and Hudson. Those who imagine that the region is losing its White and disproportionately college-educated children to the city are liable to support the many active plans to build market-rate TOD rental units and "affordable" rental housing targeted to moderate-income households at 80% to 130% of area median income, a relatively affluent group with household incomes of more than $80,000 in Nassau and Suffolk. Those Latino young adults who are already moving to the region (or aging into the young adult cohort) and have lower median incomes overall are less likely to benefit from this new stock. Since this group includes undocumented immigrants that some communities would prefer to exclude as residents, the youth drain–affordable housing discursive strategy – while seemingly race neutral and progressive on its face – may have exclusionary implications.

Suburban Immigration and Its Contentious Politics

If the class and ethnic composition of young adults is mostly avoided in the debate on the "youth drain," they come to the fore in the contested regional politics of immigrant integration and exclusion. Across the United States, immigration flows have increasingly bypassed traditional urban gateways and entered the suburbs instead (see, for instance, Katz, Ceighton, Amsterdam, & Chowkwanyun, 2010; Singer, Brettell, & Hardwick, 2008). And this story is certainly true for New York as well, but the story for the New York suburbs is also rather different. First, the suburbs of the New York metropolitan area have always been home to immigrant communities. In 1950, New York's older suburbs[3] and newer suburbs were 15% and 13% foreign-born, respectively. The immigrant share of the region's suburbs shrank from the 1950s until

1990, but that was because the rate of growth of the native born population was faster than that for immigrants, not because the immigrant populations shrank.

The second way in which the suburbanization of immigration in New York has been different from the rest of the country is that a slim majority (51%) of the foreign-born population in the region still resides within the five boroughs of New York City. For the 100 largest metropolitan areas around the country, this is not true, and more than 61% of the foreign-born population was suburban in 2013 (Wilson & Svajlenka, 2014). To some extent this is a function of late-19th-century consolidation that created the five-borough municipality. Simply put, large portions of Queens, Brooklyn, Staten Island, and even the Bronx are "suburban" in their built form, demographics, commuting patterns, and functional integration into the larger metropolitan region.

Finally, New York's individual suburban counties and the metropolitan area as a whole currently have a foreign-born population that is more complex and multinational in its countries of origin than any other in the United States. No single country of origin dominates, and there are disparate communities across the region – Koreans in Bergen County (NJ), Indians in Middlesex County (NJ), Mexicans in Westchester County (NY), Salvadorans in Long Island. There are also sizable Dominican communities throughout the region, as well as Jamaican, Haitian, and Guyanese communities (Black, in the New York metropolitan region, has long since not meant "African American" in the usual way Americans mean that term). Within counties this diversity is also evident, and the largest immigrant groups in Westchester and Nassau counties (Mexicans and Salvadorans, respectively) comprise only little more than 10% of the counties' total immigrant populations. All told, the immigrant population for the suburbs is now a bit less than 3 million people, or 21% of the suburban population.

Immigrants may not be new to the region's suburbs, but the suburbanization of immigration has certainly been transforming them in recent decades. Social conflict surrounding Latinx immigration has been particularly intense (Duncan & Duncan, 2004; Dolgon, 2005; Mahler, 1995). National discourses about illegality and criminality – in particular the animus of nativist, non-Hispanic Whites towards young Latino men – surface in local debates about migrants who are now living and working in the suburbs. But they also reflect the everyday local discomfort of native-born (primarily White) residents during daily encounters in public and private spaces. Communities with on-street, day-labourer pick-up sites and boarding houses were flash points for conflict in the 1990s and 2000s; more recently, school overcrowding has emerged as a point of tension in areas that have received unaccompanied minors. Non-Latino native-born residents have pressed local governments to police day labourers and public spaces (Duncan & Duncan, 2004; Maney & Abraham, 2008); harass businesses used by immigrants; tighten prohibitions on overcrowding and illegal apartments; and increase cooperation between local or county police departments and federal immigration authorities. The last has been widely variable across county lines and administrations, undermining attempts to build trust between police departments and new immigrant communities (c.f., Ridgley, 2008). Anti-immigrant discourse and exclusion policy has been accompanied by extralegal violence, including the fatal assault on Ecuadoran immigrant Marcelo Lucero, in Patchogue in 2008; the Lucero killing mobilized pro-immigrant organizing on Long Island, which had already seen an reactive upsurge in response to national-level threats such as the proposed (and defeated) *Border Protection, Anti-Terrorism, and Illegal Immigration Control Act*, cosponsored by Long Island representative Peter King in 2005.

Just as this type of nativist legislation disregards the United States' role in encouraging immigration and its dependence on the low-wage labour of undocumented immigrants, a parallel dissonance exists between suburban anti-immigrant discourse and immigrants' local economic importance. The benefits that immigrants bring in the form of economic growth, housing rents, and bourgeois suburban consumption collide with race and class privilege encoded in widely held beliefs about suburban "quality of life" (Hanlon & Vicino, 2015; Walker & Leitner, 2011). Such contradictions are on display in the politics of affordable housing. Localities and counties create "affordable housing opportunities" that are designated for more affluent residents or that require one year or more of residency within the jurisdiction (though such laws, e.g., in the Town of Oyster Bay, have faced legal challenges). New accessory apartment laws, which are frequently limited to the family members of the homeowner, are paired with crackdowns on illegal housing. In this sense, panics over the youth drain and day labourers appear as two sides of the same coin, though they are seldom treated as such.

Rebuilding the Postwar Suburbs

The suburban built environment poses another potential challenge to regional growth. As the postwar infrastructure aged, it became a barrier to economic growth and accumulation (Harvey, 1985): the departure of suburban manufacturing jobs has left behind brownfields, competition with newer commercial development has dampened retail trade in suburban downtowns and older malls, and smaller single-family

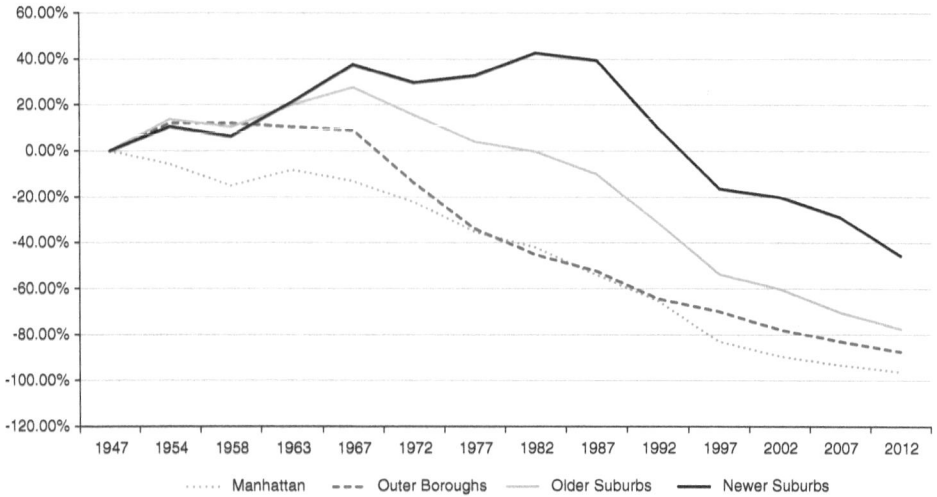

Figure 9.3 Change in manufacturing employment in New York City and its suburbs, 1947–2012. Source: Authors' analysis of Economic Census data.

homes have struggled to attract new buyers (e.g., Lucy & Phillips, 2006). Governmental fragmentation has worsened the problem: lacking the financial and coordinating functions of central cities, the tax base of older suburbs has diminished even as service needs have increased. Local governments have been unable to repair public infrastructure or meet the growing need for public services for the poor population. Economic decline and local fiscal stress have reinforced patterns of racial and class segregation, both within these older suburbs and between the older suburbs and newer "affluent job centers" (Orfield, 2002).

The New York suburbs epitomize national trends. The deindustrialization of New York's manufacturing sector has proceeded in clear geographical stages, beginning with a steady attrition of jobs from Manhattan and the outer boroughs from the 1950s onward (see figure 9.3 and table 9.2). This process accelerated with the advent of containerization, as most shipping moved to Newark, and was encouraged by real estate interests intent on capturing the waterfront (Angotti, 2008; Curran, 2007). Manufacturing jobs rose in older postwar suburbs until the 1960s, and dropped only slightly in the 1970s, but the decline accelerated in the 1980s and 1990s, and was often accompanied by losses in retail trade (Listokin & Beaton, 1983). Governmental fragmentation in New Jersey and New York has aggravated fiscal stress: in the five

Table 9.2. Manufacturing jobs, regional share of manufacturing jobs, and location quotients for employment, New York, including and excluding Manhattan

	Employment, 1947	Employment, 2012	Percentage of all regional manufacturing jobs	Location quotient	Location quotient without Manhattan
Manhattan					
New York County, NY	554,701	21,220	4.74	0.20	---
Outer Boroughs					
Queens County, NY	97,013	22,240	4.97	0.89	0.71
Kings County, NY	229,632	18,296	4.09	0.65	0.52
Bronx County, NY	41,323	6,197	1.38	0.53	0.42
Richmond County, NY	10,620	999	0.22	0.22	0.18
Older Suburbs					
Bergen County, NJ	52,508	33,434	7.47	1.62	1.29
Union County, NJ	67,033	20,790	4.64	2.09	1.66
Passaic County, NJ	78,521	18,337	4.10	2.42	1.92
Essex County, NJ	142,892	17,556	3.92	1.16	0.92
Nassau County, NY	25,845	16,580	3.70	0.62	0.50
Westchester County, NY	46,761	11,776	2.63	0.65	0.52
Hudson County, NJ	144,039	7,865	1.76	0.72	0.57
Newer Suburbs					
Suffolk County, NY	13,213	51,967	11.61	1.86	1.48
Fairfield County, CT	112,555	35,507	7.93	1.69	1.35
New Haven County, CT	110,611	31,792	7.10	2.07	1.65
Middlesex County, NJ	59,614	28,277	6.32	1.47	1.17
Morris County, NJ	12,837	14,358	3.21	1.09	0.87
Somerset County, NJ	14,217	12,329	2.75	1.24	0.99
Litchfield County, CT	18,205	9,062	2.02	3.53	2.81
Middlesex County, CT	10,544	8,920	1.99	3.04	2.42
Monmouth County, NJ	11,587	8,551	1.91	0.77	0.61
Dutchess County, NY	14,692	8,544	1.91	2.01	1.60
Rockland County, NY	7,893	8,416	1.88	1.75	1.40
Orange County, NY	13,877	7,105	1.59	1.28	1.02
Mercer County, NJ	40,422	7,070	1.58	0.82	0.65
Ocean County, NJ	824	5,069	1.13	0.81	0.65
Warren County, NJ	9,130	3,854	0.86	2.81	2.24
Ulster County, NY	7,412	3,518	0.79	1.51	1.20
Hunterdon County, NJ	4,123	3,485	0.78	1.67	1.33
Sussex County, NJ	1,396	1,932	0.43	1.25	1.00
Putnam County, NY	176	1,478	0.33	1.42	1.13
Sullivan County, NY	487	1,224	0.27	1.31	1.04

Source: U.S. Economic Census, 1947–2012

older suburban counties of New Jersey, 143 local governments served a median population of 12,754, and a median area of 7.64 square kilometres. In New York, two layers of local government have allowed tiny enclaves of the affluent – often numbering well under 1,000 residents – to avoid sharing taxes and schools with poorer suburbanites.

Yet at the same time, some New York's older suburbs may be appealing sites for accumulation through redevelopment. Waterfront residential investment and financial services jobs have migrated across the Hudson River to Hoboken and Jersey City, where residents enjoy easy transit access to Manhattan while avoiding the stratospheric prices of its hypergentrified neighbourhoods. The New York metro's extensive commuter rail systems – with by far the highest daily ridership in the United States – have provided a skeleton for redevelopment plans in more distant older suburbs like Hempstead and Wyandanch on Long Island, Yonkers and New Rochelle in Westchester, and Rahway and Perth Amboy in New Jersey. Additionally, the relatively permissive eminent domain laws in New York and New Jersey allow for the ambitious (if often controversial) downtown revitalization plans.

The reinvention of suburban downtowns has also received enthusiastic backing from state and county agencies, nonprofit planning and advocacy groups, and regional foundations. Funded by HUD, the RPA launched a Sustainable Communities Consortium in 2010, focusing on transit-oriented development across the tri-state area. On Long Island, this initiative has dovetailed with the advocacy of the Rauch Foundation, which sponsored competitions for both the bold redesign of the area's suburban landscape, as well as more modest proposals for the redevelopment of underused parking lots near Long Island Railroad stations (Williamson, 2013).

While these visions incorporate social equity goals, the negotiation between local and county governments, developers, and residents has tended to produce upmarket rental and condominium development around suburban transit hubs, often in an explicit attempt to emulate urban lifestyles popular among young adults – a "Great Reversion" of sorts. These projects present a puzzle for equity advocates: downtown suburban revitalization introduces new reinvestment, buttresses the local tax base, and may underwrite the costs of community benefits that include new affordable housing. Yet the same projects introduce displacement pressures that resemble urban gentrification and echo prior, largely forgotten rounds of "suburban renewal" (see Wiese, 2004). Here, the continued resistance of affluent suburbs to affordable housing presents an obvious but unacknowledged conundrum: including replacement affordable housing in older suburbs prevents displacement while reproducing class and racial segregation. Debates about poverty and affordable housing concentration have raged among social scientists and planners for several years (see e.g., Chapple, 2006), but are usually framed as a question of deconcentration from urban centres. The redevelopment of older suburban centres across the region – concurrent with the rapid expansion of the gentrification frontier in New York – means that the dilemmas of poverty deconcentration become ones of suburban-to-suburban or even suburban-to-urban migration.

Foreclosure Crisis in the Suburbs

The long-planned revitalization of many older suburbs was interrupted by the financial crisis, which choked off capital for downtown redevelopment in places like Yonkers. More generally, widespread foreclosures worsened conditions in majority-minority suburbs, including those that lacked transportation hubs and robust redevelopment plans. Here, the foreclosure crisis in the New York suburbs has presented an acute wave of devaluation and left community organizations scrambling to ensure accountability and prevent speculation.

Nationally, it is hard to tell whether cities, older suburbs, or newer suburbs bore the brunt of the foreclosure crisis, in part because subprime lending varied between metropolitan areas and racial/ethnic borrower groups (Anacker, 2015). Within the New York region, the geographies of subprime lending and foreclosure closely matched the settlement patterns of moderate-income African American, Latino, and Asian homeowners, but it has been more persistent in the suburbs. Suffolk, Nassau, Rockland, and Putnam counties had the highest foreclosure rates in the state, with pending foreclosures accounting for more than 2% of all units, greatly exceeding rates in New York City and upstate New York alike (New York State Office of the Comptroller, 2015). On Long Island, two older suburban areas received a torrent of high-interest mortgage capital in the mid 2000s: a cluster of majority Black but increasingly Latinx communities in central Nassau County, and another cluster of majority Latinx communities in western Suffolk County (see Niedt & Silver, 2014). Similar patterns could be observed in New Jersey's suburban cities and in Bridgeport, Connecticut.

The crisis unravelled relatively slowly at a regional level. The "first wave" of subprime lending was already apparent in in Newark, Irvington, and East Orange by the early 2000s (Newman, 2009). And the aftereffects of the bust have lingered far longer in New York and New Jersey, "judicial foreclosure" states with two of the longest foreclosure processing times in the nation. While these procedural requirements provide some protection for homeowners, the delays have created the problem of "zombie homes" that remain stalled at various points in the foreclosure process, physically deteriorate, and have negative spillover effects for neighbourhoods and local governments. At the end of 2013, New Jersey had the highest foreclosure rate in the nation, and the five cities with the highest percentage of underwater or negative equity homes were Hartford, Connecticut; followed by Newark, Elizabeth, and Paterson, New Jersey; and Detroit, Michigan (Dreier, Mollenkopf, & Swanstrom, 2014).

Again, there are plenty of opportunities for plucky capitalists to profit, and vulture investors have moved quickly to snap up bank-owned properties, usually in all-cash offers that out-competed would-be owner occupants (Niedt & Silver, 2014). Homebuyers, often trapped in underwater mortgages or with credit damaged by foreclosure proceedings, were largely unable to avail themselves of low prices and low refinancing rates. The more significant structural problem was that subprime lenders had dominated the market in majority-minority suburban communities, and when they failed (and major banks tightened lending requirements), communities found themselves starved for credit (Hanley, Maker, & Van Kerkove, 2013; Niedt & Silver, 2014). The subprime crisis made visible a hitherto invisible credit market that sliced through the suburbs, following lines of class and race segregation; even as the crisis abates, there are no signs that the affected suburbs will gain access to fair housing opportunities, at least not through the actions of lenders and governments alone.

Social Movements and Struggles for Social Justice

While the stories told thus far in this chapter have focused mostly on the actions of the elites (in different realms), there has been a vibrant set of social movements working to transform the processes and institutions that shape the suburbs in the New York region. The oppression and political marginalization that have resulted from the elite actions described above have been met with creative, important, and sometimes politically transformative social movements. Those social movements, however, are being created and forged in the context of suburbs that lack the organizational infrastructure (philanthropic and nonprofit organizations) of the metropolitan core.[4]

Despite this structural weakness, suburban social justice organizations – often working on shoestring budgets, or none at all – have sought to achieve equity and inclusion. Here we consider those focused on housing and on labour/immigration issues. Housing-justice organizations have tried to address the foreclosure crisis and the lack of affordable housing for working class and poor people in the suburbs. New York City and several of its suburban governments have explored a model originally developed in San Bernardino

County: using eminent domain, not to take real property such as foreclosed homes themselves, but to condemn, restructure, and resell the properties' underlying mortgages (Hockett, 2012). The financial sector and its trade groups have mounted a vigorous resistance to the plans wherever they have emerged, and most jurisdictions that have explored them have decided against being the first to challenge the financial sector's prerogatives. Currently, Newark, New Jersey, and Richmond, California, have moved the eminent domain plan furthest through the local legislative process, though it is unclear whether even these will withstand legal challenges and political and lender pressure.

These campaigns likely emerged in the suburbs for several reasons, including the relative absence of financial institutions as internal stakeholders. Race and class inequality also explain the viability of these plans at the local level; the suburbs that advanced them are almost without exception majority Black and Latino suburbs that have experienced significant abandonment and concentrated losses of household wealth and tax revenues during the subprime crisis. The shared experience of foreclosure across the neighbourhoods of many majority-minority suburbs – compared to the relative unevenness across larger cities such as New York – allows organizers to build popular support with a racial-justice framing that mirrors the actual racial geographies of the subprime crisis. Whether or not the plans are ultimately successful, the pursuit of these plans – and similar efforts by suburban local governments to pressure banks for mortgage modifications in Essex, Nassau, Suffolk, and Westchester counties – indicates that the hoped-for "financialization of urban politics" may in fact be more likely to occur on a suburban terrain (Fields, 2014; Niedt & Christophers, 2016).

Essex County, New Jersey, has not just seen housing organizing around the foreclosure crisis, but the formation of a countywide land trust in which the land is permanently taken off the market and controlled by a non-profit organization. This reduces the original sale price of the properties and, through a ground lease, restricts the resale price for the homeowner, thereby creating a stock of permanently affordable homes. The Essex Community Land Trust (ECLT) was created in 2011 by a set of community organizations and municipalities concerned not just with foreclosures, but also with persistent need for affordable housing (Kasen et al., 2013). Because the land-trust model creates permanent affordability, it retains the public subsidies that make affordability possible (see Davis, 2010). Given the fragmented and heterogeneous populations in the county's many and disparate municipalities,[5] the ECLT is structured around a set of partnerships with community development organizations, local homeowner and civic associations, and municipalities. The attempt, therefore, is not simply to maintain affordable housing but to bridge the divides between communities and municipalities in this suburban county.

A similar effort is underway in the older suburbs of Nassau County, where local residents and their allies formed the Uniondale Community Land Trust (U-CLT), which shares many goals with the ECLT. During the height of the housing boom (2005 to 2006), high-rate loans accounted for over one half of all originations in Uniondale, and the subsequent foreclosures had severe negative effects on local institutions, safety, and property values; the land trust seeks to rehabilitate and resell these real estate owned and abandoned "zombie" homes as affordable housing. Located in an unincorporated area of the Town of Hempstead, the U-CLT relies heavily on its relationship with the Greater Uniondale Area Action Coalition (GUAAC), an umbrella group of local civic associations. The GUAAC has incorporated the U-CLT into its visioning plan, as land-trust homes may help insulate the community from the market pressures as it pursues reinvestment, and as state and private actors redevelop a nearby office and entertainment hub.

The suburbs of New York have also been at the forefront of a set of labour organizing trends in the United States over the last 20 years. Much of this is rooted in the prominence of low-wage service-sector work in the labour market in this most globalized city region (Sassen, 1991). Social movements and community organizations have interwoven labour and immigration, engaging the immigrants who increasingly make up the suburban working class.

Immigrant activism has a deep, broad, and relatively underexplored history in the New York suburbs (but see, e.g., Dolgon, 2005). Mutual aid and advocacy organizations have long provided job training, business networks, connections to social services, and legal assistance to successive waves of immigrants. They have also served to organize celebrations of ethnic identity and culture, at times in the face of exclusion, hostility, and violence.

Activism among Latinx immigrants included early attempts to organize farm workers on the rural fringe of the region from the 1960s forward (e.g., on the East End of Long Island) and efforts to serve the growing numbers of Central and South Americans who were refugees from civil war, political repression, and economic collapse during the 1980s and 1990s. Suburban organizing in the New York area mobilized Salvadoran groups in political efforts to secure Temporary Protected Status and provide a path to citizenship for refugees (figure 9.4). In the 2000s and 2010s, local organizers supported national campaigns to provide paths to legalization for undocumented workers.

Within the New York region, a growing number of immigrant organizations have formed to defend immigrant rights. Often these

New York's Suburbs in a Globalized Metropolitan Region 191

Figure 9.4 Long Islanders demonstrating for permanent residence for Salvadoran immigrants in Washington, DC, 1995. Courtesy of the Hofstra University Special Collections, Hispanic/Latino Collection.

organizations forge labour-community alliances that erase the conventional division between residential and workplace politics, a division which is often regarded as quintessentially suburban. Hempstead's Workplace Project, and Community Resource Center and Neighbors Link in Westchester, provided varied combinations of political education and organization, job placement and hiring sites, and protections from wage theft and mistreatment (Duncan & Duncan, 2004; Fine, 2005; Gordon, 2005). Groups such as Make the Road New York, Westchester Hispanic Coalition, Long Island Wins, and the Central American Resource Center have developed a broad immigrant-rights agenda that includes ensuring language access at the local and county levels, reducing police profiling and violence, bringing attention to immigrants' economic contributions, and calling attention to exploitation within the housing and labour markets.

Two labour organizations inspired new approaches to community-labour organizing at the national level: the Workplace Project and the Stamford Organizing Project. In 1992, Jennifer Gordon, a recently minted lawyer from Harvard Law School, created the Workplace Project (also known as "Centro de Derechos Laborales"), an innovative labour-community organization that organized Latino immigrant workers to fight employer workplace violations and provided legal assistance and worker education (Gordon, 1995, 2005). Headquartered in the Village of Hempstead, it became a model for the small but significant immigrant worker centre movement nationwide (Fine, 2005).

While Latino workers were organizing in Long Island, a similar, but organizationally very different effort was emerging in Stamford, Connecticut. The Stamford Organizing Project (SOP) was created in 1998 by a set of different unions that decided to partner with each other and community organizations building an example of what has been called "community unionism" (see, for instance, Black, 2005). This effort was one of four nationwide to focus on specific geographic areas, and they came from the changed leadership in the AFL-CIO following the election of John Sweeney in 1995 (Fine, 2001). This effort focused not just on the usual rank and file union members' concerns about contracts and better wages and benefits. Instead, it addressed a whole set of issues beyond the realm of the narrower interpretations of workplace organizing, including housing and racial justice – "All the Issues in Workers' Lives," as HoSang (2000) put it. It also was not the usual labour–community coalition that amounts to labour organizers meeting with community group staff members and the members or constituency of either being more or less out of the process. Instead, it was a process that responded to the segregation of Black workers (immigrant – Haitian and Jamaican, mostly – and native born) into particular pockets of the city. Those overlaps between workers and community residents allowed the SOP to transcend the usual "trenches" (Katznelson, 1981) that divide labour from community in American politics.

Conclusion

In 2013, Bill DeBlasio won the mayoralty by promising to unite a New York City that had become riven by inequality. This was a "tale of two cities": the resurgent finance sector and affluent residents, supported by a technocratic government, enjoyed the benefits of prosperity and development, while the working class dealt with high unemployment and high rents. Once in office, the mayor moved forward with plans for universal pre-kindergarten, new affordable housing construction, and paid sick leave legislation. This legislative agenda met with mixed success, and some have noted that business community – initially alarmed a successful left-populist mayoral campaign after two decades of unabashed probusiness government – eventually found the new administration surprisingly conciliatory to corporate interests. Still, the progressive energies that enabled DeBlasio's election, rooted both in the city's radical history and in campaigns like Occupy Wall Street and #blacklivesmatter, have remained a vital counterweight to urban elite power.

Beyond the city line, there is a "tale of two suburbs." Connecticut's Gold Coast, Long Island's North Shore and Hamptons, and exclusive gated communities that dot northern New Jersey project a dreamscape of suburban wealth and excess. And the history of New York's suburbanization itself was driven by organized elite interests, even if the politics of regionalism pitted technocrats at the centre against assertive local growth machines backed by home rule. But the image of paradigmatic postwar suburb – the racially/ethnically homogenous, middle class suburb that supposedly offered security and upward mobility to its inhabitants – is becoming something of a fiction. In its place we find growing diversity, driven by immigration; competing plans for the reconstruction of "obsolescent" infrastructure; and a prolonged, intractable housing crisis.

The resulting suburban politics have been divisive and empirically intriguing. Though the New York suburbs pose unique challenges to organizing (e.g. Gordon, 2005; Ziskind, 2003), they too have been sites for progressive activism, historical and emergent, in the fields of labour, immigrant rights, and housing. The success of these movements' efforts relies in part on public and philanthropic support, which is partly conditioned by representations of the suburbs as wealthy and at least relatively untroubled. Revealing the complexity and the indeterminacy of New York's suburban futures is thus a first step in making those futures more just.

NOTES

1 While the precise boundaries of the metropolitan region of New York are debatable, for simplicity sake, we follow the definition used by the Regional Plan Association.
2 The prescriptions of metropolitan and decentralist regionalism, as Fishman notes, eventually inspired plans for urban renewal and extensive highway construction.
3 We are defining older suburban counties as currently having more than two thirds of their housing built before 1970. These are the counties of: Nassau and Westchester in New York, and Bergen, Essex, Hudson, Passaic, and Union Counties in New Jersey. The other suburban counties are considered "newer" suburbs for the purposes of this analysis.
4 There is a rapidly expanding literature on the thinness of the suburban social service net (e.g., Roth & Allard, 2015, Kneebone & Berube, 2013; Reckhow & Weir, 2011); yet, these studies seldom distinguish between direct

service provision and base-building organizations that attempt to mobilize the poor and working class around policy goals.
5 The poorest municipality in Essex County, East Orange, has a median household income that is 22% of Essex Fells, the wealthiest.

REFERENCES

Anacker, K. (2015). Analyzing census tract foreclosure risk rates in mature and developing suburbs in the United States. *Urban Geography, 36*, 1221–40.

Angotti, T. (2008). *New York for sale: Community planning confronts global real estate.* Cambridge, MA: MIT Press.

Belkin, L. (1999). *Show me a hero.* New York City: Back Bay Books.

Black, S. (2005). Community unionism: A strategy for organizing in the new economy. *New Labor Forum, 14*(3), 24–32.

Chapple, K. (2006). Overcoming mismatch: Beyond dispersal, mobility, and development strategies. *Journal of the American Planning Association, 72*(3), 322–36.

Curran, W. (2007). "From the frying pan to the oven": Gentrification and the experience of industrial displacement in Williamsburg, Brooklyn. *Urban Studies, 44*(8), 1427–40.

Davis, J. (Ed.). (2010). *The community land trust reader.* Cambridge, MA: Lincoln Institute of Land Policy.

Dolgon, C. (2005). *The end of the Hamptons: Scenes from the class struggle in America's paradise.* New York, NY: NYU Press.

Dreier, P., Mollenkopf, J. & T. Swanstrom. 2014. *Place matters: Metropolitics for the twenty-first century.* 3rd Ed. Lawrence, KS: University of Kansas Press

Duncan, J., & Duncan, N. (2004). *Landscapes of privilege: The politics of the aesthetic in an American suburb.* New York, NY: Routledge.

Ehrenhalt, A. (2012). *The great inversion and the future of the American city.* New York, NY: Knopf.

Fields, D. (2014). Finance as a new terrain for progressive urban politics. *metropolitiques.* Retrieved 10 November 2014 from http://www.metropolitiques.eu/Finance-as-a-New-Terrain-for.html

Fine, J. (2001, December 14). Building community unions: In Stamford, Connecticut, organizers are putting the movement back in labor. *The Nation.*

Fine, J. (2005). *Worker centers: Organizing communities at the edge of the dream.* Washington, DC: Economic Policy Institute.

Fishman, R. (1987). *Bourgeois utopias: The rise and fall of suburbia.* New York, NY: Basic Books.

Fishman, R. (2000). The death and life of American regional planning. In Bruce Katz (Ed.), *Reflections on regionalism* (pp. 107-23), Washington, DC: Brookings.

Fitch, R. (1993). *The Assassination of New York*. New York: Verso Books

Gordon, J. (1995). We make the road by walking: Immigrant workers, the workplace project and the struggle for social change. *Harvard Civil Rights-Civil Liberties Law Review, 30*, 407–50.

Gordon, J. (2005). *Suburban sweatshops: The fight for immigrant rights*. Cambridge, MA: Belknapp/Harvard University Press.

Hanley, M., Maker, R., & Van Kerkove, B. (2013). *The Long Island foreclosure crisis: Stabilizing the communities most impacted by foreclosures in Nassau and Suffolk Counties*. Rochester, NY: Empire Justice Center.

Hanlon, B., & Vicino, T. (2015). Local immigration legislation in new suburbs: An examination of immigration policies in farmers branch, Texas and Carpentersville, Illinois. In K. Anacker (Ed.), *The new American suburb: Poverty, race, and the economic crisis* (pp. 113–32). New York, NY: Routledge.

Harvey, D. (1985). *The urbanization of capital: Studies in the history and theory of capitalist urbanization*. Baltimore, MD: John Hopkins University Press.

Healy, J. (n.d.). Renaissance on the waterfront and beyond. *New Jersey League of Municipalities*. Retrieved from http://www.njslom.org

Hockett, R. (2012). It takes a village: Municipal condemnation proceedings and public/private partnerships for mortgage loan modification, value preservation, and local economic recovery. *Stanford Journal of Law, Business & Finance, 18*, 121–76.

HoSang, D. (2000, August 25). "All the issues in workers' lives": Labor confronts race in Stamford. *ColorLines*.

Jackson, K. (1987). *Crabgrass frontier: The suburbanization of the United States*. New York: Oxford University Press.

Kasen, R., Manieri, J., Rodriguez, C., Thibault Grof, J., & Winter, A. (2013). *Assessing fundamental issues of central server CLTs: Strategies for the Essex Community Land Trust*. Fall 2012 Community Development Studio. Bloustein School of Planning and Public Policy, Rutgers University.

Katz, M., Ceighton, M., Amsterdam, D., & Chowkwanyun, M. (2010). Immigration and the new metropolitan geography. *Journal of Urban Affairs, 32*(5), 523–57.

Katznelson, I. 1981. *City Trenches*. Chicago, IL: University of Chicago Press.

Kneebone, E., & Alan, B. (2013). *Confronting suburban poverty in America*. Washington, DC: Brookings Institution Press.

Kushner, D. (2009). *Levittown: Two families, one tycoon, and the fight for civil rights in America's legendary suburb*. New York, NY: Walker & Company.

Levine, M. (2015). *Urban politics: Cities and suburbs in a global age* (9th ed.). New York, NY: Routledge.

Linder, M., & Zacharias, L. (1999). *Of cabbages and kings county: Agriculture and the formation of modern Brooklyn*. Iowa City: University of Iowa Press.

Listokin, D., & Beaton, W.P. (1983). Revitalizing the older suburb. New Brunswick, NJ: Center for Urban Policy Research.
Lucy, W.H., & Phillips, D.L. (2006). Tomorrow's cities, tomorrow's suburbs. Chicago, IL: American Planning Association.
Mahler, S. (1995). *American dreaming: Life on the margins.* Princeton, NJ: Princeton University Press.
Maney, G.M., & Abraham, M. (2008). Whose backyard?: Boundary making in NIMBY opposition to immigrant services. *Social Justice, 35*(4), 66–82.
Massey, D., Albright, L., Casciano, R., Derickson, E., and Kinsey, D. (2013). *Climbing Mount Laurel: The struggle for affordable housing and social mobility in an American suburb.* Princeton, NJ: Princeton University Press.
New York State Office of the Comptroller (Thomas DiNapoli). (2015, August). The foreclosure predicament persists. Retrieved from https://www.osc.state.ny.us/localgov/pubs/research/snapshot/foreclosure0815.pdf
Newman, K. 2009. Post-industrial widgets: Capital flows and the production of the urban. *International Journal of Urban and Regional Research.* 33(2): 314-331
Niedt, C., & Christophers, B. (2016). Value at risk in the suburbs: Eminent domain and the geographical politics of the US foreclosure crisis. *International Journal of Urban and Regional Research,* doi:10.1111/1468-2427.12413
Niedt, C., & Silver, M. (2014). *An uneven road to recovery: Place, race, and mortgage lending on Long Island.* Hempstead, NY: National Center for Suburban Studies at Hofstra University.
Orfield, M. (2002). *American metropolitics: The new suburban reality.* Washington, DC: Brookings Institution Press.
Reckhow, S., & Weir, M. (2011). Building a stronger regional safety net: Philanthropy's role. Brookings Institution Metropolitan Policy Program, Washington, DC; Retrieved from http://www.brookings.edu/research/papers/2011/07/21-philanthropy-reckhow-weir
Ridgley, J. (2008). Cities of refuge: Immigration enforcement, police, and the insurgent genealogies of citizenship in U.S. sanctuary cities. *Urban Geography, 29*(1), 53–77.
Roberts, S. (2009, August 10). Westchester adds housing to desegregation pact. *New York Times.*
Roth, B.J., & Allard, S.W. (2015). The response of the nonprofit safety net to rising suburban poverty. In K. Anacker (Ed.), *The new American suburb: Poverty, race, and the economic crisis* (pp. 247–83). Surrey, UK: Ashgate.
Santos, F. (2007). After 27 years, Yonkers housing desegregation battle ends quietly in Manhattan Court. *New York Times.*
Sassen, S. (1991). *The global city.* Princeton, NJ: Princeton University Press.
Scobey, D. (2002). *Empire city: The making and meaning of the New York landscape.* Philadelphia: Temple University Press.

Singer, A, Brettell, C., & Hardwick, S. (Eds.). (2008). *Twenty-first-century gateways: Immigrant incorporation in suburban America*. Washington, DC: Brookings Institution Press.

Walker, K., & Leitner, H. (2011). The variegated landscape of local immigration policies in the United States. *Urban Geography, 32*(2), 156–78.

Wood, R. 1961. *1400 Governments: The Political Economy of the New York Metropolitan Region*. Cambridge, MA: Harvard University Press

Wiese, A. (2004). *Places of their own: African American suburbanization in the twentieth century*. Chicago, IL: University of Chicago Press.

Williamson, J. (2013). *Designing suburban futures: New models from build a better burb*. Washington, DC: Island Press.

Wilson, J.H. and N.P. Svajlenka. 2014. Immigrants Continue to Disperse, with Fastest Growth in the Suburbs. Brookings Institute. url: https://www.brookings.edu/research/immigrants-continue-to-disperse-with-fastest-growth-in-the-suburbs/

Ziskind, M. (2003). Labor conflict in the suburbs: Organizing retail in metropolitan New York, 1954–1958. *International Labor and Working-Class History, 64* (Fall), 55–73.

PART 3

Race, Ethnicity, and the Remaking of Suburbia

10 Diverging Racial Geographies in Phoenix's Postwar and Post–Civil Rights Suburbs

DEIRDRE PFEIFFER

Introduction

The case of Phoenix offers insight into the post–civil rights era of suburbanization, an overlooked period in the existing literature on US suburbs. Post–civil rights suburbs matured after the passage of the 1968 US *Fair Housing Act*. This was a historical context of continued large-scale, auto-oriented housing construction but legal protections against housing and employment discrimination, minority socioeconomic gains, and greater racial tolerance. Consequently, the post–civil rights suburbs are less racially stratified than the postwar suburbs and offer greater racial equity in neighbourhood conditions.

In this chapter, I first review the typology of the postwar suburb and introduce the typology of the post–civil rights suburb. Next, I turn the discussion to Phoenix. I describe its evolution after the Second World War into a suburban region and locate the post–civil rights suburbs in relation to the city of Phoenix, a postwar suburban city. I draw on recent US Census data to illustrate the distinct conditions of the post–civil rights suburbs in the region and changes wrought by the recent US Great Recession and foreclosure crisis. The chapter concludes by stressing the importance of accounting for historical context in studying racial stratification in US suburbs.

The Postwar and the Post–Civil Rights Suburbs

Postwar suburbs came of age during the two and a half decades immediately following the Second World War (1945 to 1969). They are well studied in the existing literature on US suburbanization (e.g. Nicolaides, 2002; Wiese, 2004). When the postwar suburbs developed, space was racially marked, meaning there was a strong sense of who lives where based on race. There were White postwar suburbs and African American and Latino postwar suburbs but few that became, let alone remained, racially integrated over the long term.

Postwar suburbanization happened in a context of long-standing racial intolerance and housing and employment discrimination against families of colour. In the two decades following the war, federal agencies like the Federal Housing Administration routinely refused to insure mortgage loans in racially mixed communities, and realtors, landlords, home sellers, and banks legally were able to use race as a basis for denying homes and mortgage loans to families of colour. Workers of colour were regularly denied employment and promotions on the basis of race. These structural conditions prevented families of colour from competing for housing in the postwar suburbs where Whites lived and thwarted the formation of stable, racially integrated communities, with few exceptions.

Post–civil rights suburbs emerged in a different historical and geographical context than the postwar suburbs. Families moved into them during a time of legal protections against housing and employment discrimination, minority socioeconomic gains, and increased racial tolerance, which meant that their housing markets were more diverse at their onset as compared to the postwar suburbs. Major legislation that opened up the housing market to minorities included the 1968 *Fair Housing Act*, which outlawed discrimination based on race, religion, and nationality, and the 1975 *Home Mortgage Disclosure Act* and 1977 *Community Reinvestment Act*, which encouraged banks to proactively lend to minority communities. Overt housing discrimination declined markedly for African Americans following the passage of this legislation (Schwartz, 2015), and Whites' racial tolerance for living with minorities grew (Cashin, 2001). In turn, African Americans' white-collar employment increased significantly in response to Title VII of the 1964 *Civil Rights Act*, which outlawed racial discrimination in employment (Pattillo-McCoy, 2000). These socioeconomic gains made African Americans' more competitive in the housing market. However, racial discrimination in hiring decisions, employment outcomes, mortgage lending, renting to Latinos, and real estate neighbourhood steering remained major barriers in the wake of the civil rights era.

The post–civil rights suburbs also had geographic features that predisposed them to be more nurturing of a racially diverse population. They typically were developed out of farmland and open space areas that were far from poor, racially segregated communities in the central city. This diminished the threat of social and physical decay and discouraged White flight. Space was less racially marked in these areas, meaning there was less of a sense of who lives where based on race (Pfeiffer, 2012). Families moved into post–civil rights suburbs without knowing how the communities would eventually be defined (Pfeiffer, 2012). These dynamics led the post–civil rights suburbs to be more racially integrated than the postwar suburbs.

I operationalize post–civil rights suburbs as places that 1) had 75% or more of their housing built after 1969 and 2) are located within commuting distance of at least one large city but are not the largest city in their regions. Applying these criteria, there were 1,510 post–civil rights suburbs in 88 of the 100 largest metro regions in the United States as of 2000 (Pfeiffer, 2016). Most are in the sunbelt (70%); about 40% are in the states of California, Texas, and Florida (Pfeiffer, 2016).

The typology of the post–civil rights suburb is more expansive than existing typologies of newer suburbs. These include the exurb (Nelson, 1992), edge city (Garreau, 1991), and boomburb (Lang & LeFurgy, 2007). Post–civil rights suburbs have diverse geographies, land uses, and demographic trajectories. Delineating suburbs as coming of age in the postwar or post–civil rights eras asserts the importance of historical over spatial or demographic context in assessing how suburbanization shapes social inequalities.

The demographic and environmental characteristics of the postwar and post–civil rights suburbs are dynamic. Both are increasingly home to poor people, people of colour, and immigrants (Kneebone & Lou, 2014; Suro, Wilson, & Singer, 2011). Emerging movements for social justice are taking root in both types of suburbs (Niedt, 2013). Postwar and post–civil rights suburbs also are trying to reinvent their identities as single-family tract home bedroom communities in order to thrive (Dunham-Jones & Williamson, 2011).

Yet, the postwar and post–civil rights suburbs are evolving differently in other ways. The postwar suburbs' population is aging and declining, while their infrastructure is deteriorating. They are grappling with paying for the costs of public services (Orfield, 2002). Conversely, the post–civil rights suburbs' population is young and growing. Many are still lacking adequate infrastructure and struggling to pay for the costs of growth (Orfield, 2002).

Evolution of a Suburban Region

Phoenix is the prototypical American suburban region. It largely developed after innovations in mass-produced housing and mortgage finance in the 1940s, which laid the groundwork for large-scale suburbanization in the United States. Like Los Angeles, suburbs in Phoenix

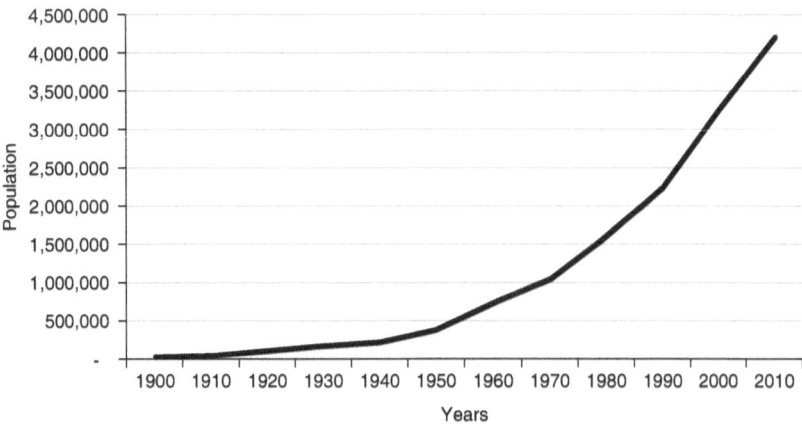

Figure 10.1 Phoenix region population growth, 1900–2010. Source: US Census (1995, 2000, 2010) and Forstall (1995).

are relatively dense. The region boasted an urbanized area population density of about 3,600 people per square mile in 2000 – a level of compactness more comparable to Washington, DC (3,400) or Chicago (3,900) than its often-cited partner in sprawl – Atlanta (1,800) (Gober, 2006).

The first known residents of Phoenix were the prehistoric Hohokam people, who more than 2,000 years ago used irrigation technology to transform portions of the Sonora desert into farmland (Gober, 2006). The Hohokam abandoned the region around 1450 A.D., partly in response to climate change and unsustainable farming practices (Gober, 2006). The area remained unpopulated until the mid-19th century, when hopeful American farmers moved in to supply the gold and silver mining boomtowns to the north and west and Fort McDowell, a military base, to the east (Gober, 2006). Prescient investors stabilized the region's water supply by renovating and expanding the Hohokam's canal network and lobbying for the construction of the Roosevelt Dam. These decisions enabled Phoenix to blossom into a major producer of citrus and cotton by the early 20th century (Gober, 2006).

Rise of the Postwar Suburban City

The Phoenix region transformed from an agricultural centre to a major metropolitan area after the Second World War. The region's population increased almost fivefold from 1940 to 1970, from 215,000 to just over one million (see figure 10.1). About 14% of the region's housing stock remains from this era (U.S. Census, 2014).

Postwar growth in Phoenix was fuelled by a combination of factors. These included 1) the establishment of military bases and war industries, 2) the invention of air conditioning, 3) a probusiness and antiunion climate that attracted employment, and 4) dedicated water rights and land availability (Konig, 1982; Luckingham, 1989). By the mid-1940s, military installations, like Luke and Williams Fields, and defence industries, like Alcoa and Goodyear Aircraft, had drawn tens of thousands of military personnel and defence industry workers to Phoenix (Konig, 1982; VanderMeer, 2010). Local boosterism, changes to the industrial tax base, and state antiunion "right to work" laws attracted close to 300 manufacturing firms between the late 1940s and 1960, which further spurred the region's population growth (Konig, 1982). Real estate developers responded to demand from settled defence industry workers and newly arrived manufacturing industry workers by taming the flat expanse of the Salt River Valley into a patchwork of single-family tract home communities. Most conformed to the 1950s suburban ideal of being homogenous, owner-occupied and auto-oriented.

Postwar suburbanization in Phoenix was populist but racially stratified, meaning that it enabled people of different races across the class spectrum to have a single-family home and backyard patio but in different neighbourhoods. Postwar Phoenix was defined by "neither the wealthy nor the poor" but "an energetic, middle-class" (Konig, 1982, p. 21). It aspired to be "a place where average people could afford to purchase a quality home" (VanderMeer, 2010, p. 192). The homebuilding industry realized (and profited mightily from) this vision by constructing tract housing communities that, with the help of prevalent federal mortgage financing and insurance, offered White migrants affordable homeownership in a more sanitized setting, far from the older, crowded, and dirty cities of the Midwest, Northeast and South. This suburban imaginary was packaged and sold to working and middle class families widely across the United States during this time (Nijman & Clery, 2015).

Postwar race relations in Phoenix resembled, but were not as strained as, those found elsewhere in the United States, particularly in the South (Whitaker, 2005). Violent racially motivated lynchings and murders did not occur in Phoenix (Whitaker, 2005). School desegregation in the wake of the *Brown v. Board of Education* decision happened without much White resistance (Whitaker, 2005). However, racial discrimination in places of public accommodation, employment, and housing was prevalent (Whitaker, 2005). Mortgage lenders routinely denied loans to Latinos and African Americans wanting to purchase homes in predominately White communities in North Phoenix, out of the belief that they would undermine property values (Konig, 1982;

Luckingham, 1994; Whitaker, 2005). Long-standing racially restrictive covenants and other discriminatory practices enacted by the White building and real estate industry discouraged integration (Luckingham, 1994; Whitaker, 2005). African Americans who had the courage and cunning to penetrate White neighbourhoods in the 1950s faced pressure from their neighbours to move and endured racial epitaphs, which were spray-painted on their houses (Whitaker, 2005). Historian Matthew Whitaker depicts the challenges faced by the Ragsdales, an affluent African American family, when they purchased a home in a "Whites only" section near the Encanto neighbourhood north of Phoenix's downtown in the early 1950s:

> When they arrived to move into their new home, Lincoln Ragsdale remembered, the realtors "wouldn't let me in." ... "Within a month of their move," three members of a neighborhood "improvement" committee rang the Ragsdales' doorbell ... they were greeted by one of their neighbors who told them that "we know you're not going to be happy here" ... the committee proceeded to offer to buy the Ragsdales' home if the family would be willing to move (Whitaker, 2005, p. 110).

In response to the threat or experience of racism, Latinos and African Americans who could afford it bought homes in subdivisions located near their long-standing communities in South Phoenix (Luckingham, 1994; Konig, 1982; Whitaker, 2005). Many of those who could not afford or did not want to buy into newly developed subdivisions lived in squalid conditions (Luckingham, 1994). By 1960, just three of the city of Phoenix's 92 census tracts housed half of its 21,000 African Americans (Luckingham, 1994). By 1970, 83% of Mexican Americans in the city of Phoenix lived in tracts with at least 400 coethnics (Luckingham, 1994).

The region's quintessential postwar suburb was John F. Long's Maryvale, a master-planned expanse of 25,000 tract homes built in southwest Phoenix in the 1950s. Long popularized new techniques in production, such as vertical integration, and marketing, such as holding open houses masked as carnivals, complete with attractions for the whole family (VanderMeer, 2010). Long also incorporated diverse amenities into Maryvale's design rather than waiting for amenities to follow housing, as was common. These included the state's then largest shopping centre, parks, libraries, schools, and medical facilities (VanderMeer, 2010). The price of buying into Maryvale was low ($7,950); demand was voracious – 156 homes were selling weekly in 1956 (VanderMeer, 2010; Ross, 2011).

The City of Phoenix responded to the growth of suburbs like Maryvale by engineering a proactive annexation policy to avoid being landlocked (VanderMeer, 2010). By 1960, Maryvale was incorporated into the City of Phoenix, along with other postwar communities such as Sunnyslope and South Phoenix (Konig, 1982; Luckingham, 1989). Phoenix expanded from 10 to 190 square miles between 1940 and 1960 (Luckingham, 1989).

These decisions transformed Phoenix from an older, compact central city to a sprawling *postwar suburban city*. Postwar suburban cities are places that matured by amalgamating postwar suburbs. Their patchwork construction leads them to have a sprawling, amorphous character. Today, visitors to Phoenix are hard-pressed to find its centre and fail to see differences among its many neighbourhoods. This fragmented orientation is reflected in the region's mass media publications such as the *East Valley Tribune,* the *Ahwatukee Foothill News,* and the *Arizona Republic,* and the City of Phoenix's planning framework, which since 1985 has encouraged polycentric living (Luckingham, 1989).

Racial Stratification in the Postwar Suburban City

As the US postwar suburbs matured in the latter decades of the 20th century, they became more *racially stratified*. Racial stratification, which is related to the concept of place stratification, is a process of increasing socioeconomic divergence among places within a region based on their racial makeup (Logan, 1978; Logan & Schneider, 1981).

Racial stratification occurred in the postwar suburbs for two reasons. First, postwar suburbs typically grew adjacent or close to central cities. These borders became more fluid over time. Overcrowding, dilapidation, high rents, and a lack of homeownership opportunities in the central city encouraged central city residents of colour to move to the postwar suburbs (Wiese, 2004). The lives of residents of adjacent central city and postwar suburban communities became increasingly intertwined over time (Cashin, 2007, 2001). White flight in response to growing racial diversity in postwar suburbs served to open up more housing opportunities for nearby central city residents of colour (Wiese, 2004). Over time, postwar suburbs located adjacent to the central city took on the demographic and socioeconomic characteristics of the older and racially stratified urban neighbourhoods that they abutted (Cashin, 2001, 2007; Pattillo-McCoy, 2000).

Racial stratification in Phoenix's postwar suburban city became more entrenched as the 20th century progressed. The spatial map of who lived where based on race remained relatively fixed among the predominately

Figure 10.2 Maryvale. Photos courtesy of author.

affluent White and poorer minority communities that were formed in the decades following the Second World War. Most of the city's African Americans and Latinos, for instance, still lived in disadvantaged South Phoenix (Whitaker, 2005). Initially middle class White postwar suburbs experienced a racial transition as they aged.

Maryvale experienced racial transition early on, when the Federal Housing Administration and Veterans Administration repossessed about 1,300 of its homes in the early 1960s (VanderMeer, 2010). Toxic chemicals generated by nearby high-tech and aerospace companies leached into the groundwater and potentially contributed to higher than average rates of leukemia and other chronic conditions, leading people with the means to live elsewhere to move out or avoid the community (Ross, 2011). Over time, working class Latinos, many of them immigrants, replaced working and middle class Whites.

Today, about 80% of Maryvale's population is Latino (U.S. Census, 2012). Maryvale's median household income averages about $37,000 among its neighbourhoods, which is lower than the region's median household income (just over $50,000) (U.S. Census, 2012). Maryvale has a reputation as being a high crime area, and is sometimes denigrated as "Scaryvale" or "Murdavale" – a far cry from its 1950s reputation of being a modern and desirable community (Urban Dictionary, 2014). Evidence of Maryvale's racial and economic transition and ongoing safety issues are evident in murals and other signage throughout its built environment (see figure 10.2). Most of the homes are still well cared for despite these challenges, and some long-standing White residents remain.

Emergence of the Post–Civil Rights Suburbs

The Phoenix region's population quadrupled from one million to more than four million during the post–civil rights era, which spanned from 1970 to the present (see figure 10.1). The vast majority of the region's housing stock (85%) was built during this era (U.S. Census, 2014). Postwar neighbourhoods in places like Tempe, Scottsdale, Chandler, and Glendale were swamped by their localities' equally aggressive annexation programs and continued growth, and eventually became smaller parts of larger post–civil rights suburbs. In turn, Phoenix continued to develop, increasing its land area to 390 square miles by 1987 (Luckingham, 1989).

Map 10.1 shows the two counties that encompass the Phoenix region, Maricopa and Pinal, and the location of the region's post–civil rights suburbs in relation to its postwar suburban city. Sixty percent and 35% of households lived in the post–civil rights suburbs and postwar suburban city in 2012 respectively (U.S. Census, 2013b). Places that do not fit either the typology of the post–civil rights or postwar suburbs also are highlighted. These include the counties' unincorporated areas and rural towns, such as Wickenburg, Gila Bend, and Coolidge. They also include unique communities located in the region's core, such as Paradise Valley, an enclave of the ultra rich, and Guadalupe, an older town of Yaqui Indians. Fewer than 5% of households lived in these places in 2012 (U.S. Census, 2013b).

The characteristics of the region's post–civil rights suburbs and postwar suburban city in 2012 are shown in table 10.1. Almost all of the housing in the post–civil rights suburbs was built since 1970 (90%) compared to 74% of the housing in the postwar suburban city. The post–civil rights suburbs are more expensive, owner occupied, and White than the postwar suburban city. They have fewer Latinos and families in poverty. They have similar proportions of African Americans and Asians, most of whom are better off than their counterparts in the postwar suburbs.

Like the postwar era of suburban growth, housing built in the post–civil rights era was relatively inexpensive, single family, and mass produced. Low housing costs resulted from the confluence of available land, low-cost labour, and abundant mortgage finance (Luckingham, 1989). Strong job growth in service, government, and high-tech industries promised wages sufficient to meet housing costs (Luckingham,

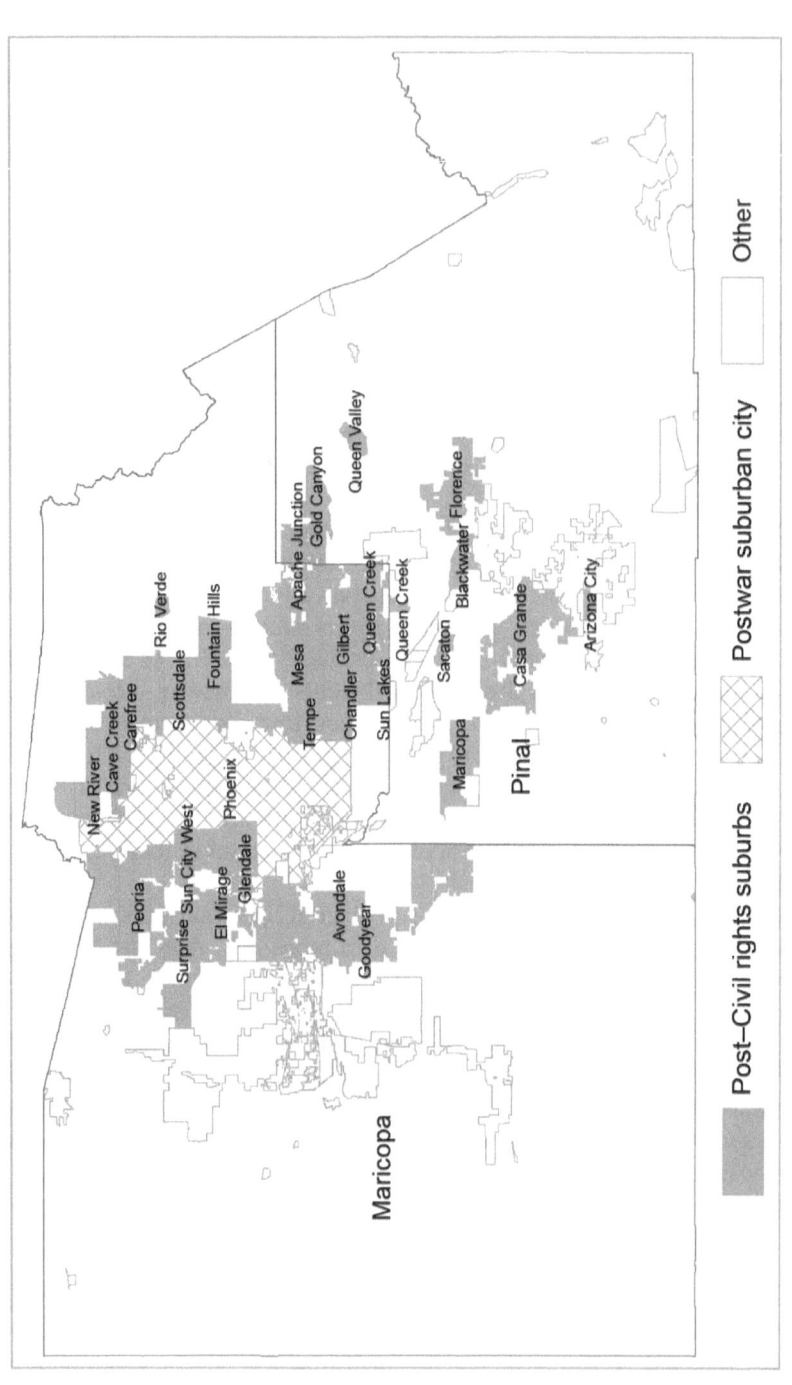

Map 10.1 Geography of Phoenix's post-civil rights suburbs and postwar suburban city. Source: author.

Table 10.1 Demographic and housing market characteristics of Phoenix's post–civil rights suburbs and postwar suburban city, 2010

Characteristics	Post–civil rights suburbs	Postwar suburban city
Housing market		
% built 1970+	90	74
Average neighbourhood housing value	$167,498	$144,170
% owner-occupied	68	57
Demographics		
% African American	4	6
% Latino	23	40
% Asian	4	3
% White	65	47
% families in poverty	9	17
Population	2,435,300	1,483.548

Source: U.S. Census (2013b).
Note: Whites are non-Hispanic. Average neighbourhood housing value is based on median neighbourhood housing value.

1989). Families of colour moved from the postwar suburban city and suburbs and central cities in California, Chicago, and other regions to Phoenix's post–civil rights suburbs, leading them to become more diverse as they matured (Aspinwall, 2006).

Housing built in the post–civil rights era was, and remains, less racially stratified (Pfeiffer, 2016). Table 10.2 shows how the neighbourhood conditions experienced by middle-income[1] African American, Latino, Asian, and White households in the post–civil rights suburbs and postwar suburban city compared in 2012. The weighted averages of neighbourhood poverty rates experienced by middle-income households in each racial group are reported.[2] The greater divergence that similar income households of different races experience in their average neighbourhood poverty rates within the postwar suburban city and post–civil rights suburbs, the greater their racial stratification. The divergence of poverty rates among Latino, African American, and White households in the post–civil rights suburbs is less than in the older postwar suburban city (see table 10.2).

The post–civil rights suburbs' lower racial stratification arises partly from their lower racial segregation. The level of segregation between Latinos and African Americans and Whites in the postwar suburban city and post–civil rights suburbs in 2012 is shown on table 10.2 through the dissimilarity index, which reports the percent of people in one racial

Table 10.2 Racial stratification and segregation in post–civil rights suburbs and in the postwar suburban city, 2012

Characteristics	Post–civil rights suburbs	Postwar suburban city
Typical neighbourhood poverty rate of middle-income households		
African American	8.3	17.7
Latino	11.2	21.2
Asian	8.1	11.0
White	7.0	10.6
Dissimilarity index		
African American/White	44.6	54.8
Latino/White	42.0	59.5
Asian/White	42.1	36.3
Income ratios		
White to African American	1.3	1.7
White to Latino	1.3	1.8
White to Asian	0.9	1.0

Source: US Census (2013b).
Note: Whites are non-Hispanic; the dissimilarity index calculates the proportion of the minority group that would have to switch neighbourhoods to be completely integrated into non-Hispanic Whites' neighbourhoods; the income ratio divides the proportion of non-Hispanic White households who are middle and high income by the proportion of households from the minority group who are middle and high income.

group that would have to switch neighbourhoods to be completely integrated into the neighbourhoods of another group, meaning that the two groups would be evenly dispersed across the region's neighbourhoods. In 2012, 42% Latinos and 45% of African Americans living in post–civil rights suburbs would have had to switch neighbourhoods to be completely integrated into the communities where Whites were living (a low to moderate level of segregation) compared to 60% and 55% of their peers living in the postwar suburban city respectively (a high level of segregation).

The post–civil rights suburbs' greater racial integration is accounted for in part by their greater *economic* homogeneity. Income ratios are reported in table 10.2 to illustrate the extent of income parity among racial groups. Latino and African American households earned incomes more similar to Whites in the post–civil rights suburbs in 2012 (income ratio of 1.3 compared to 1.8 and 1.7 respectively in the postwar suburban city).

The urban design features of the post–civil rights suburbs and the historical context in which they developed may also have contributed to their lower racial segregation, greater income parity, and lower racial stratification. Master-planned communities pioneered in Phoenix by John F. Long proliferated during the post–civil rights era, in part due to changes to local zoning regulations, federal tax incentives, and financing (Schipper, 2008; VanderMeer, 2010). By the mid-2000s, an estimated 75% of home resales and 80% of new housing permits were in master-planned communities in the region (Hedding, 2005). Master-planned communities included a greater diversity of housing types, from apartments to single-family homes, which offered greater opportunities for residency among households of varying levels of wealth. The inclusion of diverse housing types and tenures may have been critical to attracting African Americans and Latinos, who historically have had less wealth than Whites (Kochhar, Fry, & Taylor, 2011).

Housing discrimination was less pernicious during the time that the post–civil rights suburbs emerged and matured. This was partly in response to new laws such as the 1968 *Fair Housing Act* and partly due to increased racial tolerance. Local leaders of colour reported that by the 1980s, housing discrimination had become subtler, defined more by finances than skin colour (Whitaker, 2005). Long-standing middle- and upper-income families of colour in Phoenix increasingly moved to post–civil rights suburbs during this era (Whitaker, 2005). Newly arrived Asian Indian and Chinese immigrants skipped over traditional gateway destinations like Los Angeles and older neighbourhoods in the city of Phoenix for post–civil rights suburbs (Skop & Li, 2005). However, Asians had an easier time dispersing among the regions' neighbourhoods than African Americans did, an outcome potentially shaped by their higher human capital (Luckingham, 1994; Skop & Li, 2005). Table 10.2 supports this finding – unlike Latinos and African Americans, Asians, and Whites lived in similar kinds of communities in the postwar suburban city and post–civil rights suburbs in 2012, as evident by their low levels of racial segregation and stratification. African Americans were more concentrated within the region, though less so than in the postwar suburban city (Gober, 2006). Several reports in the late 1980s and early 1990s found that discrimination in the local real estate and lending industries were still major barriers for African Americans and even Latinos (Luckingham, 1994, 1989).

Two suburbs that came of age during the post–civil rights era are Avondale in the West Valley and Gilbert in the East Valley (see map 10.1 and figure 10.3). Both were originally farming communities; yet they have experienced distinct demographic trajectories. Avondale houses a population just shy of 80,000 (U.S. Census, 2015). Half of its population is Latino, followed by non-Hispanic White (34%) and African American (9%) (U.S. Census, 2015). Its median household income of $55,000 is similar to the region (U.S. Census, 2015). Gilbert is a confederation of master-planned, homeowner association-governed single-family subdivisions, boasting a population of 230,000 (U.S. Census, 2015). Although its racial diversity increased dramatically as it grew in the 2000s, it is predominately non-Hispanic White (73%) and relatively affluent (median household income of $80,000) (Aspinwall, 2006; U.S. Census, 2015).

Both Avondale and Gilbert have struggled to forge their unique identities. Avondale was one of 68 cities nationwide (together with Chandler and Tempe, also post–civil rights suburbs in the Phoenix region) to adopt a brand as an "inclusive" community in 2006, meaning that they were "welcoming diversity in race, ethnicity, and lifestyle" (El Nasser, 2006, online source). That year, the city elected its first Latina mayor, Marie Lopez Rogers, who vowed "to make sure the city is connecting with everyone who calls this place home" (Bui, 2006, online source). Today, Avondale aspires to be a "healthy city." In the early 2010s, it built a recreational complex to encourage greater physical activity and boasted a farmers market, a community garden, and a city-driven healthy-eating initiative (Let's Move Blog, n.d.). Avondale's continued rapid growth, branding as a healthy community, and struggle to pay for the costs of its growth are evident in recent images from its built environment (figure 10.3).

Like Avondale, Gilbert has tried to be an inclusive community as it has grown, though it has faced unique challenges in accomplishing this goal. In the late 1990s, Gilbert was the home of the Devil Dogs, a violent White supremacist gang, whose members wore black boots with white laces to signify their racial solidarity and made a habit of intimidating people by barking at them (Smith & Willrich, 2001). City leaders tried to heal Gilbert from its racist past and build a more inclusive identity as its population boomed from 1999 through the mid-2000s. They held town meetings for residents to reveal their racial harassment and established a Diversity Task Force, a Human Relations Commission, and a Global Village Festival, which helped to nurture and celebrate its growing diversity (Aspinwall, 2006; Smith & Willrich, 2001). Now, Gilbert aspires to be a homespun, hobby-farming community. Its most distinctive neighbourhood is Agritopia, a cluster of New Urbanist-designed homes around an orchard and garden. Gilbert was the fastest-growing boomburb in the early 2000s; yet, it stubbornly clings to its rural past by calling itself a town (Lang & LeFurgy, 2007).

The better and more racially equitable conditions that the post–civil rights suburbs offer Latinos and African Americans (and Asians) are not unique to the Phoenix region but are found to various degrees across US regions that have post–civil rights suburbs (Pfeiffer, 2016). Notably,

Figure 10.3 Avondale. Photos courtesy of author.

these conditions persisted between 2000 and 2012 (Pfeiffer, 2016). However, the exact mechanisms that lead the post–civil rights suburbs to be more racially equitable are not known and are an important topic for further research.

Post-Recession Transformations

The recent US Great Recession and foreclosure crisis had a transformative effect on the housing market in Phoenix and other sunbelt metros, such as Las Vegas, Tampa, and Atlanta. In Maricopa County, where more than 90% of Phoenix region residents live, the unemployment rate rose from 3.7% in December 2007 to 10.3% in January 2010, and hovered between 8% and 10% for much of the following 2 years (Federal Reserve Bank of St. Lo uis, 2015). Median home values fell by just under one half from 2007 through 2012 (U.S. Census, 2008, 2010, 2013a). The twin dynamics of declining home values and job loss in the context of readjusting mortgage interest rates made homeowners vulnerable to foreclosure. An estimated 220,000 homes were foreclosed on in Maricopa County from January 2004 through May 2014 (Pfeiffer & Lucio, 2015).

Foreclosures in Maricopa County during this period exhibited two dominant spatial patterns (see map 10.2). On the one hand, they were widespread across the region's neighbourhoods. It is difficult to identify a community that had not experienced at least a handful of foreclosures. On the other hand, they were more heavily concentrated in more disadvantaged communities of colour in the postwar suburban city and

216 Deirdre Pfeiffer

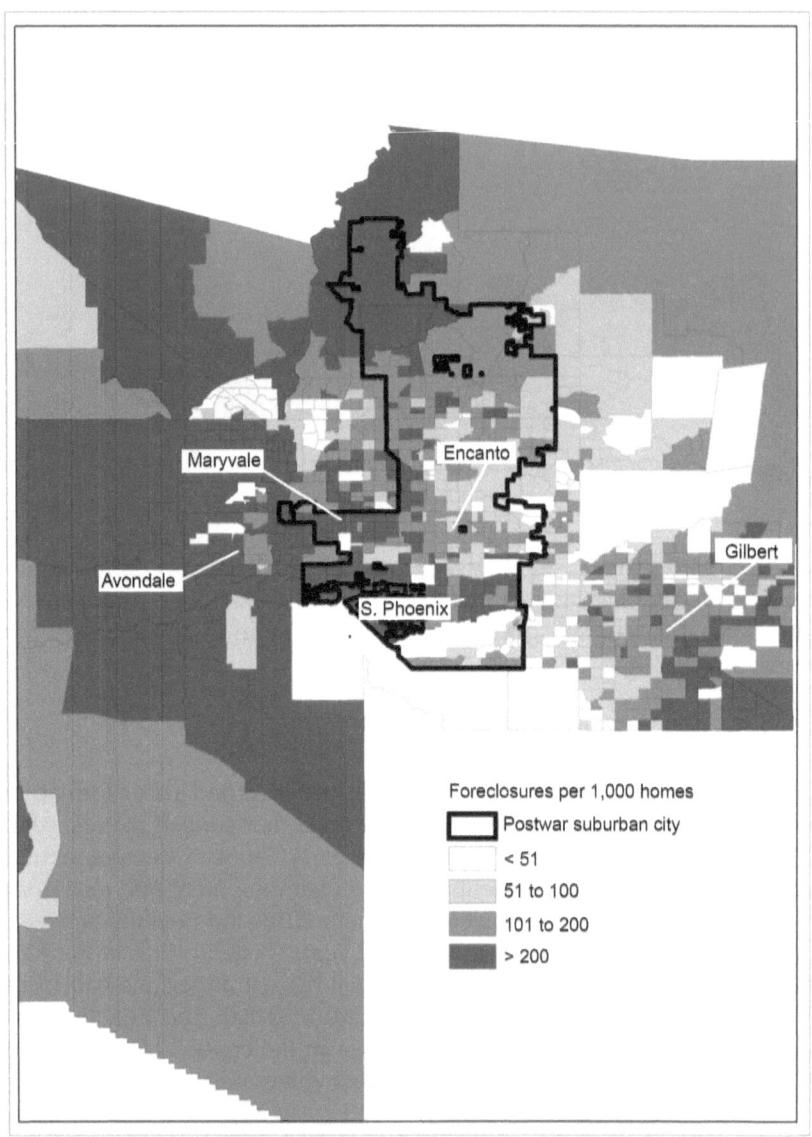

Map 10.2 The geography of foreclosures in Maricopa County, January 2004–May 2014. The spatial unit is the census tract; the very large tracts to the west, southwest, and northwest are mountain preserves and tribal communities.
Sources: Information Market (2014); US Census 2012a.

post–civil rights suburbs. In most neighbourhoods in Maryvale, South Phoenix, and Avondale, which are predominately Latino and African American, more than one in five homes experienced foreclosure (see map 10.2). This compared to slightly more than one in 20 homes in foreclosure in more affluent and White communities in Gilbert and near the Encanto neighbourhood north of downtown Phoenix, where the Ragsdales experience racial hostility in the 1950s.

Latinos' and African Americans' unequal experience of foreclosure is widely documented nationwide (e.g., Bocian, Li, & Ernst, 2010). Consensus is strong that because of racial disparities in foreclosures, the recent recession exacerbated existing nationwide patterns of residential segregation and inequality and potentially accelerated the racial stratification of the postwar suburbs (and catalyzed the racial stratification of the post–civil rights suburbs) (e.g ., Rugh & Massey, 2010; Schafran & Wegmann, 2012; Wyly, Moos, Hammel, & Kabahizi, 2009). Yet, gains in racial equity in the post–civil rights suburbs relative to postwar suburbs in Phoenix and other regions persisted between 2000 and 2012, after the official duration of the recession from 2007 to 2009 and well into the foreclosure crisis (Pfeiffer, 2016; see table 10.2). These findings put this theory into question.

An underexplored aspect of the foreclosure crisis that may help to explain enduring gains in racial equity in the post–civil rights suburbs is the transformation of neighbourhoods from majority owner- to renter-occupied. Nine percent of the Phoenix region's 972 census tracts changed from majority owner- to renter-occupied between 2000 and 2012 (U.S. Census, 2000, 2013b). Virtually all these places are located in the suburbs. Fifty-five percent are located in the post–civil rights suburbs; slightly more than 40% are located in the postwar suburban city. On average, these communities experienced about a 25-percentage point reduction in their owner-occupancy rates over the 2000s and early 2010s.

The transformation of neighbourhoods from majority owner- to renter-occupied is at once forging greater social equity and inequity in Phoenix. On the one hand, low-income renters of colour have taken advantage of this shift to move into neighbourhoods with higher performing schools, parks, and other amenities previously reserved for homeowners (Pfeiffer & Lucio, 2015). On the other hand, the opening up of the single-family home stock in the suburbs to renters disrupts traditional notions of suburbia as space of stability and household wealth accumulation through homeownership. This process may eventually increase racial stratification among the postwar and post–civil rights suburbs, but this does not seem to be occurring yet.

Whether majority renter-occupied suburban neighbourhoods are a temporary outcome of the Great Recession or a permanent shift in the Phoenix region's suburban landscape is key question. The job and housing markets in Maricopa County have slowly recovered. From January 2013 to January 2015, unemployment rates hovered at close to prerecession levels of between 5% and 7% (Federal Reserve Bank of St. Lo uis, 2015). Median home values increased 13% from 2012 to 2013 (U.S. Census, 2013a, 2014). Job stability, increases in earnings, and confidence in housing as a good investment should revive the for-sale housing market and encourage investor owners to sell their homes, potentially leading majority renter-occupied neighbourhoods to switch back to majority owner-occupied. However, the growing popularity of rent securitization among large corporate investor owners and a persistently tight mortgage-lending market may stabilize the suburban rental market for at least the immediate future (Edelman, Gordon, & Sanchez, 2014). It is difficult to draw conclusions on the effects of the Great Recession on racial stratification in Phoenix's postwar suburban city and post–civil rights suburbs until we know whether these new, majority renter-occupied neighbourhoods are a permanent feature in the region's suburban landscape. This question will be answered over the coming decades.

Conclusion

The Phoenix region at once confirms and disrupts classic notions of suburbanization in the United States. Phoenix has exhibited the mass-produced built environment that is typical of suburbanization in the rest of the United States. Yet because Phoenix came of age after the postwar era, it has a postwar suburban city at its core instead of an older central city. The case of Phoenix disrupts notions of racial stratification in the suburbs through its prolific post–civil rights suburbs, which emerged during a period of legal restrictions against housing and employment discrimination, minority socioeconomic gains, and greater racial tolerance. The recent dynamism of suburbanization also is exemplified in the Phoenix region. The mass conversion of former foreclosures to rentals in the wake of the recent recession and foreclosure crisis threatens long-standing conceptions of suburbia as a space for household stability and wealth accumulation and may eventually lead to greater racial stratification in both suburban types.

Whether or not the more racially equitable conditions in the post–civil rights suburbs reflect a fundamentally transformed suburban geography or just their relative immaturity is an important question that will be answered over the coming decades. There is fear that because of their experience of foreclosure in the wake of the recession and their dearth of middle class jobs, post–civil rights suburbs are becoming quagmires of socioeconomic stagnation and downward mobility (Schafran & Wegmann, 2012). Thus, people of colour who moved to them in search of a better life may eventually experience the same conditions that they sought to escape in the postwar suburbs and central city. It is important to investigate how people of colour are faring in the post–civil rights suburbs over time through further research. However, this chapter has shown that, at least in Phoenix, African American and Latino households are still experiencing far more racially equitable neighbourhood conditions in the post–civil rights suburbs than in the postwar suburban city, even after the official end of the recession. This suggests that the post–civil rights suburbs may be offering people of colour more lasting gains than other scholars and journalists predict, an outcome that also may apply to other regions nationwide (Pfeiffer, 2016).

NOTES

1 Middle-income refers to families earning between $50,000 and $125,000 in 2010 dollars. The Phoenix region's median household income in 2012 was about $54,000 (U.S. Census, 2013b).
2 This means that neighbourhood conditions where more middle-income households of the racial group are living are weighted more highly in determining their average neighbourhood conditions. Poorer neighbourhoods tend to have poorer performing schools, higher unemployment, and worse health and environmental conditions.

REFERENCES

Aspinwall, C. (2006, May 30). Gilbert sheds racist past as diversity grows. *Arizona Republic*. Retrieved on 12 September 2015 from http://www.azbilingualed.org/News_2006/gilbert_sheds_racist_past_as_diversity_grows.htm

Bocian, D.G., Li, W., & Ernst, K.S. (2010). *Foreclosures by race and ethnicity: The demographics of a crisis*. Durham, NC: Center for Responsible Lending. Retrieved 5 July 2013 from https://www.responsiblelending.org/mortgage-lending/research-analysis/foreclosures-by-race-and-ethnicity.pdf

Bui, L. (2006, March 29). Hispanic issues topic of Avondale forum. *Arizona Republic*. Retrieved 12 September 2015 from http://www.azbilingualed.org/News_2006/hispanic_issues_topic_of_avondale_forum.htm

Cashin, S. (2001). Middle class black suburbs and the state of integration. *Cornell Law Review, 86*, 729.

Cashin, S. (2007). Dilemma of place and suburbanization of the black middle class. In R.D. Bullard (Ed.), *The Black metropolis in the twenty-first century* (pp. 87–110). Lanham, MD: Rowman and Littlefield Publishers, Inc.

Dunham-Jones, E., & Williamson, J. (2011). *Retrofitting suburbia: Urban design solutions for redesigning suburbs*. Hoboken, NJ: Wiley.

Edelman, S., Gordon, J., & Sanchez, D. (2014). *When Wall Street buys Main Street: The implications of single-family rental bonds for tenants and housing markets*. Washington, DC: Center for American Progress.

El Nasser, H. (2006, August 4). Cities make quiet plea for tolerance. *USA Today*. Retrieved 12 September 2015 from http://usatoday30.usatoday.com/news/nation/2006-08-04-inclusive-cover_x.htm

Federal Reserve Bank of St. Louis. (2015). Unemployment rate in Maricopa County, AZ, percent, monthly, not seasonally adjusted. *FRED Graph Observations*. Economic Research Division, Federal Reserve Bank of St. Louis. Retrieved 13 May 2015 from http://research.stlouisfed.org/fred2

Forstall, R.L. (1995). *Arizona: Population of counties by decennial census: 1900 to 1990*. Washington, DC: U.S. Census Bureau.

Garreau, J. (1991). *Edge city: Life on the new frontier*. New York, NY: Anchor Books.

Gober, P. (2006). *Metropolitan Phoenix: Place making and community building in the desert*. Philadelphia: University of Pennsylvania Press.

Hedding, J. (2005). Master-planned communities: Majority of homes being built are part of a master-planned community. *About.com*. Retrieved 11 June 2015 from http://phoenix.about.com/cs/real/a/masterplanned.htm

Kneebone, E., & Lou, C. (2014, June). Suburban and poor: The changing landscape of race and poverty in the U.S. *Planning*, 17–21.

Kochhar, R., Fry, R., & Taylor, P. (2011). *Twenty to one: Wealth gaps rise to record highs between Whites, Blacks, and Hispanics*. Washington, DC: Pew Research Center.

Konig, M. (1982). Phoenix in the 1950s: Urban growth in the "Sunbelt". *Arizona and the West, 24*(1), 19–38.

Lang, R.E., & LeFurgy, J.B. (2007). *Boomburbs: The rise of America's accidental cities*. Washington, DC: Brookings Institution Press.

Let's Move Blog. (n.d.). Today, first lady Michelle Obama makes Let's Move! cities, towns and counties announcement. *Let's Move*. Retrieved 14 June 2015 from https://letsmove.obamawhitehouse.archives.gov/blog/2012/07/18/today-first-lady-michelle-obama-makes-let%E2%80%99s-move-cities-towns-and-counties-announcem

Logan, J.R. (1978). Growth, politics, and the stratification of places. *American Journal of Sociology, 84*, 404–16.

Logan, J.R., & Schneider, M. (1981). The stratification of metropolitan suburbs: 1960–1970. *American Sociological Review, 46*(2), 175–86.

Luckingham, B. (1989). *Phoenix: The history of a southwestern metropolis*. Tucson: University of Arizona Press.

Luckingham, B. (1994). *Minorities in Phoenix*. Tucson: University of Arizona Press.

Nelson, A.C. (1992). Characterizing exurbia. *Journal of Planning Literature, 6*(4), 350–68.

Nicolaides, B.M. (2002). *My blue heaven: Life and politics in the working class suburbs of Los Angeles*. Chicago, IL: University of Chicago Press.

Niedt, C. (Ed.). (2013). *Social justice in diverse suburbs: History, politics, and prospects*. Philadelphia, PA: Temple University Press.

Nijman, J., & Clery, T. (2015). The United States: Suburban imaginaries and metropolitan realities. In P. Hamel & R. Keil (Eds.), *Suburban governance: A global view* (pp. 57–79). Toronto, Canada: University of Toronto Press.

Orfield, M. (2002). *American metropolitics: The new suburban reality*. Washington, DC: Brookings Institution Press.

Pattillo-McCoy, M. (2000). The limits of out-migration for the Black middle class. *Journal of Urban Affairs, 22*(3), 225–41.

Pfeiffer, D. (2012). Has exurban growth enabled greater racial equity in neighborhood quality? Evidence from the Los Angeles region. *Journal of Urban Affairs, 34*(4), 347–71.

Pfeiffer, D. (2016). Racial equity in the post–civil rights suburbs? Evidence from U.S. regions 2000–2012. *Urban Studies, 53*(4), 799-817.

Pfeiffer, D., & Lucio, J. (2015). An unexpected geography of opportunity in the wake of the foreclosure crisis: Low-income renters in investor-purchased foreclosures in Phoenix, Arizona. *Urban Geography, 36*(8): 1197-1220.

Ross, A. (2011). *Bird on fire: Lessons from the world's least sustainable city*. New York, NY: Oxford University Press.

Rugh, J.S., & Massey, D.S. (2010). Racial segregation and the American foreclosure crisis. *American Sociological Review, 75*(5), 629–51.

Schafran, A., & Wegmann, J. (2012). Restructuring, race, and real estate: Changing home values and the New California metropolis, 1989–2010. *Urban Geography, 33*(5), 630–54.

Schipper, J. (2008). *Disappearing desert: The growth of Phoenix and the culture of sprawl*. Norman: University of Oklahoma Press.

Schwartz, A.F. (2015). *Housing policy in the United States* (3rd ed.). New York, NY: Routledge.

Skop, E., & Li, W. (2005). Asians in America's suburbs: Patterns and consequences of settlement. *The Geographical Review, 95*(2), 167–88.

Smith, P.M., & Willrich, P.L. (2001). *Diversity issues in Gilbert, Arizona: Effectiveness of human relations commission for resolving human rights*

violations (Unpublished master's thesis). Springfield College, San Diego, CA.
Suro, R., Wilson, J.H., & Singer, A. (2011). *Immigration and poverty in America's suburbs.* Washington, DC: Brookings Institution.
U.S. Census. (2000). *2000 decennial census.* Washington, DC: U.S. Census Bureau.
U.S. Census. (2008). *2007 American community survey 1-year estimates.* Washington, DC: U.S. Census Bureau.
U.S. Census. (2010). 2010 *decennial census.* Washington, DC: U.S. Census Bureau.
U.S. Census. (2012). *2011 American community survey 5-Year estimates.* Washington, DC: U.S. Census Bureau.
U.S. Census. (2013a). *2012 American community survey 1-year estimates.* Washington, DC: U.S. Census Bureau.
U.S. Census. (2013b). *2012 American community survey 5-year estimates.* Washington, DC: U.S. Census Bureau.
U.S. Census. (2014). *2013 American community survey 1-year estimates.* Washington, DC: U.S. Census Bureau.
U.S. Census. (2015). *State & county quickfacts, Gilbert (town), Arizona.* Washington, DC: U.S. Census Bureau.
Urban Dictionary. (2014). Murdavale. Retrieved 29 May 2015 from www.urbandictionary.com/define.php?term=murdavale
VanderMeer, P.R. (2010). *Desert visions and the making of Phoenix, 1860–2009.* Albuquerque: University of New Mexico Press.
Whitaker, M.C. (2005). *Race work: The rise of civil rights in the urban west.* Lincoln: University of Nebraska Press.
Wiese, A. (2004). *Places of their own: African American suburbanization in the twentieth century.* Chicago, IL: University of Chicago Press.
Wyly, E., Moos, M., Hammel, D., & Kabahizi, E. (2009). Cartographies of race and class: Mapping the class-monopoly rents of American subprime mortgage capital. *International Journal of Urban and Regional Research, 33*(2), 332–54.

11 Suburbanization and the Making of Atlanta as the "Black Mecca"

KATHERINE HANKINS AND STEVEN R. HOLLOWAY

Introduction

Between 1990 and 2010, metro Atlanta experienced a dramatic change in its suburban population: two thirds of the almost one million new Black residents of the city moved into suburbs outside the historic urban core. This change, we suggest, reflects the long-standing emphasis by the Black elite ("Black capital") on creating Black suburbs for the growing Black middle class. Deemed the "black mecca" in 1971 by *Ebony* magazine, Atlanta was portrayed as a city of Black prosperity and opportunity – a place to stay in or move to if you were part of the Black middle class in America – to take advantage of the economic opportunities and political power of the central city and the housing opportunities of the growing suburbs. This reputation was largely sustained by over four decades of Black mayors and the concentration of several of the country's elite historically Black colleges and universities (HBCUs). The emergence of Atlanta as a "black mecca" in the first place is, we argue, a fundamentally suburban phenomenon, built on the decades of active development of the suburban reaches of the city by its African American elite in the 20th century.

This chapter places the dynamics of Atlanta's African American suburbanization in historical context. It highlights key moments, such as the importance of the city's early Black-owned banks and real estate companies, negotiations that led to the city's planned Black middle class suburbs, and the striking move of African Americans to more distant suburban spaces in recent decades. We examine the institutional frameworks and the changing spatial economy of race in Atlanta as the city's role as the "black mecca" was made through suburbanization.

Before There Were Suburbs

Founded at the intersection of major railroads, Atlanta developed rapidly at the end of the 19th century and became the economic hub of the southeastern US regional economy by 1900. Of its approximately 89,000 residents, roughly 60% were categorized as White and almost 40% were African American. The residential patterns at the time reflected relatively integrated streets and neighbourhoods. As Lands (2009, p. 2) asserts "Social practices and understandings may have separated classes and races, but housing location did not" as the 19th century gave way to the 20th. Even so, African Americans were relatively concentrated in a few neighbourhoods, such as South Atlanta (then called Brownsville, a neighbourhood that grew up next to the Gammon Theological Seminary, the African American seminary founded in 1883), the neighbourhood of Mechanicsville along the rail lines south of the city, and the Old Fourth Ward, formerly called Shermantown, whose residents patronized Auburn Avenue, deemed by a *Fortune* writer as "the richest Negro street in the world" on the city's east side (Hughes, 1956). For the most part, however, African Americans lived in and among White residential sections of the city from its founding in the mid-19th century until the 1906 race riot. The race riot, in which racial tensions boiled over amid unfounded rumours of assaults on White women by Black men, resulted in the death of more than two dozen Blacks as White mob violence targeted Black residents across the city over a period of 3 days (Lands, 2009; Mixon & Kuhn, 2005; White, 1982). The riot ushered in a reordering of the city, as Black residents feared continued violence. Racialized residential segregation deepened over the next several decades (see map 11.1), but was not (successfully) formalized until the advent of zoning in the 1920s, through which municipalities designated separate land uses and regulated space based on race and class (Toll, 1969). Even then, residential patterns in the early years of the 20th century had more to do with who had access to capital (which was, of course, deeply shaped by race) than by formal land-use regulations.

The First Black Suburbs

Heman Perry, an entrepreneur whose father grew up a slave in Georgia, dramatically changed housing opportunities for Blacks in Atlanta. Perry used the reserves from Standard Life Insurance, Atlanta's first Black insurance company, founded in 1913 to fund the Service Company, an enterprise he developed to engage in a variety of commercial transactions, including the buying and selling of real estate. One of the

Suburbanization and the Making of Atlanta as the "Black Mecca" 225

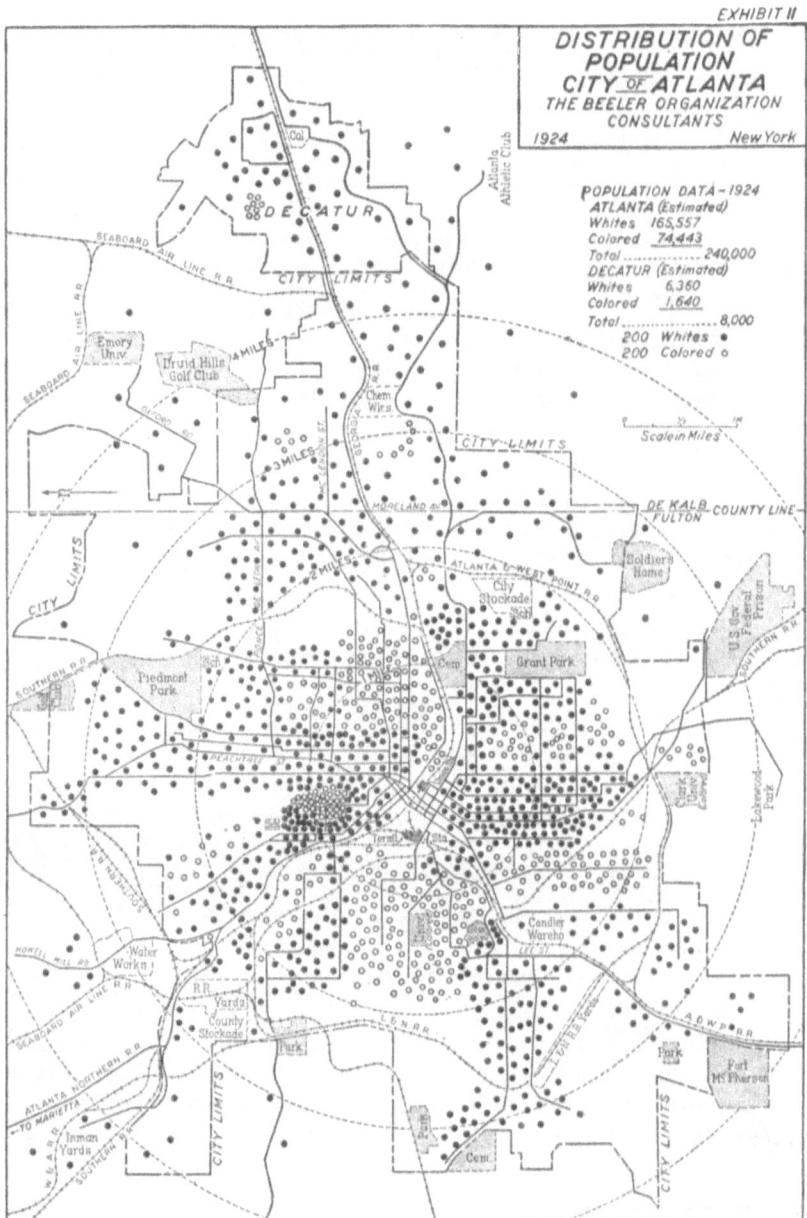

Map 11.1 Distribution of population, City of Atlanta, 1924. Source: The Beeler Organization Consultants, New York. Note: This map shows an eastern orientation of the city.

Service Company's first activities in 1917 was the purchase of 17 acres of "swamp and sagefield" on the very western edge of the city, close to Atlanta University, one of the country's premier Black universities (Henderson, 1987, p. 226). In addition, he opened Citizens Trust Bank in 1921 as a full-service bank with the expressed objective of servicing "the black community in the form of increased home ownership" (Henderson, 1987, p. 226; see also Ingham & Feldman, 1994; Wiese, 2005). The Service Realty Company, a subsidiary of the Service Company, sold a portion of its 17 acres to the Atlanta School Board, which became the site of the city's first Black high school, Booker T. Washington High School. Service Realty acquired hundreds of additional acres near the school site, which it then sold to Black homebuyers, who purchased homes built by the Service Engineering and Construction Company (Henderson, 1987). Perry pressured the city to establish a park in the area, the first for Black residents of the city (Bayor, 1996). Perry's companies were responsible for the development of some 500 homes for Black families near the first Black high school and the city's first park designated for Black families. According to a journalist from the Black newspaper the *Pittsburgh Courier*, "the opening of the Westside was the greatest single contribution to the improvement of living conditions in Atlanta for Negroes made in 25 years" ("Heman Perry ... ," n.d., in Henderson, 1987, p. 228).

The impact of Perry's enterprises, which had collapsed rather dramatically by 1925, were profound in shaping the ability of Black Atlantans to purchase homes built by a Black-owned company with loans from a Black-owned bank.[1] Washington Park, the neighbourhood that Perry created, remained an important site of Black residential space for much of the first half of the 20th century. And it inspired further development of Black residential subdivisions on the west side of the city, well beyond the city limits. As Wiese (2005, pp. 176–77) points out, "Heman Perry's vision of building self-contained black neighbourhoods on the developing edge of town became the blueprint for black housing efforts for the rest of the century." And, we suggest, it laid the *suburban* foundation for Atlanta's emergence as a city of Black wealth and opportunity. There, amid the neatly laid-out streets, middle class Blacks purchased the same kinds of bungalow homes that dominated White neighbourhoods at the time.

Perry's model was followed by another Black entrepreneur, Walter Aiken, who became the leading home builder in Atlanta by the late 1920s after building the suburban estate of middleweight boxing champion Tiger Flowers in 1927 (Wiese, 2005). His developments included 250 units for Black families in a subdivision called Fairview Terrace on the western edges of the city and his purchase of hundreds of acres beyond the formal city limits. During the postwar housing crisis, Aiken led a group of Black business leaders to form the Temporary Coordinating Committee on Housing (TCCH) to map out "Negro expansion" areas. These areas were "adjacent to a 'present Negro area' in the outer rim of the city" and "convenient to employment centers and existing black institutions, and each was sufficiently isolated to minimize white resistance" (Wiese, 2005, p. 181). This blueprint of self-contained Black suburbs came up against other forces that were also shaping the racial geography of Atlanta's housing and that were contributing to the city's suburbanization.

As scholars of the "New Suburban History" (Jackson, 1985; Sugrue, 2005[1996]; see also Kruse & Sugrue, 2006) have made clear, the New Deal policies designed to shore up mortgage lending during and following the Great Depression were expressed through several institutions, including the Home Owners' Loan Corporation (HOLC) and the Federal Housing Administration (FHA). In Atlanta, one of the early HOLC Security maps, which was used as a guide to assess the "risk" associated with insuring a mortgage, was created in 1938. It, like others created across the country, designated areas as green ("best"), blue ("still desirable"), yellow ("definitely declining"), and red ("hazardous") based on area characteristics, the age of housing stock, and the resident population, including the race and income of residents. In Atlanta's 1938 HOLC map, areas adjacent to the traditional downtown core were all designated as red or yellow, based on the age of the housing stock and the presence of African Americans. As other scholars have noted, the presence of even a few African American families in a district "redlined" it, as it was designated a "hazardous" risk for insuring mortgages (e.g., Sugrue, 2005[1996]). For example, the area around the city's cluster of HBCUs, west of downtown, was designated as "red" even though the accompanying notes in the "area description" identified it as "the best negro area in Atlanta and contains best type of negro residents and highest percentage of negro home ownership" (Form 8, 10–1-37, HOLC), thanks in no small part to the real estate development practices of Perry and Aiken.

The long-term impact of redlining on strengthening existing patterns of racial segregation has been well documented (Jackson, 1985; Kruse, 2005; Sugrue, 2005[1996]). Indeed, redlining created new layers of racial segregation, as older housing stock close to the historic downtown was identified as "hazardous" or "declining" and White, middle class families were unable to secure FHA-backed mortgages in any neighbourhoods other than those on the edges of the city or in more distant suburban areas. In other words, White suburbanization was anchored institutionally in the governmental and private-market institutions that created the post-Second World War metropolis. African American suburbanization, on the other hand, had a different set of drivers and geographic limitations.

The early suburbanization of African Americans in Atlanta occurred as the policies and impacts of the discriminatory practices of the FHA and HOLC were unfurling against the backdrop of the Jim Crow system, which maintained separate institutions and facilities, such as schools and parks, for Whites and Blacks (e.g., Bayor, 1996; Kruse, 2005). One result of the prosuburban bias of New Deal housing policies was to open up formerly White neighbourhoods to Black residents, which was common in many other cities across the country. Atlanta remained relatively unique, however, because of the continued concentration of wealth among key African American entrepreneurs, the growth and increasing significance of the Black colleges and universities clustered on the west side of the city, and the model of developing self-contained Black suburbs, which continued through the decades of the mid-20th century. Other cities, such as Chicago or New York City, were home to many industrious Black entrepreneurs (Ingham & Feldman, 1994), and Black colleges thrived in cities such as Tuskegee, Alabama, and Washington, DC. But no other city had the concentration of both Black entrepreneurs *and* elite institutions of higher education. In addition, several strong and politically active churches were important in consolidating the social cohesion and respectability of the Black elite in the city. The strategy of many of these early 20th-century Black leaders was to facilitate Black middle classness through developing single-family homes in the expanding city.

Hunter Hills, just outside the city of Atlanta's western border, was developed by the concerted effort of leaders of Citizens Trust Bank in partnership with the FHA, which agreed to work on the development for middle-income Blacks (Ingham & Feldman, 1994). The Works Progress Administration cut roads through the area, and the bank financed the construction of single-family homes, which became a subdivision of the Black elite. Indeed, the role of the Black-owned banks headquartered on Auburn Avenue significantly shaped the suburban patterns of Black expansion. As Wiese (2005) notes, the concentration of the wealth of those banks along with the strategic plans of "negro expansion" articulated by the TCCH resulted in the purchase of hundreds of acres on the western side of the city for the purpose of developing subdivisions of single-family homes to be sold to the city's emerging Black middle class households.

For the most part, the "negro expansion" areas were adjacent to existing Black communities and most were outside the city's existing borders, making them both political and functional suburbs of Atlanta. Nonetheless, the growth of the Black population within the city, and the growing political power of Black voters alarmed White elected officials by the late 1940s (Bayor, 1996; Stone, 1989). In a letter to community leaders, seeking support for the annexation of the northern White suburbs into the city limits, then Mayor Hartsfield suggested that outmigration [by White home-owning citizens] is good but that with the

> federal government insisting on political recognition of Negroes in local affairs, the time is not far distant when they will become a political force in Atlanta if our white citizens are just going to move out and give it to them. This is not intended to stir race prejudice because all of us want to deal fairly with them; but do you want to hand them political control of Atlanta? (Martin, 1978, p. 42 in Stone, 1989)

By the time Atlanta officials annexed over 50,000 acres in an attempt to maintain the White majority of the city's electorate, the city brought into its limits not only the northern White areas of Buckhead and other affluent White areas to the south and west, but also the elite Black suburbs of Hunter Hills and its surrounding Black subdivisions. The earliest Black suburbs, then, became part of the city in 1951 (see map 11.2).

Suburban Expansion

Securing housing opportunities in increasingly suburban reaches became a strategy for Black leaders to serve the growing needs of middle class Black families in light of the racist and restrictive covenants delimiting White neighbourhoods in the northern, eastern, and southern reaches of Atlanta (Wiese, 2005). In a race to secure property that was not yet in the clutches of White subdivisions, Atlanta's Urban League housing secretary Robert Thompson devised a strategy to achieve the development of hundreds of acres for Black homeownership, seeking to "quietly acquire" land at the western edge of the county (Wiese, 2005, p. 186). This involved "jumping over the whites" and buying land to block in existing (White) landowners, resulting in the eventual development of the communities of Fair Haven and Carroll Heights with modest two-bedroom, one-bath brick ranch homes, and Collier Heights, which was designed with three- and four-bedroom homes for more affluent families (see map 11.3). Collier Heights eventually included more than 1,700 homes, schools, and churches and was home to the city's

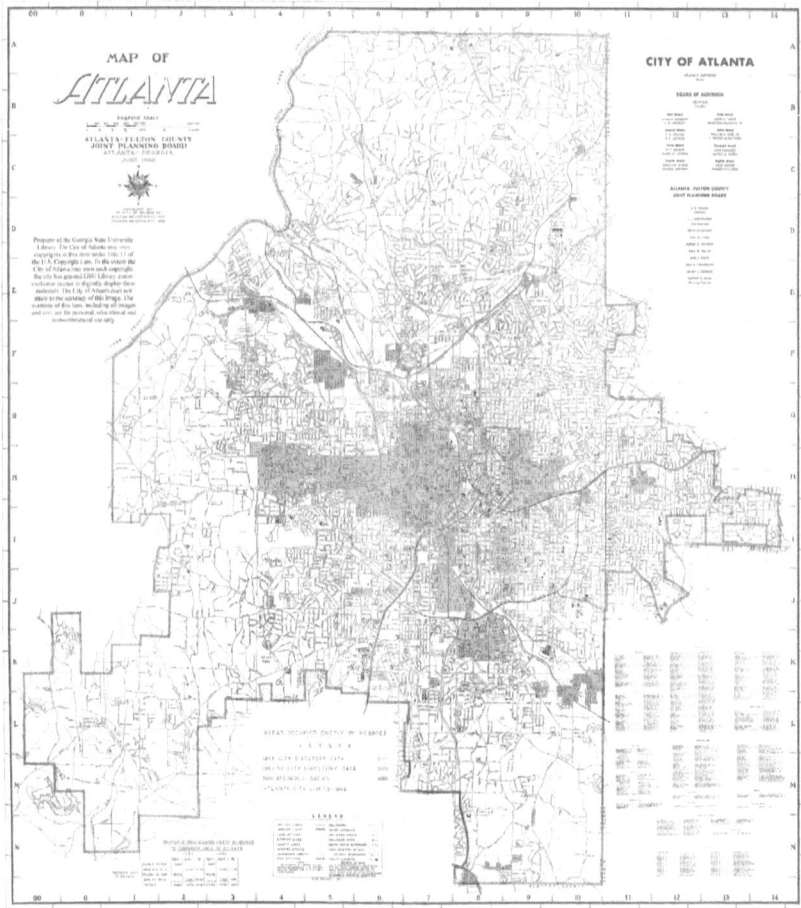

Map 11.2 Map of Atlanta areas "occupied chiefly by Negroes," 1959: Source: City of Atlanta Planning.

Black elite of the day, including Reverend Martin Luther King, Sr., and his wife, among many other prominent Black Atlanta families (Riley, 2010). It was yet another effort financed in large part by three of the nation's largest African American financial institutions: Mutual Federal Savings and Loan Association, Citizens Trust Bank, and the Atlanta Life Insurance Company (Riley, 2010). As the writer of an *Ebony* magazine article proclaiming Atlanta the "black mecca" put it "There are several of these black residential sections ... where exquisitely designed and

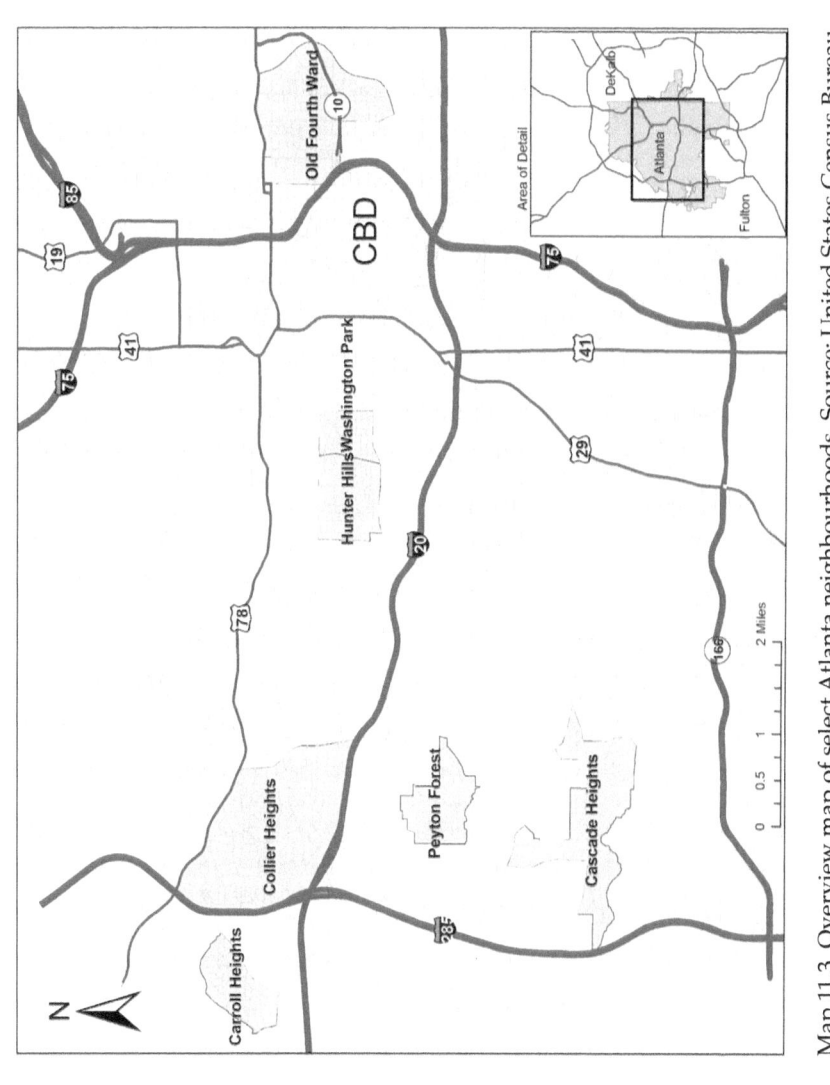

Map 11.3 Overview map of select Atlanta neighbourhoods. Source: United States Census Bureau and the Atlanta Regional Commission; map created by Dajun Dai.

spacious modern homes are set far back from the street in protective nests of towering Georgia pines. Indeed, these homes are so fabulous and numerous that visitors have difficulty believing what they're seeing" (Garland, 1971). What was so unique were the miles and miles of well-kempt, middle class *African American* neighbourhoods, the likes of which no other metropolitan region in the United States had.

This expansion westward occurred during a period between 1930 and 1960 in which the numbers of African Americans in the Atlanta area had doubled, while African Americans were being pushed out of several neighbourhoods closer to the urban core by the massive highway construction and urban renewal projects that were dramatically transforming the city (Kruse, 2005; Scott, 1961; Stone, 1989). Furthermore, the incomes of African American households increased steadily in the mid-20th century (Wiese, 2005), and new housing construction on the margins of the city and increasingly in its suburban reaches (versus converting existing White neighbourhoods) remained the primary solution to accommodate the growing numbers of middle class African Americans.

That is not to suggest that there were not racial tensions in the increasingly African American west side. In fact, one of Atlanta's more famous incidents around the breaking of racial barriers occurred in 1962 in the Cascade Heights-area subdivision of Peyton Forest, which is located south of Collier Heights, below what was, at the time, one of the city's "color barriers" (White, 1982). In an effort to appease White residents of the neighbourhood, city construction crews built roadblocks on two roads "at the precise fault line between black and white sections of the city" (Kruse, 2005, p. 3). Civil rights activists drew national and international attention to the symbolic barrier intended to stop the encroachment of Black families into the neighbourhood (Bayor, 1996). The "Peyton Wall" as it was later known was eventually dismantled after local courts ruled against the roadblocks, and within six months, all but 15 White families had sold their homes to Black buyers (Kruse, 2005). By the 1970s, Cascade Heights was one of the premier addresses for Atlanta's Black elite (and is home to the city's past and current mayors).

The central city increased modestly in population in the 1960s, and suburban expansion, particularly of White households was rather dramatic in the 1960s and 1970s. A key driver of this push, in addition to the legacy of redlining, was the reaction of many White families to the desegregation of schools, which was peacefully begun by Atlanta Public Schools in 1961. Many White families left their neighbourhood schools for private schools (the metro area experienced a dramatic increase in the number of private schools and "segregation academies" during this time) or left their neighbourhoods altogether (i.e., "White flight"). Kruse (2005) highlights the rapid change in Kirkwood, a neighbourhood east of downtown and predominantly White in the 1950s. In 1963, a group of Black parents showed up to the neighbourhood's all-White elementary school to register their children, given that the "black" schools were overcrowded. They were turned away but kept pressure on the school board, which relented, agreeing that Black children could enrol there after their upcoming holiday break. "On the Friday before the black children were to arrive, there were still 470 white boys and girls enrolled at Kirkwood Elementary... . When black students showed up the following Monday, they found only seven white children in the building" (Kruse, 2005, p. 169). Despite Atlanta's modest population growth between 1960 and 1970, student enrolment in Atlanta public schools declined from its 1967–68 peak of 113,470 (40.8% White; 59.2% Black) to its 1993–94 nadir of 59,154 (6.8% White; 91.5% Black; and 2.5% "other") (and where the population of the city had declined by around 100,000).

By 1970, a planning map of the metro region shows the spatial concentration of African Americans as a large belt across the city, with greater density in the central and eastern neighbourhoods and the extensive automobile-centred housing developments on the west side (see map 11.4). With the exception of a few clusters of Black communities to the south of the city and a few beyond the eastern edge of the city limits, the vast majority of Atlanta's Black residents lived within the city limits (although they were suburban in form in terms of the predominance of single-family homes on curvilinear streets, they relied on city services). According to the U.S. Census, approximately 48% of the city's almost 500,000 residents were White and 51% were Black, whereas the broader metro area of more than 1.7 million was 78% White.

The 1970s were a time of transition, as White flight continued. Despite Title VI of the *Civil Rights Act* of 1964 and Title VII of the *Civil Rights Act* of 1968 (*the Fair Housing Act*), which were designed to remove the barriers to free choice in the housing market, Black families encountered obstacles in obtaining homes, as documented in a special series published in 1988 by *Atlanta Constitution* journalists, who conducted studies on the illegal steering practices of real estate agents and the egregious racism of mortgage lending by area banks (see Dedman, 1988). The headline of the first instalment of the Pulitzer Prize–winning series read "Atlanta Blacks Losing in Home Loans Scramble" (Dedman, 1988), reflecting the documented discrimination among banks in Atlanta and their lending and approval rates to Black loan seekers versus White ones. The four-part series documented

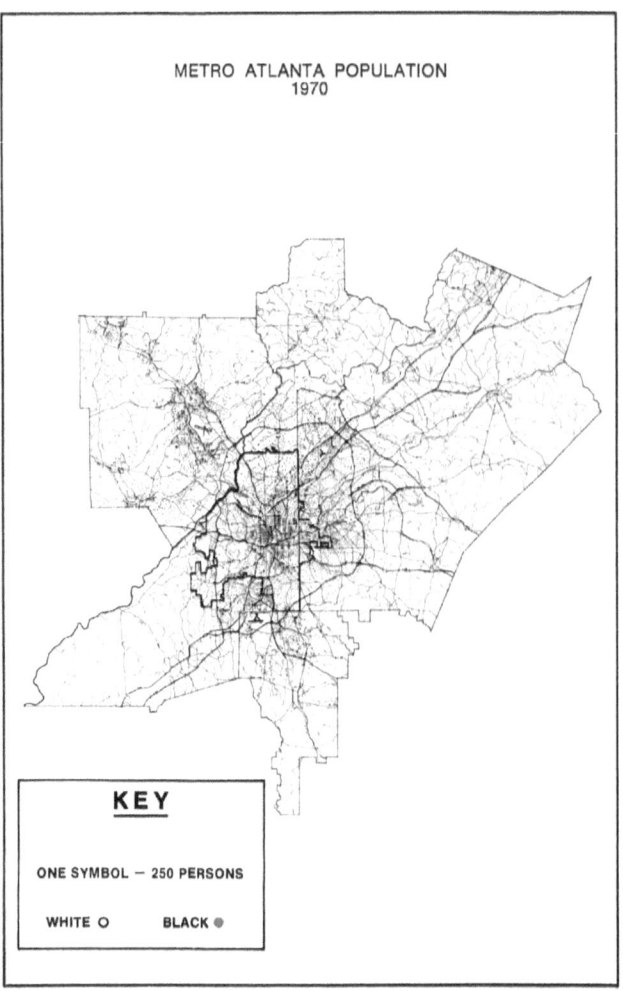

Map 11.4 Metro Atlanta population, 1970. From *Which Way Atlanta?*, March 1972, published by Research Atlanta, accessed through the Planning Atlanta digital archives.

the extensiveness of the discrimination and the persistence of Black financial institutions as the cornerstone of Black homebuying. For example, by 1985–6, 100% of loans by Citizens Trust Bank, founded by Heman Perry back in 1921, were to Black households, followed by Mutual Federal Savings and Loan, which had a 5:1 ratio in lending to Black families (based on study data that controlled for neighbourhood quality and income). In other words, Black institutions were still the foundation of homebuying for Black, middle class families in the 1970s and early 1980s. At the same time, Black residents were the majority in the city proper, and their voting power ushered in a new era of the city's politics, solidifying Atlanta's reputation as a city of opportunity for African Americans.

Atlanta as the Black Mecca

In 1974, Atlanta residents elected 35-year-old Maynard Jackson, the first Black mayor of any major city to serve the beginning of what was ultimately three terms. In addition to initiating a neighbourhood-based planning system, a major hallmark of Jackson's time in office was to increase the share of municipal contracts given to minority companies from 1% to 35% (Stone, 1989). One of Jackson's biggest projects was the expansion of the Atlanta airport, which many have suggested secured Atlanta's significance as a regional economic hub and created dozens of Black millionaires (Stone, 1989). Herman Russell, founder of what became one of the largest Black-owned construction companies in the country, was one such beneficiary of Mayor Jackson's insistence on hiring minority firms. Furthermore, Russell built his sprawling home in Cascade Heights, where many of Atlanta's Black elite lived.

While the population of African Americans within the city limits (although in automobile suburbs such as Collier Heights) continued to grow through the 1970s, the African American population of the non-central city portions of the five major counties in the Atlanta standard metropolitan statistical area tripled (Thomas, 1984; see also Jaret, 2002). Indeed, much of the new housing construction in the 1970s occurred in the suburbs with little building taking place in the central cities (Bullard, 1986, p. 60). And homeownership rates between the Black residents of the central city versus the suburbs also varied (nationally, in the 1970s, 36% of central city Blacks were homeowners, while more than 50% of suburban Blacks owned their homes (Bullard, 1986, p. 60)). As Lake (1979) points out, however, the ability of Black homeowners to realize increased value in their homes was compromised by the persistent trends in racial transition, where homes owned by White residents were sold to Black homebuyers but rarely did White homebuyers purchase properties from Black families. Instead, Black families wishing to sell often converted their properties to rentals. This dynamic effectively limited the wealth accumulation Black homeowners could extract from their homes, unlike their White counterparts (Lake, 1981). As the *Atlanta Journal-Constitution* documented, between 1975 and 1984, median sale prices of homes in Atlanta increased by 58%, but prices increased only modestly in Black neighbourhoods. For example, in Garden Hills (White), values went up by 68% but only by 20% in Cascade Heights (Black), and up 71% in Peachtree Hills (White) but only 6% in Collier Heights (Black) (Dedman, 1988). Nonetheless, Atlanta remained a significant site of Black homeownership, which increasingly moved to the more distant suburbs.

Between 1970 and 1990, the percentage of Atlanta's suburban population that was Black rose from 6.2% to 20% (U.S. Census Bureau, 1990), and approximately 53% of all Black married couples with children moved out of Atlanta's city limits between 1970 and 1990 (Goldberg, 1997). As Hartshorne and Ihlanfeldt (1993) document, in 1970, nearly 80% of African-Americans in the region lived within the city limits, but by 1990, the percentage had dropped to below 40%.

While the Black population was moving out beyond city limits, the business and political opportunities of the central city continued to attract (and form) middle class Blacks in Atlanta. For example, since 1974, Atlanta has been governed by a series of Black mayors, whom scholars have characterized as largely "pro-growth advocates" with strong support for business and commercial interests (Banks, 2000). At the conclusion of Maynard Jackson's second term in 1982, former UN Ambassador Andrew Young took the helm of Atlanta and continued to espouse business-oriented enterprises, which he termed "public purpose capitalism" (Newman, 2016). Maynard Jackson, who had developed close ties with Atlanta's White business leaders during his second four-year term (Stone, 1989), returned to the mayor's office to serve a third term, from 1990 to 1994. It was during this time that Atlanta was selected to host the 1996 Olympic Games, due in no small part to the international ties former ambassador Young had cultivated along with the support of the city's business community (Rutheiser, 1996). Bill Campbell served from 1994 to 2002 and continued the progrowth policies and the city's "unyielding commitment to African-American businesses" (Ebony, 1997, p. 72). Shirley Franklin, who had served in various capacities under both Maynard Jackson and Andrew Young, served two terms from 2002 to 2008, followed by Kasim Reed, Shirley Franklin's former campaign manager, who barely edged out his White competitor in his 2008 election ("Atlanta contest," 2009). The leadership by well-connected Black elites has shaped the perception of Atlanta by outsiders as being a city where Black leaders foster the success of Black capital.

A quarter century after designating Atlanta the country's "black mecca," *Ebony* magazine once again identified Atlanta as "the mecca for blacks" (Ebony, 1997, pp. 68–9): "A poll of the magazine's 1997 list of

100 Most Influential Black Americans found that Atlanta is overwhelmingly considered to be the best city for Blacks, the city with the most employment opportunities for Blacks, the most diverse city, and the city with the best schools and the most affordable housing for African Americans." The affordable housing for Black homebuyers was increasingly in the distant suburbs, and in fact the 1990s marked a dramatic shift in the growth of the distant Black suburbs and the increasing numbers of middle class Black Atlantans.

Atlanta's Growing Black Suburbs

Echoing the descriptions of Collier Heights from almost a half century before, an *Atlanta* magazine writer described the disbelief her relative from Detroit expressed when he visited her in 2003 in majority-Black South DeKalb County:

> Land that was once cow pastures is now crisscrossed by streets that wind up, down and around gently sloping hills, offering beautiful tree-laden vignettes at every turn. These streets meander past custom-built homes, four- or five-bedroom affairs, often professionally decorated. You spot at least one foreign-made luxury vehicle in each driveway. Lawns are manicured and perfectly reflect the strict neighborhood association codes. The landscape has become second nature to me; in the three or four hours a day I spend behind the wheel chauffeuring my kids to school, basketball practice and church, I do not pay much attention to what's outside the windshield. My uncle, however was startled by the passing landscape. "These are all black folks?" he asked incredulously. "Yes," I said. "And these? Still black?" he asked, as we left one beautiful subdivision to enter another. "Yes, as far as the eye can see," I said. We continued our tour, passing mega churches, newly built major chain grocery stores, restaurants and service businesses. "All black folks?" he asked. "Yes," I said. "Black, black, black." (Williams, 2003)

Map 11.5 shows the evolution of Black suburbs between 1990 and 2010. Areas shaded black on the maps are highly segregated majority-Black neighbourhoods. Neighbourhoods shaded dark grey are modestly segregated (majority Black), but with substantial representation of non-Black groups. The maps also show the outer boundary of the urbanized portion of the metropolitan region and do not show the racial composition of neighbourhoods in the old urban core.[2] Black suburbs in 1990 maintained the historical pattern of being located to the west, southwest, and southeast of downtown (while White suburbanization

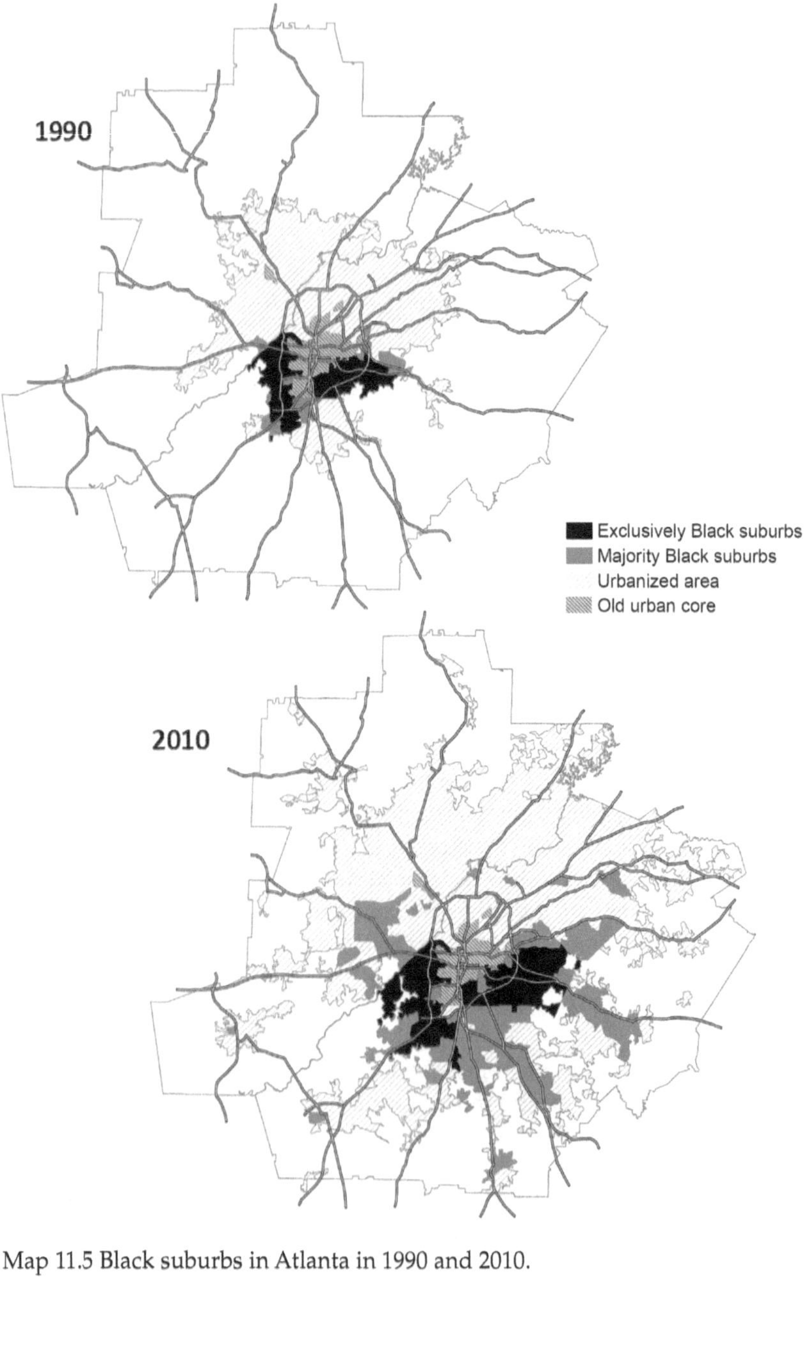

Map 11.5 Black suburbs in Atlanta in 1990 and 2010.

occurred rather consistently to the north). While most of these areas are located south of Interstate-20, a few neighbourhoods just to the north had been incorporated into the Black suburbs by 1990.

By 2010, several dramatic changes had taken place, reflecting the intensification of Black suburbanization. First, the total area of what we characterize as Black suburbs had expanded dramatically, both in highly segregated almost exclusively Black areas and in racially mixed, majority-Black areas.

Second, Black suburbs had expanded well north of I-20, especially in DeKalb County, to the east and southeast of downtown Atlanta. Many of these areas became highly segregated via racial transition – they had formerly been predominantly White. Of 146 inner suburban neighbourhoods classified as Black suburbs in 2010 (~43% of the total of 346 neighbourhoods), half had been majority White in 1990. The total population in these transitional areas grew by about 43% to just over half a million. The White population of these areas declined by 62%, while the Black population of these areas increased by almost 350%.

Third, Black suburbs – especially racially mixed, majority-Black areas – expanded long distances away from downtown, primarily along major highways. Many of these areas were developed as new Black suburbs (rather than as White suburbs that later transitioned to Black). Of 124 neighbourhoods that developed after 1990, 24% were majority Black in 2010. The total population of these neighbourhoods more than doubled, from 140,635 to 356,521. Black population in these areas increased by 440%, while White population decreased by 5%.

As has always been true of Atlanta's Black suburbs, neighbourhoods contain a mixture of affluence and economic precarity. While we have focused on the significance of suburbanization in undergirding the development of Atlanta as a Black mecca in the 20th century, affluence and opportunity have not reached all Black Atlantans, and in recent decades, the ground in wealth accumulation gained by Black households has been compromised by the Great Recession and the predatory lending practices of the early 2000s. For example, subdivisions in South DeKalb County where affluent Black professionals live, as described in the *Atlanta* magazine writer above, have lost significant value since the Great Recession. A *Washington Post* analysis revealed that "the higher a zip code's share of black residents in the Atlanta region, the worse its housing values have fared over the past turbulent housing cycle" (Badger, 2016).

Figure 11.1 shows poverty rate changes in suburban neighbourhoods classified by race for three periods, which imperfectly capture pre- and post-recession patterns. Poverty rates increased in all types

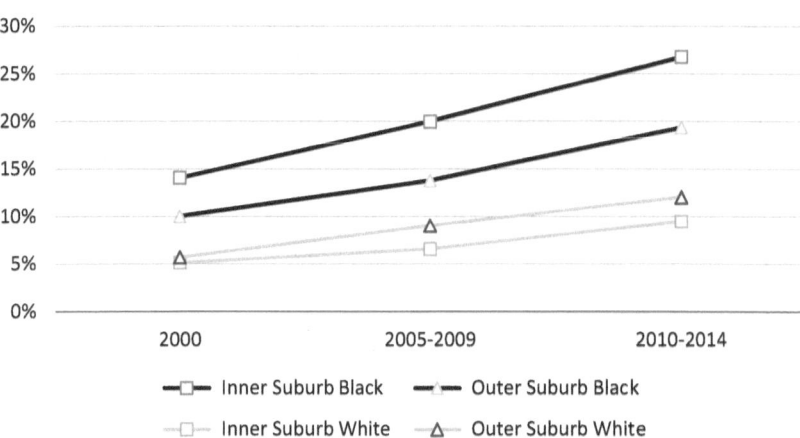

Figure 11.1 Poverty rate in Black and White suburbs of Atlanta, 2000–2014.
Source: Neighbourhood poverty rates for 2000 were derived from Census 2000 estimates, while rates for 2005–9 and 2010–14 were derived from 5-year pools of American Community Survey estimates.

of neighbourhoods in all parts of the metropolitan area, both between 2000 and the period ending 2009, and in the subsequent 5-year period. Poverty rates are substantially higher in Black suburban neighbourhoods than they are in White suburban neighbourhoods in each of the three periods, and they increased over time more prominently in predominantly Black suburbs. Poverty rates increased most dramatically for Black inner suburban neighbourhoods, from 14% in 2000 to 20% in the period ending 2009, and 27% in the period ending 2014. The poverty rate almost doubled in just 15 years! Many of these areas were hit hard by the foreclosure crisis, as these are precisely the neighbourhoods that were targeted by predatory lenders. While poverty also deepened in the newer outer suburbs, the magnitude of increase was more modest than in the Black inner suburbs, rising from 10%, to 14%, to 19%. Unsurprisingly, the impact of economic woes on White suburbs was much more muted. For majority-White suburbs, poverty rates were higher in each of the three periods in the outer suburban neighbourhoods than in the inner suburban neighbourhoods – opposite the pattern evident in Black suburbs. Our brief analysis of Black suburban poverty reveals drastic increases in material need throughout the Black suburban landscape.

One of the challenges presented by the increasing poverty rate in the suburbs involves the ability and capacity of suburban governments to provide services to their residents. Unlike in many other major cities where suburbs have incorporated and municipalities provide services, such as in Chicago or Los Angeles, Atlanta's suburbs remain largely unincorporated. Instead, comprehensive urban county units provide utility services (water, sewer, trash collection), education, police, fire, and other functions, a trend that begin in DeKalb County in the 1940s and 1950s, followed by Cobb County in the 1960s; Clayton, Douglas, Gwinnett, Fulton, and Henry counties in the 1970s; and Cherokee, Forsyth, and Coweta in the 1980s (Hartshorne & Ihlanfeldt, 1993). This governance structure, which is fragmented by numerous municipalities (including more recent municipal incorporations such as that of Sandy Springs and Johns Creek), is further splintered by various homeowner associations and other forms of private governance (Nijman & Clery, 2015), presenting even more challenges for the coordination of services to address increasing needs.

Conclusions

Atlanta's Black suburbs represent a distinct set of urban spaces. The development of Black suburbs began early in Atlanta's history, reflecting the impact of a relatively privileged segment of the Black population that sought to promote racial uplift by leveraging its capital and later its political power. Drawing from the presence of important financial, religious, and educational institutions in the city, the Black elite expanded Black residential spaces to the edges of the growing city, first to neighbourhoods adjacent to downtown Atlanta, and by the mid-20th century, miles from the central city. These suburban residential communities were sites of continued Black wealth accumulation, solidifying the significance of the city's Black elite in industry and in politics, contributing to the city's national reputation as a Black mecca.

Most recently, Atlanta's Black suburbs have grown rapidly, creating striking new patterns. As our data show, racially mixed Black suburbs now stretch far beyond Atlanta's core, following along major highway routes and transforming previously majority-White suburbs in counties where neoliberal forms of suburban governance prevail. Overall, Atlanta's Black population is now overwhelmingly suburban and increasingly living in racially mixed majority-Black suburbs, both in the older and newer portions of Atlanta's suburban spaces. Current Black suburbanization, thus, can best be understood as a continuation of the forces that first gave rise to Atlanta as "black mecca," which, we argue, was always rooted in suburban expansionism.

NOTES

1 Taken over by other Black entrepreneurs in Atlanta, Citizens Trust Bank survived.
2 The maps, charts, and tables shown in this section are based on a classification of metropolitan space into four categories, based on the period during which land cover and population density became "urban" as defined by the U.S. Census Bureau. The "old urban core" is the area that had been developed by 1950, the "inner suburbs" had been developed by 1990, the "outer suburbs" had been developed by 2010, and the "exurbs" had not achieved urban status as of the 2010 Census.

REFERENCES

Atlanta contest shows battered Black electorate. (2009, December 4). Associated Press.
Atlanta is new Mecca for Blacks. (1997, September). *Ebony*.
Badger, E. (2016, May 2). The nation's housing recovery is leaving Blacks behind. *The Washington Post*. Retrieved from https://www.washingtonpost.com/graphics/business/wonk/housing/atlanta/
Banks, J. (2000). A changing electorate in a majority Black city: The emergence of a neo-conservative Black urban regime in contemporary Atlanta. *Journal of Urban Affairs*, 22(3), 265–78.
Bayor, R. (1996). *Race and the shaping of twentieth century Atlanta*. Chapel Hill and London: University of North Carolina Press.
Bullard, R. (1986). Blacks and the American dream of housing. In J. Momeni (Ed.), *Race, ethnicity, and minority housing in the United* States (pp. 53–68). New York, NY: Greenwood Press.
Dedman, B. (1988, May 2). Southside treated like banks' stepchild? *The Atlanta-Constitution*.
Garland, P. (1971). "Atlanta: Black Mecca of the South" *Ebony*, August, 152–157.
Hartshorn, T., & Ihlanfeldt, K.R. (1993). *The dynamics of change in Atlanta: An analysis of growth over the past two decades*. Atlanta, GA: Research Atlanta, Inc.
Heman Perry started Atlanta on its home building program. (n.d.). *Pittsburgh Courier*.

Henderson, A. (1987). Heman E. Perry and Black enterprise in Atlanta, 1908–1925. *Business History Review, 61*(2), 216–42.
Hughes, E. (1956, September). The Negro's new economic life. *Fortune, 54*, 248.
Ingham, J., & L. Feldman. (1994). *African-American business leaders: A biographical dictionary.* Westport, CT: Greenwood Press.
Jackson, K. (1985). *Crabgrass frontier: The suburbanization of the United States.* New York and Oxford: Oxford University Press.
Jaret, C. (2002). Suburban expansion in Atlanta: 'The city without limits' faces some. In E. Squires (Ed.), *Urban Sprawl: Causes, consequences & policy responses.* Washington, DC: The Urban Institute Press, 165-206.
Kruse, K. (2005). *White flight: Atlanta and the making of modern conservatism.* Princeton, NJ, and Oxford: Princeton University Press.
Kruse, K., & Sugrue, T. (Eds.). (2006). *The new suburban history.* Chicago, IL and London: The University of Chicago Press.
Lake, R. (1979). Racial transition and Black home ownership in American suburbs. *Annals of the American Academy of Political and Social Sciences, 441*, 142–56.
Lake, R. (1981). *The new suburbanites: Race and housing in the suburbs.* New Brunswick, NJ: Center for Policy Research.
Lands, L. (2009). *The culture of property: Race, class, and housing landscapes in Atlanta, 1880–1950.* Athens: University of Georgia Press.
Martin, H. (1978). *William Berry Hartsfield: Mayor of Atlanta.* Athens, GA: University of Georgia Press.
Mixon, G., & Kuhn, C. (2005). Atlanta race riot of 1906. *New Georgia Encyclopedia.* Retrieved from http://www.georgiaencyclopedia.org/articles/history-archaeology/atlanta-race-riot-1906
Newman, H. (2016, October 30). Andrew Young's book on the 'making of modern Atlanta' describes the 'Atlanta way.' *Saporta Report.* Retrieved from saportareport.com/Andrew-youngs-book-making-modern-atlanta-describes-atlanta-way/
Nijman, J., & T. Clery. (2015). The United States: Suburban imaginaries and metropolitan realities. In P. Hamel & R. Keil (Eds.), *Suburban governance: A global view* (pp. 57–79). University of Toronto Press.
Riley, B. (2010, May 1). A separate peace: An iconic African American neighborhood, home to kings and Hollowells and Abernathys, makes history again. *Atlanta Magazine.* Retrieved from http://www.atlantamagazine.com/civilrights/collier-heights/
Rutheiser, C. (1996). *Imagineering Atlanta: The politics of place in the city of dreams.* London and New York, NY: Verso.
Scott, S. (1961, July 6). City planning department maps Collier Heights future growth. *Atlanta Daily World*, p. 1.

Stone, C. (1989). *Regime politics: Governing Atlanta 1946–1988*. Lawrence: University of Kansas Press.

Sugrue, T. (2005 [1996]). *The origins of the urban crisis: Race and inequality in postwar Detroit*. Princeton, NJ and Oxford: Princeton University Press.

Thomas, R. (1984). Black suburbanization and housing quality in Atlanta. *Journal of Urban Affairs*, 6(1), 17–28.

Toll, S. (1969). *Zoned American*. New York, NY: Grossman Publishers.

White, D. (1982). The Black sides of Atlanta: A geography of expansion and containment, 1970–1870. *Atlanta Historical Journal* 26(Summer/Fall), 199–225.

Wiese, A. (2005). *Places of their own: African American suburbanization in the twentieth century*. Chicago, IL and London: University of Chicago Press.

Williams, T. (2003, March). Money talks. *Atlanta*.

12 Edmonton, Mill Woods, *Amiskwaciy Waskahikan*

ROB SHIELDS, DIANNE GILLESPIE,
AND KIERAN MORAN

Introduction

The first suburb of Edmonton, the capital of the oil-rich western Canadian prairie province of Alberta, was arguably a proposal for an Indian reservation. Situated close to the settlement that had grown up around the Hudson's Bay Company's Fort Edmonton, the reservation was later eliminated as its starving populace one-by-one "took scrip" in the mid-to-late 1800s. They accepted payment to cede their Aboriginal rights to reservation land.[1] Almost a century later, around 1970, parts of the old reservation became the site of an idealistic project to create an affordable suburb, Mill Woods. The way in which Edmonton is cast as a place and region, or "spatialized," is one in which this past is "flattened, shifted, reimagined, and elided in spectacular and spectral settler imaginaries" (Baloy, 2015, pp. 19–21). The effect is to render the Aboriginal past barely present. We explore the intersection of the social ideals of the 1960s and 1970s and the subsequent influx of migrants from all over the world with colonial oppressions that are an unacknowledged legacy of settler society.

"Mill Woods will be a new urban community housing over 120,000 people, in its own right – a new city in a suburban environment" proclaimed the Edmonton City Planning Department (City of Edmonton, 1971, p. 1). No doubt Edmonton's Chief Commissioner P.F. Bargen and the Executive Director of the Alberta Housing Commission B.R. Orysiuk were proud of their plans for this new suburb. An ambitious development to encompass more than 6,000 acres, Mill Woods began as a publicly sponsored land assembly initiated in 1969, the first of its kind in North America (see figure 12.1).

Canada is a suburban country. More than two thirds of Canadians now live in suburbs, and 90% of growth in Edmonton between 2006 and 2016

Figure 12.1 Portrait of Mill Woods today. Top left: Edmonton Public Library, Seniors & Multicultural Centre. Top right: Mill Woods Town Centre Transit Hub. Bottom left: Gurdwara Siri Guru Singh Sabha Sikh Temple. Bottom right: Example of original (1970s) townhouses. Photos courtesy of Kieran Moran.

occurred in suburban developments in which people get around mostly by car (Gordon, Hindrichs, & Willms, 2018; Gordon & Janzen, 2013). Suburbs have been defined by their periurban location relative to the urban core and the rural periphery or as mainly residential developments with segregated uses. Mill Woods falls within these two definitions of suburbia since it is located between Edmonton's urban core and the rural areas of southeast Edmonton. It is mainly a residential development with segregated services. But with a hyperdiverse population of more than

80,000, Mill Woods defies the image of White, nuclear families that represent the stereotypes of the 20th-century North American suburb (see figure 12.1). Ultimately, however, Forsyth (2012) suggests that we abandon the word "suburb" altogether in understanding these spaces and focus on more context-specific definitions of space.

One of the ways of identifying suburbs is by examining the transport habits of the residents. Along these lines, Gordon and Janzen's (2013) Canadian study delimited suburbs according to the relative intensity of use of public transit, walking/cycling, and automobile commuting. First, they sort census tracts by their potential for public transit service based on population density. Second, they classify these areas by the dominant mode of transportation used to travel to work. However, we have reservations. Census data is notoriously unhelpful because it disregards seasonal differences in transportation modes, allowing for only the one mode that is used most frequently throughout the year. That is, mode sharing is often ignored. Five-month-a-year seasonal bicycle commuters who drive in the winter are counted as automobile commuters. In addition, mixed transportation options are coded as the mode with the greatest distance, so someone who drives to a "park-and ride" and then boards public transit may easily be classified as driving to work. For many cities such as Edmonton with light-rail transit communities on the one hand and widespread use of the car on the other, census transportation data are unreliable bases for defining suburbia.

Mill Woods, following Gordon and Janzen's classification scheme, is essentially a hybrid suburb (Filion, 2001). In the early planning stages of Mill Woods, a relatively dense population was intended. Not surprisingly, the older western and northern segments of Mill Woods are classified as a transit suburb, where a higher proportion of commuters use public transit. The remaining segments of Mill Woods are considered auto suburbs, where almost all people commute by car (Gordon & Janzen, 2013). The densest areas of Mill Woods, such as the Mill Woods Town Centre area (MWTC – includes a bus transit hub and is the site of a major enclosed shopping mall), is characterized not by large families but by almost 55% single-occupancy high-rise apartments. This concentration of high-rise apartments is double the city's overall average. Of the residents in this area, 73% drive to work while only 18% take public transit (compared to the city average of 78% driving and 15% taking public transit), despite the bus transit hub at their doorstep. It is possible that more people use public transit both from an economic necessity and also because Mill Woods is among the best-served suburbs for bus links to areas closer to the downtown. While a light-rail link from Mill Woods to downtown is

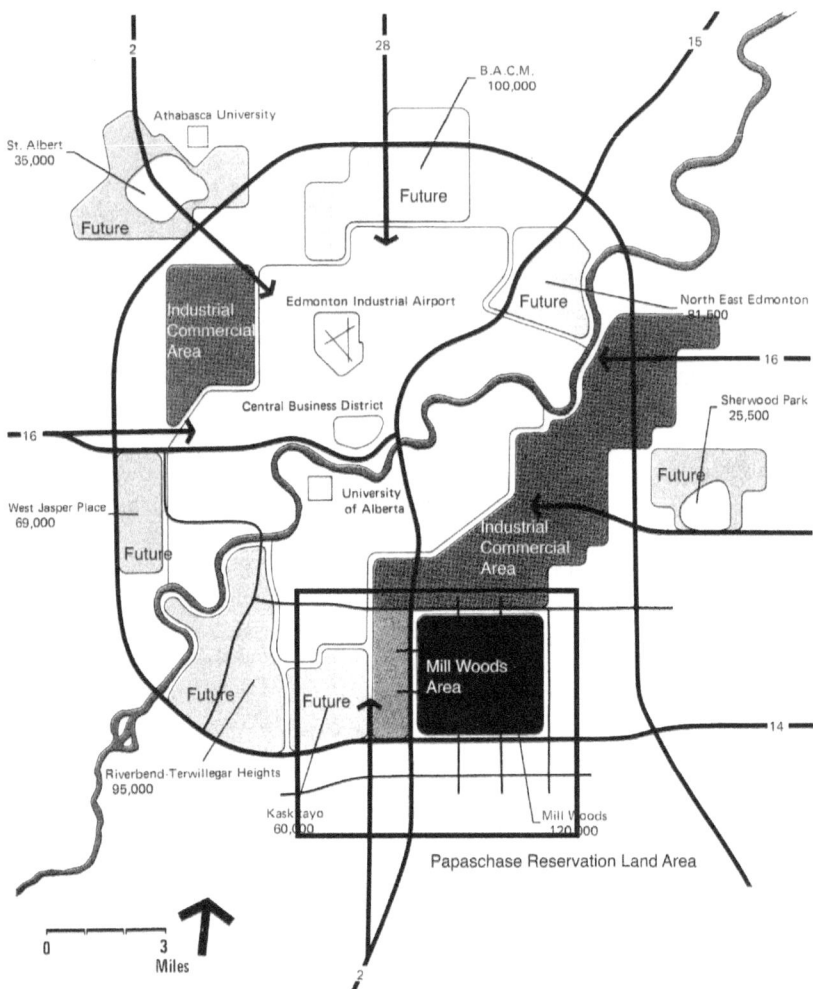

Map 12.1 The Alberta Capital Region with future growth plans showing Edmonton at the centre and proposed annexations of surrounding rural lands in the city region where rapid housing development is occurring. Adapted from City of Edmonton, 1971.

under construction, popular commuting patterns reveal destinations that are only accessible by car-based commuting: not towards downtown, but to work in light industrial suburbs to the south and northeast of Mill Woods (see map 12.1).

Edmonton's central business district white-collar employment is not the destination for the majority of Mill Woods commuters. Industrial parks, supported by Alberta's major oil and gas industry, are major employers in Edmonton, with large industrial parks bordering Mill Woods. The Leduc-Nisku Business Park for instance, located just south of Edmonton, is the second-largest industrial park in North America, employs thousands, and is largely inaccessible by public transit. As Australian studies indicate, commuting patterns often occur from one suburban location to another suburban location (Brennan-Horely, 2010). Forsyth (2012) cautioned that it is frequently problematic to think of the suburb as a singular entity. As such, even when a highly selective definition of a suburb is applied to a suburb such as Mill Woods, it has many internal differences and does not have a stereotypical city–suburb relation to the downtown core.

Edmonton and Mill Woods Today

The Edmonton Census Metropolitan Area (CMA), is now the fifth largest metropolitan area in Canada, population 1,321,426, according to the 2016 census (Statistics Canada, 2017a). In 2011 it was comparable in size to Raleigh, NC, Buffalo, NY or Hartford, Connecticut, and similar in metropolitan area to Helsinki or Liverpool. Located in central Alberta, at 53 degrees north, Edmonton is the most northerly large Canadian capital. It is currently a hub for the northwest, with an oil-and-gas services economy. Edmonton has the second largest metropolitan inmigration in Canada (Edmonton, n.d.). During the decade 2005–15, it was the third fastest-growing metropolitan area in Canada and the United States, ranking alongside sunbelt destinations including Austin, Texas, and Myrtle Beach, South Carolina (United States Census Bureau, 2015). Edmonton has a young population compared to the rest of the country (71.3% between the ages of 15 and 64 (Pratap, 2015, online; Nichols Applied Management, 2015) with more residents of working age than other cities).

Edmonton is significant among Canada's major cities because it has a high proportion of First Nations and other Aboriginal groups residing within it. More than half of Canada's Indigenous people live in urban areas. 2016 statistics indicate Edmonton has the second-largest Aboriginal population, at 76,205 people, behind Winnipeg's

Aboriginal population. Six percent of the city's population claimed an Indigenous identity (Statistics Canada, 2018). Edmonton's Aboriginal population is significantly greater than larger metropolitan areas (CMAs) in Canada, such as Vancouver or Toronto. Edmonton's Aboriginal population is a very significant differentiator from Calgary, the other large metropolitan centre in Alberta, approximately 250 kilometres south of Edmonton.

The Aboriginal population in Edmonton is also young when compared to the Canadian average. The median age of the Aboriginal population in Edmonton is 28 (Statistics Canada, 2018), compared to the median ages of 35.7 and 41.2 of the Edmonton population and the overall Canadian population respectively (Statistics Canada, 2017b; 2017c). The local Aboriginal population has doubled since 2006. This demographic trend has intensified already simmering tensions between the Aboriginal and larger Edmontonian populations. Seventy-eight percent of Aboriginal people living in Edmonton report experiencing some form of discrimination in their lifetimes (Edmonton Community Foundation and Edmonton Social Planning Council, 2013; 2015; Leger Research, 2015, p. 29). As of March 2019, the unadjusted 3-month moving average of the off-reserve Indigenous unemployment rate in Edmonton was 17.5% (Government of Alberta, 2019b, p. 8), more than twice the rate for Edmonton in general (Government of Alberta, 2019a, p. 1). By June 2018, with the oil industry in a downturn, the off-reserve Indigenous unemployment rate for Alberta was 10%. The 2016 national median personal income for First Nations people was $21,875, below Inuit, $24,502, and Métis people, $31,916 (Statistics Canada, 2018).

Mill Woods

In the context of Mill Woods, this put Aboriginal median income in the bottom 15% of the working population, compared with other residents (Alberta Treasury Board and Finance, 2018; Statistics Canada, 2018; Parsons, 2015). Although the Aboriginal population is a relatively small proportion of the total in Mill Woods, statistically, we can legitimately consider poverty as Aboriginal and Indigenized in Mill Woods specifically and in Edmonton generally.

In 2018, Mill Woods had a population of about 50,300, according to the Alberta Treasury Board (2018), of which 30% were under 25 years old and 29.6% were never legally married. Compared to other Edmonton suburbs and neighbourhoods, the population has very high levels of Canadian citizenship (84.7%). Reflecting the multiculturalism of Canada's immigrant history, 32.1% identify as having immigrated.

Filipino and South Asian backgrounds are reflected in language and visible minority statistics. After English (77.7%), the major mother tongues were Tagalog (Filipino) at 21.2%, Punjabi at 3.5%, and Cantonese at 3.1%. However, 7.9% gave multiple responses for home language, which may have inflated the English figure. 14.8% identified themselves as a South Asian visible minority, but only 7.3% as Filipino, followed by 6.7% as Black, 1.2% as Southeast Asian, and 2.9% as Chinese (this second set of figures more closely corresponds with the linguistic figures). These figures may reflect residents' anxiety to present themselves as Canadian, not "Filipino," for example. This hypothesis is supported by the high percentage of residents who actually hold a Canadian passport.

A caveat is necessary, however, for there is a lack of geographically specific data for suburbs. In particular for the period 2001–16, census questions were reduced. Thus the census for this period yields less information and is less comparable to other censuses. In a boom–bust regional hydrocarbon resource economy, the population across Edmonton's metropolitan city region expanded by almost one third just between 2005 and 2015.

"Aboriginal urbanization is largely a function of historical conditions" (Place, 2012). Because Edmonton is located where Aboriginal people had historically settled and gathered (Fromhold, 2015), Aboriginal urbanization patterns, for First Nations as well as Métis, are linked to colonial actions that removed Indigenous people from urban areas (Peters, 2004). Many Indigenous people living in Edmonton are in fact residing in their traditional territories (Peters, 2004). Ultimately, patterns of urbanization for Indigenous people differ from migration patterns of other people who have come from other parts of Canada or from abroad (Browne, McDonald, & Elliott, 2009) as colonial settlers or new, "arrivant," immigrants. While more than 2,800 Métis Edmontonians live in Mill Woods (Statistics Canada 2017d, p. 3), there are fewer First Nations living there. Indeed, most First Nations Edmontonians live near the city centre and are in turn stereotyped as dependent on the welfare services that are concentrated there.

Is there a legacy of the Canadian federal government's treatment of Aboriginal people that is reflected in the nature of all Canadian cities and the way that suburbs emerged as settler enclaves? Colonial strategies of displacement of Aboriginal peoples have contributed to the spaces of Canadian suburbia, in which few Indigenous people reside. Such spaces may not be the stereotypical White suburbs of the post-Second World War era; however, Mill Woods is a spatial expression of what has come to be called "settler society" (Frew, 2013). Aboriginal responses in the form of assertion of First Nations title to land and civil unrest reflect the ongoing "discursive management" of the "indigene" or Aboriginal people by settler cultures (Goldie, 1989). In recent years Western Vancouver, Edmonton and Central Canadian suburbs such as Caledonia (west of Toronto) and Oka (near Montreal) have all been the subject of Indigenous title assertion in the courts as well as barricades on their streets and golf courses, resulting in disruption as well as death (Doucette, 2014). The entrenchment of Indigenous title struggles in Canadian culture indicates an ongoing challenge to the legitimacy of settler-colonial society and of the dispossession, oppression, and effacement of First Nations Canadians (Goldie, 1989).

1800s *Amiskwaciy Waskahikan*

Amiskwaciy means "Beaver Hills," and "Beaver Hills House" was the Cree name for the late-18th-century Hudson's Bay Company Fort Edmonton (Fromhold, 2015). The "flats" in the river valley below the fort were a popular meeting place and river-crossing site. As a trade route, the river route brought Edmonton a role in the commodity trade in furs.[2] Later, in 1891, Edmonton became the northern terminus of the Edmonton, Yukon, and Pacific Railway, a city in 1904 and the capital of the new province of Alberta in 1905. The lands to the southeast of Edmonton, however, were home to the Papaschase Cree people. The "Papaschase Indian Reserve No. 136," proposed in the 1880s, lay just outside the boundaries of the Hudson's Bay Company settlement (see map 12.2). Although initially established, the Papaschase reservation was annulled, and the members of the band were induced or forced to surrender their rights and the land. Under pressure from local settlers, the reserve was sold off by the Government of Canada. In January 1887, the Indian agent and assistant commissioner were directed to remove any remaining Papaschase without their consent and obtain a formal surrender of the reserve. The descendants of the Papaschase argue that

> the purported surrender is invalid and void ab nitio because it did not comply with the strict procedures governing the surrender of Indian reserves as set out in section 39 of the *Indian Act*, R.S.C. 1886.
> According to the terms of the surrender instrument, all lands within IR 136 were surrendered in trust to the Crown to be disposed of "upon such terms as the Government of the Dominion of Canada may deem most conducive to our welfare and that of our people." The Govt. also undertook to collect all monies received from the sale or lease of IR 136 lands and to deposit the net proceeds after deducting management expenses into

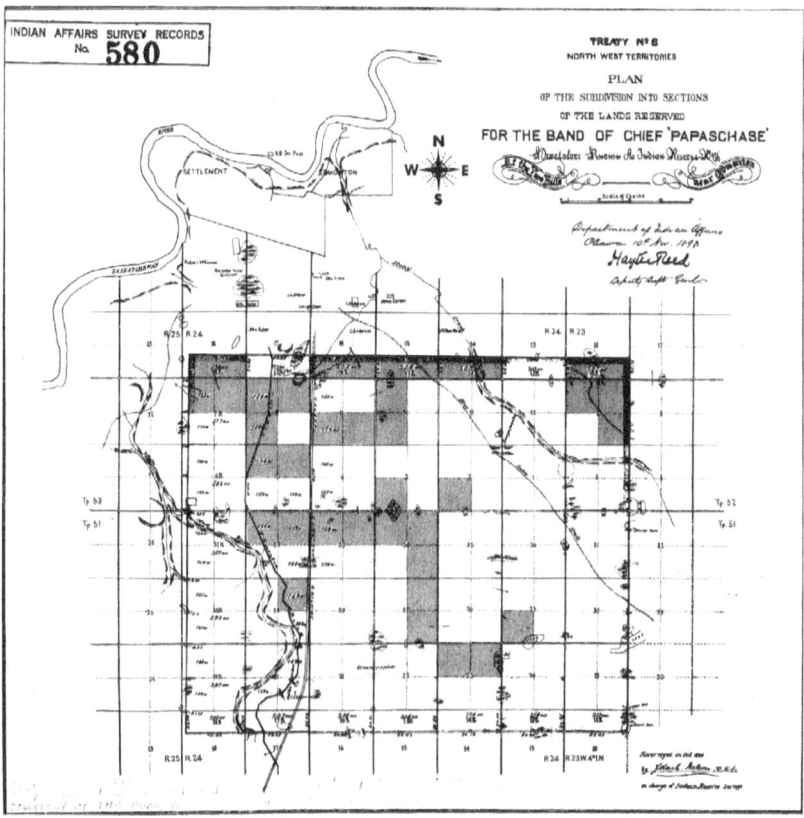

Map 12.2 Papaschase First Nations Reserve as proposed and later subdivided. Adapted from Indian Affairs Survey Records No. 580. Source: Government of Canada. (1899). Plan of the subdivision into sections of the lands reserved for the Band of Chief "Papaschase" Heretofore known as Indian Reserve No. 136 at the Two Hills near Edmonton. Image: Library and Archives Canada.

an interest bearing account to be held in trust. The surrender instrument expressly states that only the interest accruing from such monies shall be paid annually or semi-annually to the Papaschase Band and to "our descendants forever."

From 1890 to 1930, the Govt. of Canada sold all of IR 136 lands to third parties and received monies as a trustee and fiduciary on behalf of the Papaschase Band and their descendants as per the terms of the surrender instrument. The Papaschase descendants allege that Canada has acted contrary to the express terms of surrender and its trust and fiduciary obligations to the Papaschase Band by:

a) failing to hold the principal amount collected on account of the sale of IR 136 lands in trust for the exclusive benefit of the Papaschase Band and their descendants forever,
b) failing to distribute the interest generated from the sale of IR 136 land to the Papaschase Band and their descendants on an annual or semi-annual basis and,
c) distributing any portion of the proceeds of the sale, whether it is principal or interest, to any person who was not a member of the Papaschase Band or a direct descendant. (Bruneau, 2001)

The Papaschase Reserve presages the changing spatial and social relationships between "inner" and "outer" urbanisms. Suburbia is not only a "mix of different classes, races, and ethnicities" (Teaford in Hanlon, 2009, p. 221) located at ever-increasing distances from the urban core. The fate of the Papaschase Reserve shows us that there is a root and a core to a place that remains while the identity of this place changes over time. History books focus on the story of the Hudson's Bay Company, but the Papaschase and other Indigenous people were recognized by the Crown as earlier occupants in 1763. In this sense, they are founding architects of the place identity of the area. Despite the shifting identity of the area up to and including its development as the planned community of Mill Woods in the 1970s, these earlier identities return in court cases, in the names of streets, and in media stories. The effect is one of erasure (Shields, 1991), a palimpsest (Seed, 2001) or *pentimento*:

> *Pentimento* is a concept borrowed from the study of painting that I have chosen as a metaphor for the problem of historicism. The history of Aboriginal people before and after contact with Europeans has been "painted over" by mainstream interpretations of official history. In that sense, we can say that an attempt was made to displace or replace Aboriginal history and memory (as the history of Canada) with a new "painting" of a new civilization. The Aboriginal "painting" was not considered to be a useful or viable portrayal of the new brand of Canadian society that was emerging. However, Aboriginal history and memory have begun to show through in the official history of Canada, conceptual holes in the historical narratives have become obvious, and this has caused many to look more closely to see what has been missed. (Donald, 2004, p. 23)

Histories of previous Aboriginal occupation do feature in some urban histories (Doucette, 2014). However, in contrast to Australian studies (Flew, 2011; Morris, 2014; O'Reilly, 2012; Potter, 2012), the original Aboriginal presence is largely absent in Canadian and American suburban literature. The story of the Papaschase Cree began with Chief Papaschase, his brothers, and their families, who moved to the Edmonton area in the late 1850s from the Lesser Slave Lake region to the north. This small band of Cree people hunted in the Fort Edmonton, Fort Assiniboine, and Lesser Slave Lake areas for some time before making the Edmonton area their home. In 1876, commissioners for Canada met with Chiefs and Headmen at Fort Carlton and Fort Pitt to negotiate the terms for Treaty 6 (Indian Reserves – Western Canada, 2007). Chief Papaschase and his brother Tahkoots signed an adhesion to Treaty 6 in August of 1877 subdividing "into sections of the lands reserved for the band of Chief Papaschase, heretofore known as Indian Reserve No. 136, at the Two Hills near Edmonton" (Library and Archives Canada, 2007).

Ambitious White settlers such as local newspaper publisher and later Canadian Minister of the Interior Frank Oliver vociferously denounced the Papaschase reservation proposal:

> As at presently surveyed, a lasting injury will be done to this settlement, without any corresponding benefit accruing to them. Now is the time for the Government to declare the Reserve open and show whether this country is to be run in the interests of the settlers or the Indians. (Oliver, *The Bulletin*, Edmonton, Jan. 17, 1881, cited in Donald, 2004, p. 38)

The logic of the reserves was neither benign nor beneficial to Indigenous people. The reservation system imposed a spatial logic of domination (Weizman, 2007, p. 196) and assimilation that would today be categorized as cultural genocide (Woolford et al., 2014), a term also used officially by the federal government–sponsored Truth and Reconciliation Commission (2015). By 1885 the remaining Papaschase people had either settled on other reserves or accepted scrip payment. This ended their legal entitlements as Indians and their assertion of land title (Daschuk, 2013). Thus, the 1880s and 1890s saw the steady erosion of the Papaschase Reserve, as settlers from Eastern Europe and Britain purchased large allotments of the land:

> Under Section 41 of the *Indian Act* they (the Papaschase) be given the sale of the lands comprised in the reserve without any restrictions as regards to occupation and cultivation and the terms of payment to be one-tenth of the purchase money at the date of sale, and the balance in nine equal annual instalments with six percent interest. (cited in Papaschase Historical Society, 1984)

On November 19, 1888, the remaining Papaschase Indian Reserve, comprising 40 sections of land, was "surrendered" for sale (Papaschase Historical Society, 1984):

> This reserve, (signed by the Crown) which is five miles south of the flourishing Town of Edmonton, contains some fine agricultural land: the soil throughout is rich in herbage, and in a greater portion of it there is a plentiful supply of wood and water. (Cited in Papaschase Historical Society, 1984)

"The immigration boom of the early twentieth century increased pressure on the newly-founded Indian reserves and the government began to actively encourage Indian land surrenders and moved to make 'excess' Indian reserve land available for non-Indian settlement" (Donald, 2004, p. 38). Most of the land was sold at that time to speculators, who then resold the land as the value increased.

In 1949, oil was struck at the Leduc No. 1 Well, southwest of Edmonton, sparking an oil boom that transformed the largely agrarian economy of Alberta into one focused on the rapid growth of the hydrocarbon resource sector (Dant, 1954). To accommodate unprecedented growth, Edmonton city council adopted a growth strategy that included Clarence Perry's "neighborhood unit" design, which incorporated an elementary school, parks, and nearby shopping. First proposed in 1929, Perry's design produced a "harmonious interplay of these three functions" (Patricios, 2002, p. 4). Perry described four urban areas where the concept could be applied: new sites in the suburbs, vacant sites in the central area, apartment districts, and areas in need of revitalization (Perry, 1933). Moreover, a decentralized pattern of growth emerged that accommodated "industrial growth in outlying areas within a radius of 20 miles from Edmonton" (Gertler, 1955, p. 151). Devon, a new town southwest of the city, was founded to accommodate oil workers. "In the mid-1960's the supply of serviced land for suburban housing was declining and the cost of land was increasing dramatically" (City of Edmonton, 1971, p. 19).

Overall, Edmonton is low density and horizontal in its land development and built form. This is certainly true of most of Mill Woods (see figure 12.1). At 7,252 people per square mile, Edmonton is relatively low density – a sprawling city is one image the city has of itself. However, Edmonton is actually closer to the density of Washington, Winnipeg, or Miami than to Houston or Dallas (4553 p/sq.m. and 4641 p/sq.m., respectively), with which it is often compared

(Shankar, 2016). The North Saskatchewan River valley cuts through the city, resulting in a significant allocation of natural space within the city and even in the downtown core. Both French and English dominant cultures in Canada have specified a strict binary divide between the "civilized" and the frontier; the city and the country. These "founding" Canadian cultures were uneasy about the vision of a settled and urbanized quasi-nature offered by suburban development (Laforest, 2013). The growth of suburbs from the 1920s onward deviated from a more civilized urban lifestyle and intensified this "uneasiness." Low-density, sprawling cities such as Edmonton failed to fit into the ideal "dual" narrative of Canadian urbanization as civilization. Edmonton, then, remains improper, nonconforming, and disdained. It is treated as a relatively insignificant Canadian city despite its size, economic importance, attraction of migrants, rate of growth, and alarming boom–bust cycles related to the petrochemical economy. Mill Woods is a suburb that epitomizes this abject status.

The city region is seen by many as lacking charm, despite the initial quality of its prairie environment. It is a region where its government and high-tech research does not quite belong (Gow, 2007). It is important not to underestimate the impact on both residents and investors of media representations of the place and of rising or falling global economics and distant military coalitions. These external factors contribute to a sense of unsustainability, especially given ambivalent place images and the negative place myth that has developed (Shields, 1991): that of Edmonton as a sprawled, blue-collar "winter city." The conflict between the media's portrayal of Edmonton and residents' positive sense of livability may cause a relatively low sense of investment in the place. This was reported as the lowest among Canada's six largest metro areas.

1970s' Utopian Planning of Mill Woods

"Historically, Edmonton's growth pattern has followed the direction of the North Saskatchewan River Valley in a north easterly, south westerly orientation" (City of Edmonton, 1971; see map 12.1). Heavy industry sites were established in the northwest and northeast sectors of the city that have further influenced and emphasized residential growth patterns along the river valley. Mill Woods represented a major departure from this trend, "offering citizens the opportunity to reside in the southeast sector of the metropolitan area" (City of Edmonton, 1971).

Early ad hoc, low-density rural development of the Papaschase Reserve land followed the ousting of the Indigenous entitlements. Twentieth-century cycles of rapid population growth and a concomitant lack of affordable housing pressured the City of Edmonton to undertake a huge project to transform the former Papaschase area into what would become the suburb of Mill Woods.

Land banking was used as a mechanism to assemble the land for a relatively self-contained suburb for which residential land would be released over time at below-market prices using a lottery system. Although it is considered rare in North America, Eidelman (2014) notes a complex pattern of public land ownership and disposal. In land banking, municipalities first acquired land, then utilities and at least major roadways were installed to prepare land for private sector-led residential construction.

The Mill Woods Plan represented the state of the art in academic and progressive planning at the end of the 1960s. Ironically, given the exclusion of the Aboriginal population in Edmonton, Mill Woods was an attempt to diversify homeownership. This was seen as the form that full membership in modern society took. Mill Woods was both a comprehensively planned and self-contained suburb. This attempt to lower land prices was intended to provide affordable housing. By releasing lots by lottery at below-market prices, it was intended to house low- to middle-income families attracted by the oil industry. They might self-build or construct houses among each other or among family members. Edmonton faced cycles of resource-industry-driven boom and bust since the discovery of oil locally in the mid-1950s, which drove up housing demand and housing prices (Basford, 1973, p. 2258). Demographic booms routinely contributed to rising house prices, which were further inflated by speculative developers.

Following the recommendations of the Hellyer Commission, a national task force that studied the positive experience of Saskatoon, a city to the east of Edmonton, amendments were made to the *National Housing Act* (R.C.S.1973) for the express purpose of supporting the assembly of lands (Task Force on Housing and Urban Development 1969; Hellyer, 1977). Residential construction in Mill Woods officially began in 1972. In the years after the Mill Woods land bank was established, in particular from 1974 to 1976, average land costs across Canada increased by 46.5%, almost five times the average annual inflation rate. This accompanied strong waves of housing starts, which declined almost 20% annually in 1973–4, then jumped almost 15% in 1974–6, before slowly declining again by 6.4% over the next five years from 1976 to 1981 (CMHC, 1983).[3]

However, since this time, continual national divestment of public assets and public land holdings, has transferred assets from various orders of government (municipal, provincial, and federal) to private speculators rather than citizens (Eidelman, 2014; Hellyer, 1977). Nonetheless, in other northern towns and cities surrounded by Crown land, governments continue to act as primary land developers, controlling the release and availability of land (Shields, 2012).

Mill Woods: Next Wave

Today Mill Woods's population of more than 80,000 people is now hyperdiverse, but still reflects the traces of dispossession and exclusion. Mill Woods residents represent 85% of the world's cultures and languages (Kuban, 2005). With ethnic origins predominantly in India, Pakistan, the Philippines, and China, Mill Woods represents an "ethnoburbia" (Li, 1998), but one that is even more diverse than this term suggests. Canadian ethnoburbias emerged with the federal government's merit-based immigration policy in 1967 and the subsequent waves of non-European immigration. The White European immigrants who "settled" first in the Mill Woods suburbia have been replaced by an increasingly ethnically diverse immigrant population, who, in turn, have cemented Mill Woods's character as an arrivant immigrant community. Mill Woods epitomizes mobility in terms of diasporic "shifts," and it embodies a spirit of "migrancy" (Chambers, 1994).

The 1972 expulsion of Ugandans of Indian origin, the 1973 coup in Chile, the emigration of Sikhs from Punjab, and of other South Asians from Pakistan in the early 1970s, and the arrival of ethnic Chinese "boat people" from Vietnam in 1979–80 combined to shape the neighbourhood. Some landed in Mill Woods because of the affordable housing available; others followed friends and countrymen. By 1989, the population of Mill Woods was 30% people from visible minorities (as cited in Mill Woods Living Heritage n.d., "Cultural Diversity").

While wave after wave of immigrants came to Mill Woods, the Indigenous peoples that were the first inhabitants in the area are no longer a significant population. As with many other Canadian cities, there are very few areas outside of low-rent inner city neighbourhoods that are understood to be "Aboriginal" today (see de Costa & Clark, 2015).

Despite a rapidly growing Indigenous population in Edmonton, as of 2018, 5.1% of the Mill Woods population identified as Aboriginal (about the same as in 2011), with roughly half being Indigenous (First Nations), half being Métis, and 0.4% Inuit (see Statistics Canada, 2012).

A study commissioned by the City of Edmonton based on 2006 census data showed that the Aboriginal population of Mill Woods accounted for only 7.9% of the total Aboriginal population in Edmonton. This ranked the Mill Woods district as similar to other suburbs of the city in terms of the proportion of Aboriginals in its population. It has the third highest Aboriginal population outside of the downtown core.

Yet looking forward, 27% of Edmonton households with children under age 15 are Aboriginal. Almost 40% of those youth reside in Mill Woods. Relative to other parts of the city, Mill Woods has a concentration of Métis and First Nations children. This is significant, because it is not clear that there are more culturally appropriate services for these youth in Mill Woods than in any other Edmonton suburb (Andersen, 2009, p. 3). Notwithstanding the growth that occurred from 2011 to 2016, the Aboriginal population of Mill Woods remains a minority segment of the population, and an Indigenous presence is largely missing in the popular conception of the suburb as an immigrant community. The Papaschase presence is effectively erased, persisting only in a few street names and the occasional historical exhibit in the mall. How will the Aboriginal population, drawn from mixed Indigenous groups and traditions, manifest in the future? How will new waves of migration and ethnically diverse youth populations intermix and inflect Indigenous identity in these places? Such questions put peripheral Canadian suburbs such as Mill Woods at the forefront and centre of grassroots social and cultural change in North America.

The land and its place names tell a story of transformation that does connect New Canadians to the First Nations. Whereas Donald states that "the history of Aboriginal people before and after contact with Europeans has been 'painted over' by mainstream interpretations of official history" (2004, p. 23), there remains some resonance of First Nations' experiences through the evolving experience of living in the place. Specifically, the process of attachment to place imbues subsequent inhabitants with at least an appreciation, if not acknowledgment, of the experiences of people who came before them. Numerous factors influence attachment to place (Lokocz, Ryan, & Sadler, 2011), including personal memory of a place (Measham, 2006). The legacy of the Papaschase and other Indigenous people of the area is not only articulated in the enshrinement of traditional paths as roadways or in signage and place names, but also the courts have returned it to the collected memory. The recent revival of Indigenous-title court cases reasserts the Papaschase spatialization of *Amiskwaciy* as a river-crossing and meeting place. Of course, the colonial spatialization of the Papaschase Reserve influences the material contours of Mill Woods today. Moreover, while people's homes and localities may be biographical products of their creation of space (Knox, 2005), the past contextualizes this spatialization despite its suburban architecture and the veneer of 20th-century suburban planning and material culture.

Veracini states that suburbia "re-enacts settlement" by mirroring an anxious "escape" from threatening environments (2012, pp. 341–2). Combined with the forced "standardizing ideals of whiteness, masculinity, and Britishness" (Frew, 2013, p. 281), the settlement process of suburbia takes on the spatial and social forms of separation. Consequently, the sociocultural "status" of people in Canada, forced into three broad categories, "Canadians, Indians and Immigrants" (Thobani, 2007, pp. 24–8), is played out in suburbia as "settlers" move in and displace previous groups. Can we speculate that "ethnoburbia" be seen as a recent expression and manifestation of settler society in an ongoing historical pattern? In this case, is suburban "ethnoburbia" not only a residential pattern but also a "domesticating" consumption environment that initiates immigrant New Canadians into the relations and lessons of settler society as echoes of colonial relations? Postcolonial relationships between transnational migrants, Indigenous First Nations, and a dominant Canadian culture moulded in the image of its British colonial past can be detected in the material landscape, planning history, and the statistical trends over time. However, this latest state of suburban "over-painting" finds neocolonial, rather than postcolonial, accents to the suburban process.

The Mill Woods case suggests that the neglect of these aspects of the North American suburb contributes to a neocolonial pattern of settlement and ultimately creates barriers to cross-cultural and self-understanding. The history of the Papaschase Cree, who once lived on the lands that are now Mill Woods, helps to reimagine the notion of the "suburb" and the analysis of suburbia in more social and historical terms. We focused on this suburb of a sprawling "suburban" city to highlight the significance of Indigeneity, migrancy, and ethnicity for Canadian suburbs as social spaces. Our contention is that this is poorly captured by the transportation, density, and infrastructural categories conventionally used as indicators of "suburbaneity." Rather than being "new" and free of history, the case of Mill Woods suggests that researchers should pay more attention to the temporalities and historical legacies that underlie the dominance of the "new" as suburbs continually expand. This expansion makes suburbia literally a moving target that needs qualification.

Ethnicity and Indigeneity are not well understood through the transportation categories most often used by suburban historians and theorists, such as the "streetcar suburb" or "transit suburb." This perpetuates a blindness to diversity in urban and suburban research and planning historically. Planners and developers have had little sense of the ethical needs of a diverse society, and modernist ideology privileged the separation of races and cultures just as much as it separated functions. This leads to the neglect of the force that previous occupations instil. Rather than buried layers, previous occupations and settlements intersect with and peek through the arrangements of the present – even *pentimento* oversimplifies the situation. While it is not likely to create change on its own (Kim, 2014), the acceptance by the Government of Canada of the 2015 Truth and Reconciliation Commission's report may combine with the changing environmental and energy logic of urban society to change the presuppositions underlying periurban land development in the country. Cultural differences make suburbs more complex than a mere playing out of the logic of the economy and infrastructure of a given time.

ACKNOWLEDGMENTS

Our thanks to Ashley Lee and the assistance of the copy editor in proofreading this chapter.

NOTES

1 For the purposes of this chapter, "Aboriginal" denotes those who fall under treaties and administrative categories of entitlement and are thus defined with Canada's 19th-century *Indian Act*. "Indigenous" covers a wider group, including all those with a claim to an Aboriginal heritage. While "Native" is colloquially used, it is imprecise and is confused and contested by those born in or with long family histories in the area, as in "a native of Edmonton." "Indian" refers to a colonial and governmental discourse, such as in the Federal Government of Canada's *Indian Act* (R.S.C. 1876, 1985 , 2015; see also R.S.C. 1857). For a more complete history see Dickason (2002).
2 In effect, the place has long been seized by colonial and global empires and today participates in contemporary military adventures globally: less visible but significant is a military economy. Today, nearby bases are staging points and training bases for NATO air and troop deployments. The University of Alberta is part of an unacknowledged colonial legacy, established as the first act of the new Province by its English settler-elite who saw European

pedagogy as essential. Despite its distance from other centres, both global oil prices and global politics, including the invasion of Afghanistan and the coalition action against Islamic States in Iraq and Syria, have impacted the city region viscerally over the last two decades.

3 During the postwar period, average annual changes in housing starts saw a strong jump in 1952–3 (26.6%), followed by steady increase of 10.3% over five years until a slump in 1958–60 (-18.5%), which recovered over the next four years (+18.6%), increasingly steadily over seven years until the 1973–4 slump (-19.9%) followed by the 1974–6 rise (14.8%) (CMHC, 1983).

REFERENCES

Alberta Treasury Board and Finance. (2018). Alberta provincial electoral divisions: Edmonton-Mill Woods. Edmonton, AB: Office of Statistics and Information, Government of Alberta (July). Retrieved from https://open.alberta.ca/dataset/8260071f-b931-464f-8c73-97b872e32a31/resource/8a7ed4b0-7fc9-4900-be67-07aa8d812ef5/download/edmonton-mill-woods_profile.pdf

Andersen, C. (2009). *Aboriginal Edmonton: A statistical story – 2009*. Edmonton, AB: Aboriginal Relations Office, City of Edmonton.

Baloy, N.J.K. (2015). Spectacles and spectres: Settler colonial spaces in Vancouver. *Settler Colonial Studies, 6*(3), 1–26.

Basford, R., Rt. Hon. (1973, January 29). Speech to the annual conference of the Housing and Urban Development Association of Canada, Toronto.

Brennan-Horley, C. (2010). Multiple work sites and city-wide networks: A topological approach to understanding creative work. *Australian Geographer, 41*(1), 39–56.

Browne, A.J., McDonald, H., & Elliott, D. (2009). First Nations urban Aboriginal health research discussion paper. A Report for the First Nations Centre, National Aboriginal Health Organization.

Bruneau, C. (2001). *Papaschase Proposal for the Settlement of the Unlawful Surrender of Papaschase I.R. #136*. Retrieved from https://www.papaschase.ca/

Chambers, I. (1994). *Migrancy, culture, identity*. London: Routledge.

City of Edmonton. (1971). Mill Woods: A development concept report prepared on behalf of the civic administration by the City Planning Department, March. Edmonton, AB: City of Edmonton.

CMHC, Canada Mortgage and Housing Commission. (1983). *Canadian housing statistics, 1983*, Ottawa: Government of Canada.

Dant, N. (1954). Edmonton: Practical results of planning measures since 1950. *Community planning review. Revue canadienne d'urbanisme* (1951). Ottawa, Community Planning Association of Canada, 4, pp. 31–40.

Daschuk, J.W. (2013). Clearing the plains: Disease, politics of starvation, and the loss of Aboriginal life. Regina, SK, CAN: University of Regina Press. Retrieved from http://www.ebrary.com

de Costa, R., & Clark, T. (2016). On the responsibility to engage: Non-Indigenous peoples in settler states. *Settler Colonial Studies, 6*(3), 191–208.

Dikason, O. (2002). *Canada's first nations: A history of founding people from the earliest times*. Oxford: Oxford University Press.

Donald, Dwayne. (2004, Spring). Edmonton *Pentimento*: Re-reading history in the case of the Papaschase Cree. *Journal of the Canadian Association for Curriculum Studies, 2*(1), 21–54.

Doucette, J. (2014). Pigs, flowers, and bricks: A history of Leslieville. Retrieved from leslievillehistory.com

Edmonton Community Foundation and Edmonton Social Planning Council. (2013). *Edmonton vital signs 2013*. Retrieved from http://edmontonsocial planning.ca/index.php/resources/digital-resources/n-food-security/48-vital-signs-a-report-on-food-security-in-edmonton/file

Edmonton Community Foundation and Edmonton Social Planning Council. (2015). *Vital signs – 2015*. Retrieved from http://www.ecfoundation.org/uploads/FINAL-VS2015.pdf

Edmonton. (n.d.). Retrieved from https://en.wikipedia.org/wiki/Edmonton

Eidelman, G. (2014). Rethinking public land ownership. Presented at the Canadian Political Science Association, St. Catherines, ON. Retrieved 1 October 2016 from ftp://host-209-183-10-27.static.dsl.primus.ca/cpsa-acsp/2014event/Eidelman.pdf

Filion, P. (2001). Suburban mixed-use centres and urban dispersion: What difference do they make? *Environment and Planning A, 33*(1), 141–60.

Flew, T. (2011). Right to the city, desire for the suburb? *M/C Journal, 14*(4). Retrieved from http://www.journal.media-culture.org.au/index.php/mcjournal/article/view/368

Forsyth, A. (2012). Defining suburbs. *Journal of Planning Literature, 27*(3), 270–81.

Frew, L. (2013). Settler nationalism and the foreign: The representation of the Exogene in Ernest Thompson Seton's "Two little savages" and Dionne Brand's "What we all long for." *University of Toronto Quarterly, 82*(2), 278–97.

Fromhold, J. (2015). *A researcher's guide to historic trading posts and forts of the west*. Canada: First Nations Publishing.

Gertler, Leonard. (1955). Why control the growth of cities? *Community planning review. Revue canadienne d'urbanisme, 5*, 151–4.

Goldie, T. (1989). *Fear and temptation: The image of the Indigene in Canadian, Australian, and New Zealand literatures*. Montréal: McGill-Queen's University Press.

Gordon, D., Hindrichs, L., & Willms, C. (2018). *Still suburban? Growth in Canadian suburbs, 2006-2016* (Working Paper No. 2). Brampton ON: Council for Canadian Urbanism and School of Urban and Regional Planning Department of Geography and Planning Queen's University.

Gordon, D., & Janzen, M. (2013). Suburban nation? Estimating the size of Canada's suburban population. *Journal of Architectural and Planning Research*, 30(3), 197–220.

Government of Alberta. (2019a). *Alberta labour force statistics March, 2019.* Retrieved from Open Government, Government of Alberta website: https://open.alberta.ca/dataset/b754ca87-2e9b-4a80-b7b2-2cfef8e53ff4/resource/5f01c28a-d327-48f6-9a33-97e5886bf357/download/public-package-2019-03.pdf

Government of Alberta. (2019b). Labour force statistics: March 2019 Alberta Indigenous people living off-reserve package. Retrieved from https://open.alberta.ca/dataset/e8afe142-162f-4656-946e-1b3472c9ba83/resource/26477100-d632-4cb7-8515-2581d26a9285/download/indigenous-package-2019-03.pdf

Government of Canada. (1899). Plan of the subdivision into sections of the lands reserved for the Band of Chief "Papaschase" Heretofore known as Indian Reserve No. 136 at the Two Hills near Edmonton. Ottawa: Library and Archives Canada. Retrieved from http://www.collectionscanada.gc.ca/lac-bac/results/mages?form=image&lang=eng&FormName=Image+Search&PageNum=1&SortSpec=score+desc&HighLightFields=title%2Cname&Language=eng&QueryParser=lac_mikan&Sources=mikan&Archives=&ShowForm=show&SearchIn_1=&SearchInText_1=papaschase&Operator_1=AND&SearchIn_2=&SearchInText_2=&Operator_2=AND&SearchIn_3=&SearchInText_3=&Media%5B%5D=&Level=&MaterialDateOperator=after&MaterialDate=&MaterialDate=&DigitalImages=1&Source=&cainInd=&ResultCount=50

Gow, B.A.S. (2007). Creating a past for the Canadian petroleum industry's technology: The drilling rig. *Material Culture Review* 65 (Spring), 89–92.

Hanlon, B. (2009). A typology of inner-ring suburbs: Class, race, and ethnicity in U.S. suburbia. *City & Community* 8(3): 221–46.

Hellyer, P. (1977, December). How they killed public land banking: A political memoir. *City Magazine*, 3(2), 31–8.

Kim, J.J. (2014). *They made us unrecognizable to each other: Human rights, truth, and reconciliation in Canada* (PhD diss.). University of Tennessee.

Knox, P.L. (2005). Creating ordinary places: Slow cities in a fast world. *Journal of Urban Design*, 10(1), 1–11.

Kuban, R. (2005). Edmonton's urban villages: The community league movement. Edmonton: University of Alberta Press.

Laforest, D. (2013). La Banlieue dans l'imaginaire Québécois: Problèmes originels et avenir critique. *Temps Zéro*, (6). Retrieved from http://temps zero.contemporain.info/document945

Leger Research. (2015). *Survey of Edmontonians 2015*. Edmonton: Edmonton Community Foundation. Retrieved from http://www.ecfoundation.org /uploads/Report-ECF-July-6-2015.pdf

Li, W. (1998). Anatomy of a new ethnic settlement: The Chinese ethnoburb in Los Angeles. *Urban Studies, 35*(3), 479–501.

Library and Archives Canada. (2007, January 17). Indian reserves – Western Canada. Retrieved from http://www.lac-bac.gc.ca/databases/indian -reserves/001004-119.01-e.php?isn_id_nbr=1902&interval=50&PHPSESSID =t517cjd0faa8lvonqqvi6kdnf6

Lokocz, E., Ryan, R.L., & Sadler, A.J. (2011). Motivations for land protection and stewardship: Exploring place attachment and rural landscape character in Massachusetts. *Landscape and Urban Planning, 99*(2), 65–76.

Measham, T.G. (2006). Learning about environments: The significance of primal landscapes. *Environmental Management, 38*(3), 426–34.

Mill Woods Living Heritage. (n.d.). Retrieved from http://www.millwoods history.org/.

Morris, B. (2014). *Protests, land rights, and riots: Postcolonial struggles in Australia in the 1980s*. Berghahn Books.

Nichols Applied Management. (2014). *City of Edmonton growth study*. Edmonton.

O'Reilly, N. (2012). *Exploring suburbia: The suburbs in the contemporary Australian novel*. Sydney: Teneo Press.

Papaschase Historical Society. (Ed.). (1984). *South Edmonton saga* (1st ed.). Edmonton, AB: Papaschase Historical Society.

Parsons, P. (2015). Edmonton living wage $17.36, according to report. Retrieved 10 August 2015 from https://edmontonjournal.com/news/ local-news/edmonton-living-wage-17-36-according-to-report

Patricios, N. (2002). The neighborhood concept: A retrospective of physical design and social interaction. *Journal of Architectural and Planning Research, 19*(1, Spring), 70–90.

Perry, C.A. (1933). *The rebuilding of blighted areas: A study of the neighborhood unit in replanning and plot assemblage*. New York, NY: Regional Plan Association.

Peters, E.J. (2004). Three myths about Aboriginals in cities: Breakfast on the Hill Presentation, Canadian Federation for the Humanities and Social Sciences. Retrieved 4 November 2015 from http://caepr.anu.edu.au/sites /default/files/Seminars/presentations/Three%20Myths.pdf

Place, J. (2012). *The health of aboriginal people residing in urban areas*. Prince George, BC: National Collaborating Centre for Aboriginal Health.

Potter, E. (2012). Introduction: Making Indigenous place in the Australian city. *Postcolonial Studies, 15*(2), 131–42.

Pratap, V. (2015, February 11). Edmonton region has Canada's 2nd-highest population growth: Stats. *Global News*. Retrieved from globalnews.ca

R.S.C. (1857). *Act for the gradual civilization of Indian tribes in the Canada*. 20 Victoria Ch. 2.

R.S.C. (1876). *An Act to amend and consolidate the laws respecting Indians*. 39 Victoria Ch. 18

R.S.C. (1973). *National Housing Act Amendment*.

R.S.C. (1985). *Indian Act* c. I-5.

R.S.C. (2015). *Indian Act. An Act Respecting Indians* RSC 1985 C I-5. Government of Canada. http://laws-lois.justice.gc.ca/eng/acts/I-5/

Seed, P. 2001. *American pentimento: The invention of Indians and the pursuit of riches*. Minneapolis, MN: University of Minnesota Press.

Shankar, N.D. (2016, June 27). City weighted population density. Retrieved 7 April 2019 from Comparing the densities of Australian, European, Canadian, and New Zealand cities | Charting Transport website: https://chartingtransport.com/2015/11/26/comparing-the-densities-of-australian-and-european-cities/#comment-33741

Shields, R. (1991). Imaginary sites. *Between Views* Exhibition Catalogue. Banff, AB: Walter Phillips Gallery, Banff Centre for the Arts, 22–26.

Shields, R. (2012). Feral suburbs. *International Journal of Cultural Studies, 15*(3), 205–15.

Statistics Canada. (2012). Population and dwelling counts, 2011 Census. Catalogue no. 98–310-XWE2011002. Ottawa: Statistics Canada. Released 8 February 2012. Retrieved from https://sttc.ent.sirsidynix.net/client/en_GB/default/search/detailnonmodal/ent:$002f$002fSD_ILS$002f0$002fSD_ILS:10201/ada?qu=2012&qu=%22Population+and+dwelling+counts%22&te=ILS

Statistics Canada. (2017a). *Edmonton [Census metropolitan area], Alberta and Alberta [Province]* (table). *Census Profile*. 2016 Census. Statistics Canada Catalogue no. 98-316-X2016001. Ottawa. Released 29 November 2017. Retrieved 8 September 2019 from https://www12.statcan.gc.ca/census-recensement/2016/dp-pd/prof/index.cfm?Lang=E

Statistics Canada. (2017b). *Edmonton, CY [Census subdivision], Alberta and Division No. 11, CDR [Census division], Alberta* (table). *Census Profile*. 2016 Census. Statistics Canada Catalogue no. 98-316-X2016001. Ottawa. Released 29 November 2017. https://www12.statcan.gc.ca/census-recensement/2016/dp-pd/prof/details/page.cfm?Lang=E&Geo1=CSD&Code1=4811061&Geo2=CD&Code2=4811&SearchText=edmonton&SearchType=Begins&SearchPR=01&B1=Population&TABID=1&type=0

Statistics Canada. (2017c). *Canada [Country] and Canada [Country]* (table). *Census Profile.* 2016 Census. Statistics Canada Catalogue no. 98-316-X2016001. Ottawa. Released 29 November 2017. https://www12.statcan.gc.ca/census-recensement/2016/dp-pd/prof/details/page.cfm?Lang=E&Geo1=PR&Code1=01&Geo2=&Code2=&SearchText=Canada&SearchType=Begins&SearchPR=01&B1=All&TABID=1&type=0

Statistics Canada. (2017d). *Edmonton Mill Woods [Federal electoral district], Alberta and Edmonton [Population centre], Alberta* (table). *Census Profile.* 2016 Census. Statistics Canada Catalogue no. 98-316-X2016001. Ottawa. Released 29 November 2017. Retrieved 8 September 2019 from https://www12.statcan.gc.ca/census-recensement/2016/dp-pd/prof/index.cfm?Lang=E

Statistics Canada. 2018. *Canada [Country]* (table). *Aboriginal Population Profile.* 2016 Census. Statistics Canada Catalogue no. 98-510-X2016001. Ottawa. Released 18 July 2018. https://www12.statcan.gc.ca/census-recensement/2016/dp-pd/abpopprof/details/page.cfm?Lang=E&Geo1=PR&Code1=01&Data=Count&SearchText=Canada&SearchType=Begins&B1=Income&C1=All&SEX_ID=1&AGE_ID=1&RESGEO_ID=1

Task Force on Housing and Urban Development. (1969). *Report of the Federal Task Force on Housing and Urban Development.* Ottawa: Task Force on Housing and Urban Development.

Thobani, S. (2007). *Exalted subjects: Studies in the making of race and nation in Canada.* Toronto: University of Toronto Press.

Truth and Reconciliation Commission of Canada. (2015). *Honouring the truth, reconciling for the future: Summary of the final report of the Truth and Reconciliation Commission of Canada.* Retrieved 8 February 2016 from http://epe.lac-bac.gc.ca/100/201/301/weekly_acquisition_lists/2015/w15-24-F-E.html/collections/collection_2015/trc/IR4-7-2015-eng.pdf

United States Census Bureau. (2015). New census bureau population estimates reveal metro areas and counties that propelled growth in Florida and the nation. Release CB15–56 (March 26). Retrieved 8 February 2016 from https://www.census.gov/newsroom/press-releases/2015/cb15-56.html

Veracini, L. (2012). Suburbia, settler colonialism and the world turned inside out. *Housing, Theory and Society, 29*(4), 339–57.

Weizman, E. (2007). *Hollow land: Israel's architecture of occupation.* London: Verso.

Woolford, A., Benvenuto, J., & Hinton, A.L. (Eds.). (2014). *Colonial genocide in Indigenous North America.* Durham, NC: Duke University Press.

13 Economic Development and the New Immigrant Segregationist Politics in Suburban Chicago

DAVID WILSON

Introduction

The winds of change now ominously blow through many of America's aged inner-ring suburbs. The forces of globalization, transnationalism, and neoliberalism foster new trends: accelerated immigration, economic growth dominating in low-wage occupations, crippled unions, shrunken social welfare nets, growing homelessness, and retrenched government (Gallagher, 2013). The days of simple, monochrome inner-ring suburbs, if they ever existed, are clearly over: new kinds of communities take shape that increasingly mirror their nearby city brethren. While it is unclear what these suburbs will become, it is extremely clear that they are in a rapid state of becoming something new. We would expect such communities to change over time: this is not surprising, but change here has recently been fast, dizzying, and unanticipated (Keil, 2013).

This chapter focuses on Chicago's suburbs and chronicles a central change in these communities: the rise of a new immigrant segregationist politics tied to a changed economic development. It is a politics I term "the new immigrant exclusion politics" that is rife with tensions and contradictions. The heart of the matter: inner ring suburban governances in Chicago now scramble to build a new economy, one around the co-drive to enhance low-wage services and downtown physical and cultural revamping, and struggle with the dual imperatives of whether to promote class-race inclusion or race-class exclusion. Political programs and sentiments that work to both include and exclude these immigrants rub abrasively against one another. This politics takes the form of diverse and contradictory renditions of immigrants that end up challenging more deeply and ominously their civic worth, moral character, cultural content, and political rights. Let me elaborate.

On the one hand, the drive to exclude mounts with pressure to contain flows of new in-migrants in efforts to economically renew downtowns and nearby enclaves. Immigrant presences in these locations are imagined as a kind of toxin that could subvert new redevelopment hotspots. Governances thus strive to intensify this population's segregation and ghettoization (working through community-speak that emphasizes preserving community integrity, civic orderliness, property values, and property investment opportunities). On the other hand, an equally visible drive to include this population in the communities is also offered. Immigrant tolerance is purportedly important as governing agencies seek to both bolster their low-wage economic sectors with essential workers and appear as socially progressive, inclusive places. Immigrant identities and human rights become talking points for mayors, council people, and the media around which a human inclusiveness is projected. "A people" are momentarily endowed with positive cultural and social habits and traditions and located in supposedly open, welcoming communities.

Economic development, it follows, is deeply contradiction ridden, as governances desire to revamp downtowns for affluent restructuring and consumption that conflicts with their aspirations to rebuild economic bases around low-wage services. Its boldest evidence: a new ambiguous immigrant segregationist politics in these communities whose fault line is the contradiction between the governance's sense of these places' ideal spatial form and their crucial labour needs. The resolution at the moment: attract these poor into places and then rationalize their systematic exclusion and segregation. A politics emerges that is a delicate balancing act: assign to them – in abstract terms – a positive notion of citizenship and identity but eviscerate this at a moment's notice. Immigrants become free-floating signifiers caught in a political tug-of-war whose ascribed character is momentary and depends on the precise political task of governing agencies at hand. Economic development becomes awkwardly located in both assertions of worker unalienable rights and their problematic citizenship.

To be sure, governances desiring a balkanized and segregated community morphology to facilitate downtown economic growth is not new to these communities. But now these downtowns for the first time are to be dramatically "post-industrialized" and elaborately fortressed from low-income immigrant presences. Sassen's (2015) notion of a new visual obsession in economic development (what can be observed and what it means) has taken hold here, and takes the form that poor immigrant bodies are to be totally erased from view.

Moreover, for decades these communities have welcomed low-wage jobs and employers; this too is not new. Yet today this has changed: this economic subsector is trumpeted and planned for as the centre of economic growth in these communities, and workers are aggressively sought to bolster it. This economic subsector, it seems, must be made to work. The result today is an economic development politics of indecision and tension around the immigrant presence that lurks beneath seemingly confident declarations of community economic vision.

This chapter chronicles the new politics in an assessment of two innerring suburbs in Chicagoland, Berwyn, and Cicero (map 13.1).[1] These increasingly heterogeneous communities of 56,693 and 84,103 people, respectively, now feel the full impact of globalization, transnationalism, and neoliberalism (table 13.1). Today, their economies thoroughly root in low-wage services as reflected in the dramatic growth of hotels, fast-food restaurants, bars and pubs, and car dealerships along their major arterials. Once dominant industrial bases – in 1950 more than 67% of total employment in both places – have faded into obscurity. At the same time, new waves of immigrants have moved into the places to transform their geographies. More than 83,000 Latino people, mainly from Mexico and secondarily from Puerto Rico, currently live in Cicero (59% of the population) and Berwyn (77% of the population). More than 11,000 people from Latin America have moved in within the last 3 years seeking decent jobs and decent lives, and constitute these communities' most vulnerable and poorest population. Singled out as "the new immigrants" in these communities, they have become the focus of the new immigrant segregationist politics.

One Side of the Dialectic: The Drive to Include

Cicero and Berwyn now increasingly receive immigrants, particularly people from Mexico, who come to pursue waged work and decent lives. Mexico's current political and economic instability, along with a perception of better job opportunities in the United States, has fueled the movement of more than 250,000 migrants from Mexico to Chicago's suburbs in the last five years: the municipalities of Berwyn and Cicero have been major destinations (U.S. Census Bureau, 2013). These communities have been eyed by many as wonderful places of refuge compared to existing places in Mexico (Greene, Bouman, & Grammenos, 2006). "We came here because of the promise of how things could be," one recently arrived women from Mexico in Berwyn told me.

272 David Wilson

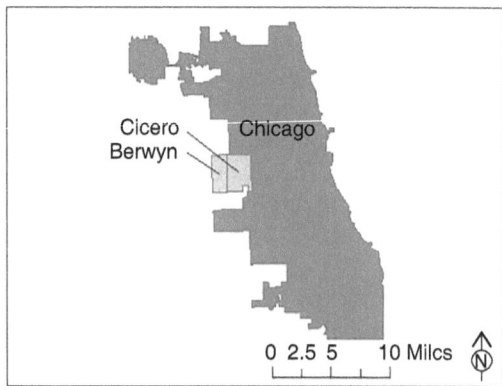

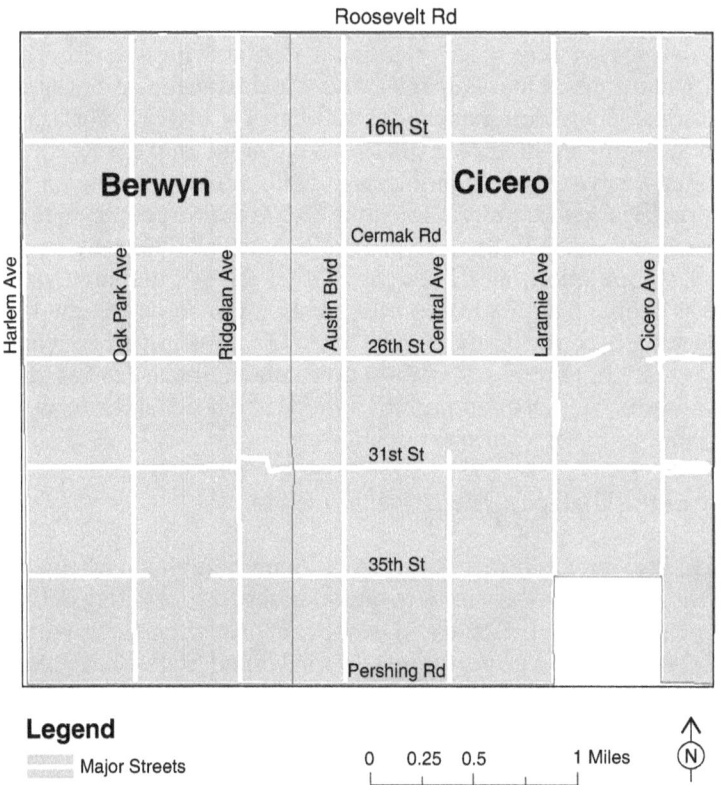

Map 13.1 Cicero and Berwyn in Greater Chicago.

Economic Development and Segregationist Politics in Chicago 273

Table 13.1 Statistics for Cicero and Berwyn

	Cicero	Berwyn
Population	84,103	56,693
Latin population	66,299 (77.4%)	33,675 (59.4%)
Management, professional, and related occupations	10,091 (29.1%)	6,492 (26.2%)
Sales and office jobs	9,779 (28.2%)	7,439 (30.0%)
Construction and maintenance jobs	3,502 (10.1%)	2,386 (9.6%)
Other service jobs	5,063 (14.6%)	3,482 (14.1%)

Source: United States Bureau of Labor Statistics, 2013

"We thought good-paying jobs would be plentiful and the country would welcome us open-arms … it's proven not that simple … dreams … and more dreams … that's what carried us."

But the new arrivals quickly find themselves starkly segregated in these places, occupying the oldest and most dilapidated part of Cicero (north side) and Berwyn (east side). Vacant housing, dilapidated structures, poor schools, and unemployment and underemployment characterize many of their new blocks (map 13.1). These are the community's most stigmatized areas, widely referenced by locals as "dead zones" and "long-term ghettos" and purportedly scarred by what Cook County Planner A. Henry (2014) called "too much social strife" and "considerable social disorganization that hurts both kids and adults." The decades-long reproduction of these residential spaces – their new environments – reflects the local governance's power to dictate spatial forms and perpetuate the existence of poor disadvantaged neighbourhoods (Consuala, 2014). These sustained residential "islands," mirroring community-wide socioeconomic stratifications, have allowed realtors, banks, developers, and speculators to extract maximum profit in everyday land and housing transactions across Berwyn and Cicero (particularly in their downtowns) (Ruggles, 2014).

Yet the immigrant presence has helped staff the burgeoning low-wage economies in these communities, and this has been recognized by residents and political officials. Needed in the new economies, this population engages in restaurant work, supermarket labour, hotel cleaning, home-maintenance jobs, and agricultural harvesting (now more than 55% of the local workforces in Cicero and Berwyn). Wages across these occupations rarely exceed $10.00 per hour (Consuala, 2014). The workers also toil in the area's three day-labour sites, in Berwyn, Cicero, and the nearby Little Village neighbourhood in Chicago. Daily, hundreds of workers gather in Home Depot parking lots and chase down drive-by vans in pursuit of just-in-time employment (Wilson, Beck, & Bailey, 2009). The day-labour sites deliver reliable stocks of inexpensive warm bodies willing to work hard for paltry wages. Workers do mainly construction, warehouse, assembly-line, and manual work around the Cicero and Berwyn areas. Work sites are often in back rooms, back yards, plan floors, and tucked-away factory sheds. These immigrant workers typically flitter precariously between poverty and material subsistence.

In this context, planners and politicians have anointed local retailers such as Long Horn Steakhouse, White Castle, Lalo's Mexican Restaurant, Green Acres Food Mart, and Jewel-Osco as "an important plank in current economic development of these communities ... that requires hard work from locals ... " (Henry, 2014). Cicero and Berwyn's two largest arterials, Cermak Road and Roosevelt Road, are now dominated by this economy. Moreover, flanking the more visible grocery stores, restaurants, motels, and food marts are the seamier payday lenders, pawn shops, and temporary labour agencies that profit off this low-wage workforce. These establishments advertise themselves as one-stop shops that provide fast access to resources and money that this population especially needs. As this low-income labour force grows, it is expected that this type of corporate opportunism will proliferate. As never before, a low-wage economy in its two tiers (formal retailers, neighbourhood predatory retailers) booms in these two suburbs.

Not surprisingly, a revised economic development rhetoric has emerged in Cicero and Berwyn. The rhetoric incorporates the desirability of having immigrant populations within their boundaries and including them within a purportedly growing economy. "Economic growth in Cicero is too include everyone as participants, including our recently arrived Mexican citizens ... there is a role to be played by all" (Henry, 2014). Multiculturalism, inclusion, and ethnic diversity are proclaimed and celebrated in the new community economic-speak: planners, politicians, and the local media repetitiously chirp the new line. Berwyn now purportedly "opens itself up to racial and ethnic diversity as today's Illinois and the world continue to change demographically and economically" (Smith, 2013). "Community after community across America has been able to absorb new immigrants and bring them into their economies," said Planner Smith (2013), "Berwyn should be no exception to this."

Cicero and Berwyn give the rhetoric a happy face, annually celebrating the culture and ways of Mexican-ness. A yearly Latino Music Festival, a Multicultural Brew Fest, and a Latino Film Festival are all packaged as building community inclusiveness, community diversity, and new ethnic economies. Racism and exclusion from community opportunities and economic endeavours are purportedly a thing of the past. It is supposedly "a new cultural and social day, tolerance now reigns supreme" (Henry, 2014). Magically, institutions and economies have become race blind and class blind. "Not to trivialize racism," says Henry, "but we effectively don't see race and ethnicity any more in our governing … yes isolated cases of this occur in our town, but its institutional forms have gone south." According to Henry (2014), "what we see now is something else, an evolution to a degree of civility, social stability, and potential contribution of all to an evolving economy that leads us forward."

The publicized extolling of blocks on Berwyn's North-East Side has been the leading edge of this new inclusive-speak (map 13.1). In the mid 2000s, governance actors began discussing this historic 15-block immigrant catchment and control zone as important to Berwyn's fabric. For decades, this area physically eroded by intense institutional neglect that the City totally ignored (except in its policing). Then, a rich and authentic ethnicity – "Mexicanness" – was found alive and well and "rediscovered" here. Such inclusive-speak was generally good community politics, but also Berwyn's low-wage subsector had multiplied, such jobs were rife in the area, numerous employers had complained to the City that they needed more workers to staff their operations, and the governance eyed the North-East Side as amenable to being more intensively mobilized in the local economy (Henry, 2014; Smith, 2013). Galvanizing these workers and perhaps bringing in even more workable bodies became the order of the day.

The drive continues to this day and features proclamations of something compelling: a rich Mexican history here. Currently, the area is widely proclaimed as more culturally iconographic and historically meaningful then previously realized. The North-East End now supposedly jumps with vibrancy, distinctiveness, and ethnic fervour (after years of being off the governance's radar). This "rediscovery" highlights notions of Mexican culture, a people's authentic music, and exotic Mexican traditions as objects of bourgeois fascination. Beneath bouts of social trouble and struggle, it is declared, is an industrious and sturdy area whose history and traditions add spice to a diversifying community. Now a neighbourhood and "a people" are to be known in one way: as simple, raw ethnic objects that contain fantastic cultural uniqueness and exotic social ways.

In this context, Berwyn's mayor talks frequently with City planners and City officials about possibilities to declare the area a historic district. "History is a key part of Berwyn," Planner A. Plane (2014) noted, "and the North-East Side has it, a legacy of rich traditions and values, it is that simple." Also, talk had previously veered to the possibility of using tax-increment financing to create the historic district. This tool, with humble beginnings in nearby Chicago when Jane Byrne applied it to a rundown block in the Central Loop in 1981, ended up not being used in Berwyn. There was also talk at this time of applying for federal historic preservation funds to achieve the designation. The application was never written: talk of using tax-increment financing and federal funds petered out shortly thereafter.

At the core of this recasting of the North-East Side, governance actors work through an imaginative frame of raw raciality. On the one hand, media voices and politicians idealize an object – "the authentic Mexican barrio" – and reference the benefits of creating a tame civic barrio that can diversify Berwyn. A served-up cast of characters, crucial in this social constructing, is illuminated by playing to established stereotypes: colourful but civic youth, hard-working and self-sacrificing men, deeply nurturing family women. These constructions gain meaning in a melodrama of people and community fighting to subsist in new social and economic realities. Thus, Planner Henry (2014) spoke of a "once Mexican ghetto and its men and women that are now becoming important community markers for diversity and multiculturalism." This racializing identifies two reductionist symbols in the area: "sturdy, worker homes" and "Mexican restaurants and eateries" (Henry, 2014). The first one references "a people's own dwellings"; the second, "a community-enhancing cultural resource that is unique and distinctive" (Henry, 2014). Yet, this celebrating of "Mexican-ness," widely discussed, stays as a one-dimensional performance of inner city poor minorities.

On the other hand, not all is presented as so positive: Mexican "street ways" are said to still linger in the area and are in need of being identified and eradicated. Berwyn's North-East Side ostensibly still contains a dangerous and counter-civic side. Celebrating this diverse community element (a space, its people) is thus set in a somber, realistic take of current realities. This realism, one more time, roots in ethnic stereotype: Mexicans dwelling in their own cultural worlds, kids carrying on mindlessly and drawn to the street, adults struggling to subsist in complex American communities. As Planner A. Smith (2014) summed it up, "The North-East Side [Berwyn] excites and fascinates people, and this area has a lot to offer the local economy. But it's socially rooted problems are there and cannot be ignored." A supposed Mexican poverty and set of ways purportedly need reengineering for the good of Berwyn, even as the space is said to "be a potentially major contributor to Berwyn's evolving economy and social fabric" (Henry, 2014).

In this context, Berwyn's North-East Side today is proclaimed as the latest example of an ethnic area that helps diversify a place. Many sections have (it is claimed) lost their ghetto roughness, now exude some of the best qualities of true Mexican barrios, and have been substantially tamed and integrated into a place's supportive milieu. And its politicians are allegedly on the forefront of culturally renewing the area. To make this case, a simple and unambiguous politician filled with benevolence and hope is served up. This highly stylized construction assigns to one caricatured being human benevolence, economic pragmatism, and community economic concerns. The leading figure, City councilperson Ray Rhodes, is cast as a caring and progressive being who boldly strikes out to integrate the North-East Side into the local economy. "Ray's out there imploring kids here to make a difference, to assume their proper role in your household and the community ... no kid or adult can elude Ray. He tirelessly energizes people to work and to work hard on behalf of them, their families, and Berwyn" (Smith, 2014).

This North-East Side, so coded, has been mobilized to assist a fundamental undertaking: entrepreneurializing Berwyn. The North-East Side becomes imaginatively transported to a fuzzy and feel-good realm by governance actors and assumes an important functional orientation: a community economic development aide. The area is to be seen as community serving and economically propelling; there is seemingly no other reasonable way to see and understand this area. In short order, Berwyn needs this area to flourish: the community's economy and cultural base depends on it. In insinuation, workers in the new low-wage economy can be found and efficiently settled here with public support. The hoopla of community diversity and tolerance is probably to be seen as containing some hype, but also undergirded by a functional logic and importance that gives this fluff-speak a kind of credibility.

The Dialectic's Other Side: The Drive to Exclude

At the same time, Cicero and Berwyn's planning and business amalgams invoke another community rationale that also marks many suburbs and cities in America: to sociospatially peripheralize recent immigrants, particularly those from the Latino world. The dominant theme in Cicero and Berwyn: these places are now in new global times and must be re-entrepreneurialized to be economically competitive. That means downtowns and select social and residential spaces are to be carefully cultivated as vibrant, aesthetic terrains (with the task at hand to attract key presences and repulse undesired populations and land uses). "It is new global days," local Planner Henry (2014) said, "and we can't get around the reality that our downtown especially needs to be attractive to business and investment. Economic development needs a robust downtown."

A central rhetoric drives this, which I have elsewhere termed "the global trope." It invokes a reality of new ominous global times and suggests communities are easily discarded as places of investment, production, and business, and this must not happen. Here, once economically robust and confident places have recently become porous and leaky landscapes rife with a potential for a dramatic economic hemorrhaging. A kind of haunting accumulation of uncertainty now hangs over them, as a new reality that must be recognized and faced. As Planner P. Jones (2014) notes, "Globalization now frames our community just like everywhere else ... it is the new world order that we as planners need to see and work through." To Jones, "Times change, and it is undeniable that these are new global times ... Berwyn will be shaped by these times, but we can seek to aggressively navigate this reality to the best of our abilities."

Yet these communities are also presented as historically resilient locales that have the capacity to act innovatively and survive. The time to act is now; there can be no other alternative. Identified signs of this ominous possibility – municipal fiscal depletion; aging physical infrastructure; decommodified residential, commercial, and production spaces – are the purported visible proof of what this bleak future may hold without proper action. Absent a strong community response, Berwyn and Cicero stand to "lose their economic and cultural heart ... be gutted and lose the bulk of their cultural and economic resources that remain" (Evans, 2014). Through this rhetoric, a kind of shock treatment of regulating and disciplining the spatial order of these communities is grounded and rationalized. Community plight is to be ultimately understood through "emergency time": the need to act immediately to build an aesthetically attractive, economically propulsive downtown and a complimentary set of spaces.

Compounding this offer of loss is the deft working through of something strategic: nostalgia and a coveted past. Thematically, these are not just any communities under the new threat of globalization, these are a serenely historic Cicero and a beloved Berwyn. Nostalgia and history are used as the signposts to know what could be lost: cherished traditions, rich cultural legacies, proud resident ways. For example, Cicero has "its traditions of people working hard and together, Berwyn is a storied bungalow community of neat and tidy homes and people. " As Planner Evans (2014) noted, "Berwyn today ... stands to lose a lot in these retrenching global times: what we have been about and what we have been ... Berwyn has always been about hard-working folk, a sturdy economy, and gracious people ... it's what we stand to lose ... and we all have to realize that." Such loss would seemingly gut these communities of what they most covet. A kind of fantasy is served up, one that illuminates glorious and revered things. Offers of imagined community and imagined monolithic population, this way, become tools to coerce.

The recent "Berwyn Comprehensive Plan" and "Cicero Growth Plan" map out the changes desired by the governances. Both call for restructuring downtowns, to be complemented by building new megaparks, condominium complexes, office parks, and enhancing dominant thoroughfares with new restaurants, clothing stores, coffee bars, theatres, and music clubs. These parts would be connected and aesthetically enticing for new investors and investment, and would communicate Cicero's and Berwyn's commitment to a progressive redevelopment. A core, concentrated growth would ostensibly mean growth for the entire community as a powerful ripple would pour across the entirety of these places. "It may be a dream," Evans (2014) noted, "but maybe it isn't ... a downtown revitalization would mean more jobs, energy, clout that would move our entire community forward ... one thing is for sure, this is the kind of approach both of these communities need. " Berwyn and Cicero, as places of becoming, would "show their resiliency and illustrate how global realities can be tussled with and fought successfully ... we just need to concentrate our energies on the downtown." Here is Kevin Cox's (1993) shift in governance activity away from "a politics of redistribution" to a "politics of resource attraction" par excellence.

A cast of elaborately narrated characters and processes are crucial to this projecting of burdened but resilient communities. The cornerstone of this has been an aggressively narrated rise of a new heroic community leader, writer Dean Mossiman's (2002) "new entrepreneurial mayor." Discussions of these mayors, a strategic politics, identify civic visionaries who should forcefully lead the needed economic development planning. They are cast as emboldened and all-seeing neoliberal beings: Mikhael Bakhtin's (1981) "eye of reason." These politicians, made resonant symbols of and articulate mouthpieces for best economic development strategies, are positioned as key narrators and truth-tellers. Mayors Robert Lovero (Berwyn) and Larry Dominick (Cicero) are the anointed truth-tellers. Lovero, mayor since 2009, is "the steady voice of reason and rationality that will steadfastly guide Berwyn into a promising future" (Johnson, 2014). Dominick, town president since 2011, "has the eye and acumen to make Cicero work in an era rife with potential economic pitfalls" (Jones, 2014). Through them, the public is to see rational, omniscient voices that poignantly pinpoint municipal ills and best solutions.

In narration, a knowledge and groundedness in local life anchor these mayors. Lovero and Dominick are widely proclaimed as attuned, intuitive neoliberalists who know the pulse of local affairs and beliefs. Bearing the local in their souls sets them up as important policy and planning voices. "Dominick is Midwest born and raised, with extensive understanding of Cicero's people and neighborhoods," and, to Planner Jones (2014), "he is perfect for dissecting what our [Cicero's] problems are and what we need." For Jones, "The business mentality that cities and communities need today to flourish ... Dominick is cut from that cloth ... he's ambitious and innovative and we're lucky to have him." These politicians ultimately gain legitimacy and voice via this framing, which renders them fellow foot soldiers to all that hold these communities dear.

For the recently arrived from Mexico, Puerto Rico, and Latin America, this vision spells trouble. In the new economic development, they cut a deeply ambivalent figure. They properly service the local grunge economies but ostensibly add little to the community's need to "glisten, sparkle, and attract the goodies and resources that Cicero and Berwyn now cry out for" (Plane, 2014). In an abstract sense, they are welcomed into the communities, and civic festivals honour their presence. But in the community morphology, their place is clear: confinement within existing peripheral spaces. Community form now must be starkly entrepreneurialized to ensure a taut business space. Their presence cannot be allowed outside these spaces for fear that community economic progress will be compromised. In the rhetoric's suggestiveness, it is not racism or discrimination at work, but rather simple economic practicality.

At the core of these deeply ambivalent figures is a pervasive but powerful reality: a dubiously constructed citizenship of "a group." Deeply rooted in Cicero's and Berwyn's mainstream imagining of this population are caricatures of "a people" viscerally felt to be ominous, impulsive, and threats to existing spatial-social orders. It appears that a wisdom of "localness" and "down-to-earth" neoliberal values, coursing through the everyday life and sensibilities of these places, transports these ideas to mainstream populations. These mainstreams now imagine a Latino Cicero and Latino Berwyn as a subset of sociospaces that need to be carefully monitored and contained. These spaces and their populations are ultimately treated in a curious way: compassion and abstract acceptance into communities are advocated but to ominous and potentially transgressive beings that seem more wild and woolly than deprived. A people suspected of being routinely and ritually anticivic is best steered to its own kinds of neighbourhoods. These people and spaces, amid provision of detail, remain caricatured and abstract phenomena: what Liam Kennedy (2000) calls a sociopsychological fantasy.

The new economic development rhetoric builds on this to suggest "a struggling people, one new to Berwyn and Cicero and still learning our ways and values ... that can collide with what they have been led to believe and know" (Johnson, 2014). "Our community needs to be refined as a place of economic development," Cicero Planner P. Jones (2014) notes, "and we need to realize the economic and cultural limitations of recent immigrant arrives [sic] ... and at the same time pay special attention to [cultivate] those who could boldly propel us forward." Here, following Rancière (2013), a racialized subaltern is less visibly repressed than "distributed" into spatial, moral, and economic oblivion (by allocations of places, people, names, functions, authorities, activities). The senses of a logically struggling and pervasively anticivic "group of immigrants" takes on resonance as Rancière's "patterns of allocations" relentlessly situate and marginalize them.

In this context, patterns of socioracial segregation in these communities proliferate. Both Cicero and Berwyn concentrate on renewing their downtowns and reculturalizing retail nodes along major thoroughfares (and to a lesser extent restabilizing middle class and working class neighbourhoods). In the shadows of these purported growth poles, poor Latino neighbourhoods fester. Here housing stocks continue to erode, streets and thoroughfares decline, and job bases languish. Poor areas mired in poverty since their inception more than two decades ago have grown in size and population. Little attention to these needs marks the "new economic development realism," as uneven development is provided a new logic rooted in the supposed need to enact a politics of resource attraction. Offers of global haunts, the rhetorical spark plug for this, have profound material consequences that build on a sordid past of racism and segregation.

Conclusion

Inner-ring suburbs in Chicago flourished as manufacturing sites nearly a century ago but have since suffered persistent economic decline accompanied by increased rates of poverty and deprivation. Some of these communities have successfully reinvented their economies in current neoliberal and transnational times; others continue to struggle to accomplish this. Today, the widely chosen place-rehabilitation option is to bolster low-wage economies, re-entrepreneurialize landscapes, and transform community demographics. Currently, Chicagoland's two quintessential inner-ring suburbs, Cicero and Berwyn, struggle along this path to reinvention. The current mantra: create spaces that tether notions of desired culture, tax base vibrancy, and economic possibility to the promise of postindustrial-service economic development.

In current Cicero and Berwyn a new immigrant segregationist politics flanks this economic development drive. It emerges out of the rapid in-movement especially of immigrants from Mexico, who are treated as important inputs into local labour markets but detrimental inputs into community redevelopment. These governances thus simultaneously accept and spurn this population: they entice them inward but drive to isolate and invisibilize their presence in these communities. This is no small accomplishment: a narrative must be crafted that identifies a population's rights and desires but also rationalizes the need to socio-spatially banish them to a community's furthest physical and social corners. Contradictions abound: these governances ultimately vacillate in praising "a group's" cultural virtues, flag their negative cultural realities, identify their material and emotive needs, impugn their material and emotive needs, and provide this population an ever-changing and ambiguous notion of citizenship.

Such politics could be seen as soft revanchist or benevolent neoliberal, but a better and more direct way to describe the phenomenon is as a pragmatic racist politics. People designated as rightful citizens of the United States and these communities are also coloured, originized, and raced to be logical targets for social and spatial marginalization. Their labour is needed in these communities, and everyone knows this. Governments exoticize them to bring them in and provide a level of inclusion. Like the historic preservation craze that currently sweeps across America's inner cities (Wilson, 2004, 2007), they are proclaimed compelling for their social and cultural otherness: as race-authentic and culturally iconographic amalgams. Yet they are to remain second-class citizens, be objects of suspicion and distrust capable of subverting crucial redevelopment which these communities supposedly need. By their mere presence in the wrong locations, in the communicating, economic redevelopment efforts could go awry.

The newness of this segregationist politics lies in its mix of ocular-centric privileging, use of global fears, and protecting a new economic vision: low-wage, dead-end labour combined with downtown upgrade economies. The visual (what is to be seen) has always been important in drives to economically develop these two communities and others across suburban America (Sassen, 2015). Today, the visual has become an obsession. Raced and classed bodies, now a potent text in suburban America for communicating who occupies and colonizes spaces (Anderson, 2010), are to be elaborately controlled and contained as geographically mobile things. In the rhetoric, a Mexican youth's whisk through a downtown park can destroy a community's slowly repairing sociospatial fabric and is fair game for being problematized.

At the same time, deployments of the new post-1990 community villain – globalization – reach new heights. The global trope has become the central fear-invoking drama through which community economic development is to be understood. This offer imposes webs of meanings, that like symbolic cages, build bars around senses of community reality and places gazes within discrete and confining visions. One reality of community and world, in rich emotive and provocative detail, is advanced while alternative visions are purged. Here is Mikhael Bakhtin's (1981) implicit dialogue with one and other points of view, the simultaneity of asserting one vision of community and world and eradicating alternatives. The global trope ultimately serves up a supposed frank and blunt package of truths about community realities and needs that can no longer be suppressed. In assertion, its pleas correspond to nothing less than core truths. As this trope has resisted and beaten back competitive visions of community and societal realities, it grows stronger in these two Chicagoland suburbs.

Finally, the new segregationist politics aligns with something unprecedented: the full-fledged fostering of low-wage and downtown-upgrade economies. These economies are now the anointed recipe for community economic and fiscal reversal that need to flower. By staunchly promoting communities built around low-wage economic operations and downtown upscaling, the rise of this new segregationist politics make sense and gains resonance. Normalized is the producing and maintaining of a shadow population whose labour can be easily tapped but whose lives are to be conducted in isolation. The dark margins of community become functional, a disinvested and neglected terrain that nevertheless provides the economy something important. This politics ultimately serves to cultivate something specific and new in these communities: a warmed-over first-world sweatshop that is to be the base of local economies. These shops are to hum and be active but not be seen by the nonpoor.

Yet, a final point of optimism must be made: we must realize that the future of this new segregationist politics is open and uncertain. It is already running into problems in these communities, which is not surprising. Many recently arrived migrants are profoundly disillusioned with their present lives. Proclaimed in abstract terms as welcome in their places, a punishing segregation and isolation gnaws at them. Frustration is exacerbated by the thin and debilitating array of economic possibilities open to them as the new low-wage service economy is aggressively pushed by the governances. The exclusion is deeply felt: "I know I should feel grateful for being in America," a 32-year-old recently arrived from Mexico women noted, "but I didn't expect this … we're really nowhere … there's almost no economic opportunity for me and my family, and we're really pushed into a corner here … so for now it's survival time." She could have been speaking for the more than 82,000 Latinos in Cicero and Berwyn today whose lives are dallied with in the new segregationist politics. Not surprisingly, immigrant rights groups, immigrant day-labour organizing, and immigrant housing movements are now flowering in these two communities. Recent arrivals have had enough and are institutionalizing their frustration. What will happen next is unclear – but stay tuned as this politics becomes increasingly identified as an object of anger and scorn.

NOTE

1 Interviewees and key actors in the communities have been provided pseudonyms to ensure frank and credible data acquisition.

REFERENCES

Anderson, M.B. (2010). The discursive regime of the 'American dream' and the new suburban frontier: The case of Kendall County, Illinois. *Urban Geography, 31*(8), 1080–99.
Bakhtin, M. (1981). *The dialogic imagination*. Austin: University of Texas.
Consuala, L. (2014, October 9). Discussion with political activist, Town of Berwyn.
Cox, K. (1993). The local and the global in the new urban politics: A critical view. *Environment and Planning D: Society and Space, 11*(4), 433–48.
Evans, B. (2014, April 29). Discussion with Planner, Town of Cicero.
Gallagher, L. (2013). *The end of the suburbs: Where the American dream is moving*. New York, NY: Penguin.
Greene, R., Bouman, M., & Grammenos, D. (2006). *Chicago's geographies: Metropolis for the 21st Century*. Washington, DC: Association of American Geographers.
Henry, A. (2014, May 11). Discussion with Cook County Planner, Town of Berwyn.

Johnson, C. (2014, July 14). Discussion with Councilperson, Town of Berwyn.
Jones, P. (2014, May 25). Discussion with Planner, Town of Cicero.
Keil, R. (2013). *Suburban constellations*. Berlin: Jovis.
Kennedy, L. (2000). *Race and urban space in contemporary American culture*. Edinburgh: University of Edinburgh.
Mossiman, D. (2002, June 13). Mayor get boost from corporations. *Wisconsin State Journal*, p. 1.
Plane, A. (2014, April 23). Discussion with Planner, Town of Berwyn.
Rancière, J. (2013). *The Politics of Aesthetics*. London: Bloomsbury.
Ruggles, D. (2014, May 19). Discussion with Planner, Town of Berwyn.
Sassen, S. (2015). The global street comes to Wall Street. In *Global crises and the challenges of the 21st century*. Retrieved from https://books.google.com/books?id=0dfOCgAAQBAJ&pg=PT16&lpg=PT16&dq=global+crisis+and+the+challenges+of+the+21st+century+book&source
Smith, A. (2013, June 12). Discussion with Planner, Town of Berwyn.
United States Census Bureau. (2013). Housing and population statistics, Department of Commerce.
Wilson, D. (2004). Toward a contingent urban neoliberalism. *Urban Geography*, 25(8), 771–83.
Wilson, D. (2007). *Cities and race: America's new Black ghetto*. London: Routledge.
Wilson, D. Beck, D., & Bailey, A. (2009). Neoliberal-parasitic economies and space building: Chicago's southwest side. *Annals of the Association of American Geographers*, 99(3), 604–26.

PART 4

Contested Suburbs

14 Governance, Politics, and Suburbanization in Los Angeles

ROGER KEIL AND DEREK BRUNELLE

Introduction

Long considered an outlier in terms of urban form among American cities, in the last half century, Los Angeles has shown similarities with changing suburbanized landscapes elsewhere on the North American continent: dramatic industrial restructuring, demographic change, and economic shifts, especially after the 2008 recession. While some recent authors have created the rather problematic impression that North America has reached "the end of the suburbs" (Gallagher, 2013), more serious scholarship has emphasized the suburban in North America as an enduring landscape that is undergoing rapid and often convulsive changes (Anacker, 2015; Forsyth, 2012; Hanlon, Short, & Vicino, 2010; Knox, 2008; Nijman, 2013; Nijman & Clery, 2015; Peck, 2015; Walks, 2012). These changes include various "involutions" and "inversions." Deploying the concept of "involution," by which they mean "a complexification and folding in of suburbanizing cultures and rationalities," Peck, Siemiatycki, and Wyly (2014, p. 389) challenge the perceived linearity of the suburbanization process. This pattern can also be named "inversion," a term first introduced by Robin Bloch (1994) with reference explicitly to Los Angeles. More recently, inversion came to be associated with major demographic change: "the poor and the newcomers are living on the outskirts. The people who live near the centre are those, some of them black or Hispanic, but most of them white, who can afford to do so" (Ehrenhalt, 2012, p. 4). Los Angeles has been showing features of peripheral poverty and diversity for a while. Before the 1970s, "segregation of minorities had been built into the structure of the city" (Fishman, 1993, p. xviii). In the 1990s, already almost one third of all new immigrants coming to the region settled directly in the suburbs (Marcelli, 2004, p. 142), and the area produced

North America's first "ethnoburbs," where (typically non-European) immigrants live in a new type of middle class suburb that is tied into both local ethnic economies and cultures and into world market based networks (Li, 1998). In more recent decades, the urban region became a classical "arrival city" (Saunders, 2010, p. 81).

At the same time, its downtown has been the site of an aggressive gentrification process often celebrated in the media. One observer wrote even that "downtown Los Angeles has, on the whole, made the most impressive recovery of any American central city in the 21st century" (Marshall, 2015). The "inversion" that has been thrown into stark relief by the crisis in Ferguson, Missouri (Kneebone, 2014), has been a defining fact of the Southern California landscape for decades, particularly as the mechanism of "privatizing with class" (Hoch, 1985) became the organizing principle of sociospatial and governmental segregation in Los Angeles County after the introduction of the Lakewood Plan in 1954, which allowed the newly incorporated municipality by that name to contract municipal service delivery from Los Angeles County and which laid the roots for subsequent waves of incorporated autonomy of suburbs with weak tax bases (Keil, 1998, p. 174).

Los Angeles has also been the inspiration for new geographies of theory (Roy, 2009). Representatives of the so-called Los Angeles School have long defied the classical centre-periphery dichotomy of much (eastern) urban scholarship as well as the presumed linearity of the urban–suburban continuum. Instead, they have used Los Angeles as the basis for a fundamental rethinking of urban form in general (Dear & Dahman, 2011, p. 74). Postsuburbia's "composite" character (Wu & Phelps, 2008) – with its global manifestations, divergences and mixing of land uses, less predictable geographic forms, new politics, new work–residence relations, and discordant land use – describes well the complexity of form, structure, and politics we find in Southern California. It is the inverted city.

The inverted landscape of suburban Los Angeles is the subject of this chapter. We will discuss, in particular, the constraints and opportunities of this landscape for governance, politics, and social activism in the Los Angeles region.

Trajectories of Suburbanization in Southern California

The beginnings of suburban Southern California were different. Los Angeles grew outward from the secured and palisaded compound of an inland Mission. By North American standards, it shows similarities in this way to Quebec City or Boston before it. It was not a city on a precipice (or hill) like those, but an outpost on Indigenous land in the western wilderness of a yet-to-be conquered continent. The position of Los Angeles as a settler colonial mission was unmasked, and its remarkable regional spread in the more than two hundred years since followed this trajectory of colonization systematically. Its suburbanization pointed inland, a reverse manifest destiny ultimately pushing east from the Pacific Ocean. Unlike the reality displayed in the pastoral images one finds in textbooks of subdivisions moving concentrically or in checkerboard patterns across vacant and verdant agrarian land, Los Angeles faced a more inimical nature of shifting sands, crumbling mountains, and fault lines. Hedged in between a vast ocean and forbiddingly uninhabitable (so it seemed) mountains, the Southland presented a volatile landscape of floods, droughts, and earthquakes. Initially, Los Angeles did not look like it might be an easy place to live but rather foreshadowed those later forms of urban settlement we now find in cities of the Global South – where squatters squeeze wedges of land out of surroundings that are less than welcoming.

Water was central. The city started on the river. Around it, the Spanish colonial water irrigation system contributed a modicum of order to the political economy of land in Los Angeles early on. The Army Corps of Engineers concreted the river in the middle of the 20th century. Built and refined over decades, this system has insured the region against the worst forms of flood-related risks. In combination with insurance issued through the Federal Emergency Management Agency for suburban homeowners, the hard spine of the river was the best foundation for the full development of the basin where the suburban expansion had massed people for almost a century. While floodwaters were designed to be rushed out to sea in the most efficient way, fresh water was shipped in long distance through an elaborate and monumental infrastructure that piped the expensive good from the Central Californian Owens Valley through the suburban (and at that time still rural) San Fernando Valley and into the Los Angeles basin. In recent decades, the Los Angeles River has become a discursive and metabolic band along which societal relationships with nature keep being rearranged (Desfor & Keil, 2004). The San Fernando Valley and its particular form of settlement at the rural frontier also serves as a reminder that suburbanization continues to grow as an expression of a semirural urbanization that retains Whiteness as identity in Southern California (Barraclough, 2011).

Outside of the settled region (map 14.1), in the peripheral suburbs, the colonization of nature continues to date as exurban developments are still pushing into the desert frontier. The persistent attempts to

292 Roger Keil and Derek Brunelle

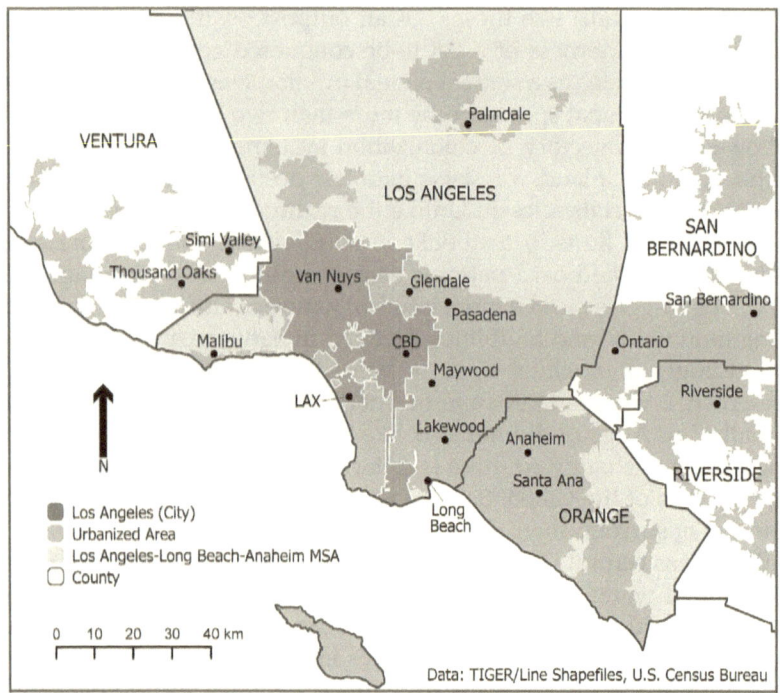

Map 14.1 Overview map of the Los Angeles urban region. Created by Rob Fiedler.

regulate air pollution on a superregional basis speak to the ongoing renegotiation of urban metabolisms across the suburbanizing region.

During the 20th century, suburban Los Angeles took shape through an amalgamation of the martial metropolis, the Keynesian conurbation, the post-Fordist region, and the global city. It has been the site of rampant and constant restructuring and never a quiet picket-fenced environment (Keil, 1998, pp. 54–75; Soja, 1989). In fact, suburbia in Los Angeles has been a dynamic maelstrom of spatial fixes, economic restructurings, and demographic shifts. Los Angeles has been both the site of fine-grain immigrant-built suburbanization (Baer, 1986) *and* home to some of the largest builders in the nation (Hise, 1997). Southern California seemed to be ahead of the rest of the United States by most measures related to suburbanization: large builders, huge subdivisions, high automobilization rates, the scale of suburban industrialization, the embrace of the freeway as the ordering infrastructure of the suburban landscape. Today, Los Angeles, while it continues to be enigmatic, equally confirms the major trends of North American and perhaps global suburbanization. One recurrent theme around L.A.'s role in redefining what cities are in the 21st century has been its polycentricity and overlapping job-housing mismatches (Modarres, 2010).

The city's fragmentation itself has been subject to much study. Already at the turn of the 20th century, the typical fragmented landscape of suburban Los Angeles could be observed (Fogelson, 1993). A class-stratified privatization of space at the basis of the social ecology of suburbanization (Hoch, 1985) coincided with the rise of big industry (from oil to consumer durables) and the interventionist force of the Keynesian welfare state. During this time, Los Angeles was both a city like many others *and* quite unique as a laboratory of Fordist modes of regulation. After the crisis of the Fordist regime, this pattern also laid the basis for continuing neoliberal, post-Fordist, and global suburbanization. Distinctive patterns of suburbanization were always part of the story (Davis, 1990; Keil, 1998; Soja, 1989).

It was never just about residential housing-tract subdivisions – perhaps the most visible form of metamorphosis from soil to suburb. Los Angeles has also been a poster child for the development of a particular kind of suburbanization that saw industrial development as the driver of a peculiar kind of massive sprawl that has known few counterparts across the United States. Capital accumulation through employment lands has been a major driver in the suburbanization of Los Angeles: from the oil patches that produced the Black Gold Suburbs of the early 20th century (Viehe, 1981) through the Cold War–fuelled expansion of the military–industrial complex, to the technopolitan-hub formations of the digital age (Fishman, 1987; Hise, 1997; Keil, 1998; Parson, 2005).

Los Angeles has been known for its horizontality, a suburban landscape devoid of gleaming towers and citadels – if we abstract, for a moment from the city's erstwhile symbols of City Hall and the Watts Towers. A "suburban metropolis," which, as Robert Fishman observed, "proved to be a contradiction in terms," it defied the original promise of the suburban ideal and lost the exclusive nature of its settlement principle (1987, pp. 15, 181). Its horizontality is dense and remarkably non-polarized in morphological terms: "Los Angeles lacks a super-dense core like Manhattan. But it also lacks a very low-density suburban periphery," leading to the description of its suburbanization as "dense sprawl" (Eidlin, 2014). Los Angeles has always been suburban, At least since its first significant property boom in the 1880s, the city sprawled and leapfrogged in ways that scholars and other observers would eventually name "suburban" (Warner, 1972) but which remained unusual for decades. In the 1920s, the city already defied common definitions of the urban boundary. Richard Neutra recalled that fact when, as the International Congress of Modern Architecture met in Brussels in 1930, all cities were mapped at the same scale. For Los Angeles "the chosen scale produced a monstrously oversized chart ... A map produced according to this established set of rules became a huge and strange jungle of misunderstandings, not possible of interpretation even by connoisseurs and experts" (Neutra, 1941, p. 189). The scale of suburban growth during the 20th century in Los Angeles was unprecedented (Wagner, 1935), although some of today's megacities have outpaced the range and speed of Southern California's urban expansion. But that it has always been suburban is really captured in perhaps the most crucial period of its existence when it became known less as an aberration and more as a model of future urban form and social structure following the Second World War (Keil, 1998).

The increasingly negative vision of Los Angeles as a significant, if problematic model of sub/urban form was held most prominently in the 1960s, when the smog was unbearable (but first measures were taken to deal with it), when the nation and the world learned about a riotous suburb called Watts, and when Charles Manson's gang murdered Sharon Tate and others in the Hollywood Hills and Los Feliz. This was also the time when the Italian director Michelangelo Antonioni made the film *Zabriskie Point* (1970), which introduced the limitless urban region through the depiction of "a vicious circle in which the escape from Los Angeles will always lead us back to the city" (Keil, 1998, p. 40). The best known outsider who noticed that Los Angeles was becoming more than an oddity but rather a leader in urban form and function was Reyner Banham. His *Los Angeles: The Architecture of Four Ecologies* (1970) revolutionized the way the suburban landscape could be viewed: disdain was replaced by curiosity and explanation. Driving on a freeway became a more normalized urban activity, and living in a suburb was part of the overall metropolitan experience rather than a social pathology. Banham's masterpiece found its match only 20 years later in Mike Davis's *City of Quartz* (1990), which painted a prohibitive and dystopian picture of Los Angeles, especially of its homeowners' association–dominated suburbs. The growth of self-governing *cities* in Southern Californian suburbia represented a form of *peripheral urbanization*, which provides the main narrative of the metropolitanization and beyond in the Southland and includes the concentration of jobs, factories, offices, and commercial areas in addition to residential expansion. Los Angeles became "exopolis," in defiance of "conventional terms of urban, suburban, exurban, and nonurban to describe divisions within contemporary metropolitan areas" (Soja,

Figure 14.1 Foreclosure Center, Palmdale, California. Photo courtesy of authors.

1996, p. 436). More recently, it was in the Southern California suburbs where many of the region's foreclosures took place during the 2008 crisis (SCAG, 2008) (figure 14.1). Even the often cited and incited desert frontier has now become postsuburban. In the empty ruins of the Great Recession, in places as far out as Rialto, the push has come to a halt, and the inner reworkings of the suburbs have begun (Martin, Meisterlin, & Kenoff, 2011). Not surprisingly, it is in the rapidly changing Los Angeles suburbs that social service needs have become particularly high (Allard & Roth, 2008).

Postsuburban Politics

The notion of postsuburbanization raises questions of governance (Ekers, Hamel, & Keil, 2012). The suburban landscape was produced and governed to a large degree by federal authorities (especially in the military area) and the dominance of the state. The federal government was instrumental in building large-scale structures such as the flood-control system that was indispensable for suburban expansion. The State of California, with its powerful legislative instruments, controlled land use. Yet the suburban municipal landscape was ostensibly built on local autonomy and self-determination and by "a shrewd alliance between private profit and public resources" (Fishman, 1993, p. xxii). The city's landed property capital – from citrus farmers and cattle ranchers to railway and oil barons, and more recently international investments – was always guided in its flow into residential, commercial, and industrial suburbs by concerted and often quite concentrated state action, especially in the area of infrastructure. This was complemented by the region's notorious "private authoritarianism," expressed most visibly in pervasive gating (Davis, 1990; Flusty, 1994; Low, 2008).

Yet, imbricated with the narratives of state, market, and privatism we find multiple everyday suburbanisms that have shaped the ethnically mixed (and sometimes segregated) landscape of suburban California in often unpredictable ways (Cheng, 2014; Waldie, 1996). Social and political struggles have been inherently connected to its suburban form. Examples range from the Van Nuys GM closure in the 1980s to the Bus Riders Movement of the early 2000s. These struggles have not necessarily been a reaction against the form but rather to the conditions that have emerged through spaces of oppression (Keil, 1998).

While American labour activists have discovered the suburbs as an organizing ground proposing to adapt union strategies to the "suburbanization of the US working class" (Narefsky, 2014), the Los Angeles region has traditionally been already the home of some of the largest industrial facilities in the United States with a lively working class culture since at least the 1920s. With similar ideals, largely a backlash against the typical urban life of other cities, working class suburbs closely represented the broader trend of suburban development during the period. As with developments for the wealthy, industrial suburbs reflected the encouragement of industrial dispersal (Nicolaides, 2001). The region boasted 524,000 manufacturing workers in 2015, to make it the centre of industrial production in the country (Morath & Van Dam, 2015). Yet some of the manufacturing suburbs, especially in the deindustrializing belt along the Los Angeles River, such as Cudahy, Bell Gardens, and Huntington Park, have historically also been among the country's poorest suburbs (Keil, 1998, pp. 194–201).

Progressive movements were not the sole political outcome of the vast suburban diversity of Southern California. The region has seen its share of conservative splintering urbanist movements. Based on previous antibusing movements and buttressed by the semiprivate homeowner political culture, a powerful secession movement emerged in the 1990s in the San Fernando Valley in the north as well as other parts of Los Angeles, including the Wilmington–San Pedro area in the South (Boudreau & Keil, 2001).

Governance, Politics, and Suburbanization in Los Angeles 297

Figure 14.2 Postsuburban landscape in Southern California. Photo courtesy of authors.

Beyond the challenges of horizontality, "the *altered geographies of postmodern urbanism are redefining the meaning and practice of urban politics*," potentially leading to "the subordination of the local state to plutocratic privatism" (Dear & Dahman, 2011, pp. 71–2). Yet sprawl also enables certain autonomies, creative local economies, and ecological politics (Dear & Dahman, 2011, p. 75). What we see in Los Angeles, perhaps more than elsewhere in the United States or anywhere, is a struggle between the tendency for "suburbia [to] become a strategically significant nexus for open-ended, deregulatory experimentation, systematically favouring more decentralized, voluntaristic, privatized, and market-oriented approaches" (Peck, 2015, p. 145). Arguably, this countermovement to neoliberalization as exemplified by the complex postsuburban politics of social and spatial justice has its origin in the vastly diverse landscapes of the Southland (figure 14.2).

Social Justice and Suburban Governance

If we treat suburban Los Angeles as a new and dynamically changing terrain, there are three interrelated areas where suburban-based and themed politics have been emerging as important features of the political arena: the environment, transportation, and labour/social justice. In environmental policymaking, Los Angeles suburban-based politics has been known to be leading nationally or even internationally for decades. The struggle of suburban communities of colour, for example, to push for justice demands in the context of air-pollution control is well documented (Carpio, Irazabal, & Pulido, 2011). The Los Angeles River, likewise, has been remade into a suburban green axis mostly due to the strength of suburban-based political activists who have imposed their socioecological demands on a debate once ruled by a narrative of flood control (Desfor & Keil, 2004). Similarly, the historic victory of the Bus Riders Union (BRU) in having their demands for more equitable distribution of transit investment recognized by a consent degree in the 1990s is well known. Suffice it here to mention that it was the demands of the low-income suburbanites of colour for better bus service that superseded the traditional favouring of prime network spaces – linking the downtown to wealthy suburbs by rail. One arena where Los Angeles has more recently led the way has been social justice victories through political intervention on a regional scale. We will have a closer look at this arena in the remainder of this chapter.

A crucial aspect of Los Angeles's postsuburban form is the new regional lens of social justice struggles and the ways in which this has influenced other jurisdictions. This line of thinking had already entered the political conversation in Southern California during the struggles against plant closings in the 1980s and 1990s, as the diverse and scattered suburban working class in the Los Angeles area "transcended the sociospatial patchwork of Southern California rather than *reproducing* it" (Keil, 1998, p. 209). Labour and community strategists fighting to keep the General Motors assembly plant in Van Nuys open were keenly aware of the suburban location of the plant itself but also of the residential geography of the workers who commuted in from African American and Latino suburbs south and east of Los Angeles (Mann, 1987, pp. 98–108). The outcome was a form of "economic planning from the bottom up" (Mann, 1987, p. 379) as the workers in their plants and communities pushed the agenda of their struggle beyond the boundaries of the disparate location of their individual residential or employment suburbs and created a regional alliance. In the words of leading organizer Eric Mann at the time, "In every city, regardless of the specific objectives and demands, the ability of workers, many of whom are themselves black, Latino or Asian, to ally with minority communities and organizations is critical" (Mann, 1987, pp. 376–7).

Pastor and Benner write about three important factors informing regional equity in Los Angeles: fragmentation of the business class, new labour organizing, and the adoption of a regional lens by a wide array of community organizations (2011, p. 333). The aftermath of the 1992 rebellion has generally been seen as a rebirth of political and civic life in Los Angeles as metrowide civic organizations began to seriously cooperate and challenge the traditional political divisions of region (Nicholls, 2003). Add Pastor and Benner, "With the seeming anger of a disenfranchised population evident – and evidently channeled to the singularly unhelpful strategy of burning down one's own neighborhood – social justice organizations found themselves grappling with the need for a new approach to organizing. They soon began to think regionally; in the words of one activist, 'If you want to help south L.A. you can't talk about south L.A. apart from the region'" (2011, p. 334).

Since the late 1990s, L.A. has been the site of numerous victories and strategies that have connected urban development to community interests. The first major instance of this kind of community-led action around employment needs was the Alameda Corridor Jobs Coalition (ACJC), formed in 1997 in response to the 21-mile rail corridor project between the port of Long Beach and downtown L.A. The Alameda Corridor project was unique because it crossed multiple suburban jurisdictions and required the support of community groups to push for local hiring and living-wage agreements at the regional level. While AJCJ coleader Mary Ochs asserts that the focus was on creating jobs – not on creating a regional approach to organizing – this cross-jurisdictional approach was nevertheless the only option for achieving the hiring and job-training goals set by the organization (Ochs, 2013). The coalition, through research, partnerships, and early action, acquired expertise on the project before entering into any agreements. The ACJC was pioneering by the way it recognized the need to focus on the many issues that inhibit individuals in different locations from finding work – access to transportation, criminal records, and education (Ochs, 2013). By addressing such barriers to employment, the coalition was successful in achieving a 30% local-hiring target along with job training commitments (Ranghelli, 2002, p. 4). Beyond Southern California, such organizing has set precedents for other jurisdictions to pursue local hiring, living-wage, and diversity targets in government-funded projects. In 2013, the Toronto Community Benefits Network actively sought a local-hiring agreement based on the success of work done by the ACJC and the Los Angeles Alliance for a New Economy (LAANE) (Maytree, 2014).

In and around Los Angeles, an extensive network of organizing has emerged through the implementation of community benefits agreements (CBAs) from proposed developments. Such agreements have been created around large-scale infrastructure projects that receive government funding. In 2001, the first successful CBA was implemented with the L.A. Live development near the Figueroa Corridor in the city's downtown (Saito, 2012). Determined to secure living-wage employment and job-training opportunities, the LAANE leveraged power through publicity and successfully fought for meaningful community representation from construction to completion. Groups like the nationally based Right to the City Alliance have also begun to address the value in preserving areas for employment (UCLA Community Scholars Program, 2007). In conjunction with the Strategic Actions for a Just Economy (SAJE), the Right to the City Alliance has focused on addressing the need for skills training to accompany new employment opportunities. Such community unionism has generated greater attention to the horizontal labour strategies of the West Coast.

Yet, despite the regional framework of equity-based organizations like LAANE, SAJE, and other community action networks, their initial efforts on policy were focused within the City of Los Angeles. This was reflected in some of the strategies employed by community coalitions, especially with the availability of incentives for new developments (Pastor, Benner, & Matsuoka, 2009, p. 136). Recognizing the potential to win economic concessions at the development application stage has helped establish new organizing methods in low-income communities across the region (Saito, 2012). This view suggests the complex nature of polycentric form with organizing around new developments. Many large-scale community organizations have remarked that the major source of power – financial and political – remains in the City of Los Angeles, "the arena where regional agendas are negotiated" (Pastor, Benner, & Matsuoka, 2009, p. 137).

Conflicts around regional governance have erupted in a number of ways, but most saliently in terms of transit needs. Soja (2010, p. xviii) draws attention to the transit needs of low-income communities as reflected in the 1996 court victory between the BRU and the City of Los Angeles. This grassroots movement demonstrated the connection between transit options and quality of life. Advocates looked beyond the immediate transit issues to recognize the influence of long-standing spatial marginalization. The result of the BRU legal battle was significant by the way it exposed the general public, as Soja argues, "to the significant racial, class, and geographical biases that are embedded in all forms of public planning" (2010, p. xiii).

High-profile victories such as the BRU and CBAs such as L.A. Live have raised awareness of the Southland's suburbs as major centres for labour movement revitalization (see also Hauptmeier & Turner, 2007, p. 132). Los Angeles still retains a major industrial complex, and organizations have begun to recognize ways to connect economic development initiatives to community development. In many ways, advocacy work in Los Angeles and its suburbs has set the stage for new developments to *expect* to partner with communities for local benefits, with the ultimate goal of removing barriers faced by marginalized groups. The connection between economic development and community needs reflects an assertive role of local advocates who have begun to think regionally and have accepted the postsuburban inverted city as their terrain of action.

Conclusion

The suburban landscape of Los Angeles was envisioned and has been analysed as an expression of individualism, class difference, and segregation. In its postsuburban phase, it has become something quite different: the terrain for social change. As the downtown makes its transition from redeveloped corporate citadel surrounded by skid-row poverty and sweatshop industrial district to a gentrified urban playground, the diverse suburban expanses of the region gain significance as sites of everyday life and innovative civic politics as well as sites for urban social-movement activism.

Suburbanization in Los Angeles has changed the trajectory and narrative of peripheral urban development and has produced a new terrain for social action. The suburban histories of Los Angeles traverse the region's social and natural ecologies, social geographies, and political institutions and territories. The dense sprawl of the Southland has not been a scene of depoliticization but instead a fertile ground for change. Narratives of privatism, corporatism, and state policy seem to have shaped suburbia in the Los Angeles region earlier, in more profound ways, and in more diverse forms than elsewhere. But suburbanization and the diffusion of suburban ways of life have also produced an intriguing mix of political action around social and environmental justice and rights-based demands. The self-reliance of suburbanized communities, their power-defiant political action and persistent insistence on the right to the suburb (and ultimately the city) have become emblematic as the impressive confluence of activism, movement organization, and spatial-justice militancy, a beacon of hope for many outside the Southland. These examples of unrest and dissatisfaction belie the usual myth of the pacified suburb. They stand as defiant symbols of the possibility of autonomy and self-determination at best and class assertion at worst in a sub/urban landscape overdetermined by state, capital, and privatism. They might be a demonstration to the political prospects of the sub/urban century in North America.

REFERENCES

Allard, S., & Roth, B. (2008). Strained suburbs: The social service challenges of rising suburban poverty. *Brookings Institute*. Retrieved 2 July 2013 from http://www.brookings.edu/research/reports/2010/10/07-suburban-poverty-allard-roth

Anacker, K.B. (Ed.). (2015). *The new American suburb: Poverty, race and the economic crisis*. Farnham, UK: Ashgate.

Baer, W.C. (1986). Housing in an internationalizing region: Housing stock dynamics in Southern California and the dilemmas of fair share. *Environment and Planning D: Society and Space*, 4(3), 337–49.

Banham, R. (1971). *Los Angeles: The architecture of four ecologies*. Harmondsworth: Pelican.

Barraclough, L.R. (2011). *Making the San Fernando Valley: Rural landscapes, urban development, and White privilege*. Athens: University of Georgia Press.

Bloch, R. (1994). *The metropolis inverted: The rise and shift to the periphery and the remaking of the contemporary City* (PhD dissertation). UCLA, Los Angeles.

Boudreau, J.-A., & Keil, R. (2001). Seceding from responsibility? Secession movements in Los Angeles. *Urban Studies*, 38(10), 1701–31.

Carpio, G., Irazabal, C., & Pulido, L. (2011). Right to the suburb? Rethinking Lefebvre and immigrant activism. *Journal of Urban Affairs*, 33(2), 185–208.

Cheng, W. (2014). *The changes next door to the Díazes: Remapping race in suburban California*. Minneapolis: University of Minnesota Press.

Davis, M. (1990). *City of Quartz: Excavating the future in Los Angeles*. London: Verso.

Dear, M., & Dahman, N. (2011). Urban politics and the Los Angeles school of urbanism. In D.R. Judd & D. Simpson (Eds.), *The city revisited: Urban theory from Chicago, Los Angeles, New York* (pp. 65–78). Minneapolis: University of Minnesota Press.

Desfor, G., & . and Keil, R. (2004). *Nature and the city: Making environmental policy in Toronto and Los Angeles*. Tucson: University of Arizona Press.

Ehrenhalt, A. (2012). *The great inversion and the future of the American city*. New York, NY: Alfred A. Knopf.

Eidlin, E. (2014). What density doesn't tell us about sprawl. *Access*. Retrieved 29 May 2015 from http://www.accessmagazine.org/articles/fall-2010/density-doesnt-tell-us-sprawl/

Ekers, M., Hamel, P., & Keil, R. (2012). Governing suburbia: Modalities and mechanisms of suburban governance. *Regional Studies, 46*(3), 405–22.

Fishman, R. (1987). *Bourgeois utopias: The rise and fall of suburbia.* New York, NY: Basic Books.

Fishman, R. (1993). Foreword. In R.M. Fogelson (Ed.), *The fragmented metropolis* (pp. xv–xvii). Cambridge, MA: Harvard University Press.

Flusty, S. (1994). *Building paranoia: The proliferation of interdictory space and the erosion of spatial justice.* West Hollywood, CA: Los Angeles Forum for Architecture and Urban Design.

Fogelson, R. (1993 [1967]). *The fragmented metropolis.* Cambridge, MA: Harvard University Press.

Forsyth, A. (2012). Defining suburbs. *Journal of Planning Literature August, 27*(3), 270–81.

Gallagher, L. (2013). *The end of the suburbs: Where the American dream is moving.* New York, NY: Portfolio/Penguin.

Hanlon, B., Short, J.R., & Vicino, T.J. (2010). *Cities and suburbs: New metropolitan realities in the US.* London: Routledge.

Hauptmeier, M., & Turner, L. (2007). The politics of labor coalitions in New York City and Los Angeles. In L. Turner & D.B. Cornfield (Eds.), *Labor in the new urban battlegrounds: Local solidarity in a global economy* (pp. 129–43). Ithaca, NY: Cornell University Press.

Hise, G. (1997). *Magnetic Los Angeles.* Baltimore, MD: Johns Hopkins University Press.

Hoch, C. (1985). Municipal contracting in California: Privatizing with class. *UAQ, 20,* 303–23.

Keil, R. (1998). *Los Angeles: Globalization, urbanization and social struggles.* Chichester: Wiley.

Kneebone, E. (2014, August 15). Ferguson, Mo. Emblematic of growing suburban poverty. *The Avenue.* Retrieved 27 May 2015 from http://www.brookings.edu/blogs/the-avenue/posts/2014/08/15-ferguson-suburban-poverty

Knox, P.L. (2008). *Metroburbia, USA.* New Brunswick, NJ: Rutgers University Press.

Li, W. (1998). Anatomy of a new ethnic settlement: The Chinese *Ethnoburb* in Los Angeles. *Urban Studies, 35*(3), 479–501.

Low, S. (2008). Incorporation and gated communities in the greater metro-Los Angeles region as a model of privatization of residential communities. *Home Cultures, 5*(1), 85–108.

Mann, E. (1987). *Taking on general motors: A case study of the UAW campaign to keep GM Van Nuys open.* Los Angeles: Center for Labor Research and Education, Institute of Industrial Relations, UCLA.

Marcelli, E.A. (2004). From the barrio to the 'Burbs: Immigration and the dynamics of suburbanization. In J. Wolch, M. Pastor, Jr., & P. Dreier (Eds.), *Up against the sprawl: Public policy and the making of Southern California* (pp. 123–50). Minneapolis: University of Minnesota Press.

Marshall, C. (2015, March 5). The gentrification of Skid Row: A story that will decide the future of Los Angeles. *The Guardian Cities*. Retrieved 26 May 2015 from http://www.theguardian.com/cities/2015/mar/05/gentrification-skid-row-los-angeles-homeless?CMP=share_btn_tw

Martin, R., Meisterlin, L., & Kenoff, A. (2011). *The Buell hypothesis: Rehousing the American dream*. New York, NY: The Temple Hoyne Buell Center for the Study of American Architecture; Columbia University Graduate School of Architecture, Planning and Preservation.

Maytree. (2014). Community benefits agreements: A new tool to reduce poverty and inequality. Retrieved 22 July 2015 from http://maytree.com/policy-and-insights/opinion/community-benefit-agreements-new-tool-reduce-poverty-inequality.html

Modarres, A. (2010). Polycentricity, commuting pattern, urban form. *International Journal of Urban and Regional Research, 35*(6), 1193–211.

Morath, E., & Van Dam, A. (2015, July 15). Where are the most U.S. manufacturing workers? Los Angeles. *Wall Street Journal*. Retrieved 21 July 2015 from http://blogs.wsj.com/economics/2015/07/15/where-are-the-most-u-s-manufacturing-workers-los-angeles/

Narefsky, K. (2014). The suburbanization of the US working class. *Jacobin Magazine*, 15/16. Retrieved 17 July 2015 from https://www.jacobinmag.com/2014/10/the-suburbanization-of-the-us-working-class/

Neutra, R.J. (1941). Homes and housing. In G.W. Robbins & L. Deming Tilton (Eds.), *Los Angeles: Preface to a masterplan* (pp. 189–201). Los Angeles: The Pacific Southwest Academy.

Nicholls, W. (2003). Forging a "new" organizational infrastructure for Los Angeles' progressive community. *International Journal of Urban and Regional Research, 27*(4), 881–96.

Nicolaides, B.M. (2001). The quest for independence: Workers in the suburbs. In T. Sitton (Ed.), *Metropolis in the making* (pp. 77–95). Los Angeles: University of California Press.

Nijman, J. (2013). The American suburb as utopian constellation. In R. Keil (Ed.), *Suburban constellations* (pp. 161–9). Berlin: Jovis.

Nijman, J., & Clery, T. (2015). The United States: Suburban imaginaries and metropolitan realities. In: P. Hamel & R. Keil (Eds.), *Suburban governance: A global view* (pp. 57–79). Toronto: University of Toronto Press.

Ochs, M. (2013). Interview conducted by Derek Brunelle on April 7, 2013. Downey, CA.

Parson, D. (2005). *Making a better world: Public housing, the red scare, and the direction of modern Los Angeles*. Minneapolis: University of Minnesota Press.

Pastor, M., & Benner, C. (2011). Moving on up? Regions, megaregions, and the changing geography of social equity organizing. *Urban Affairs Review*, 47(3), 315–48.

Pastor, M., Benner, C., & Matsuoka, M. (2009). *This could be the start of something big*. Ithaca, NY: Cornell University Press.

Peck, J. (2015). Chicago-School Suburbanism. In P. Hamel & R. Keil (Eds.), *Suburban governance: A global view*. Toronto: University of Toronto Press.

Peck, J., Siemiatycki, E., & Wyly, E. (2014). Vancouver's suburban involution. *City: Analysis of Urban Trends, Culture, Theory, Policy, Action*, 18(4–5), 386–415. doi:10.1080/13604813.2014.939464

Ranghelli, L. (2002). Replicating success: The Alameda corridor job training and employment program. Retrieved 22 July 2015 from http://www.campusactivism.org/server-new/uploads/acjc%20replication%20manual.pdf

Roy, A. (2009). The 21st-century metropolis: New geographies of theory. *Regional Studies*, 43(6), 819–30. doi:10.1080/00343400701809665

Saito, L. (2012). How low income residents can benefit from urban development: The LA Live community benefits agreement. *City and Community*, 11(2), 129–50.

Saunders, D. (2010). *Arrival city: The final migration and our next world*. Toronto: Alfred A. Knopf Canada.

SCAG. (2008). Foreclosure outlook for the gateway subregion. Retrieved 1 April 2013 from http://www.scag.ca.gov/Housing/pdfs/trends/ForeclosureOutlook_GatewayCities100108.pdf

Soja, E. (1989). *Postmodern geographies*. London: Verso.

Soja, E. (1996). Los Angeles, 1965–1992. In E. Soja & A. Scott (Eds.), *The city: Los Angeles and urban theory at the end of the twentieth century* (pp. 426–62). Berkeley: University of California Press.

Soja, E. (2010). *Seeking spatial justice*. Minneapolis: University of Minnesota Press.

UCLA Community Scholars Program. (2007). Fighting for a right to the city: Collaborative research to support community organizing in L.A. Retrieved 22 July 2015 from http://164.67.121.27/files/UP/Community%20Scholars//Right_to_the_City.pdf

Viehe, F. (1981). "Black gold suburbs": The influence of the extractive industry on the suburbanization of Los Angeles, 1890–1930. *Journal of Urban History*, 8(1), 3–26.

Wagner, A. (1935). *Los Angeles: Werden, Leben und Gestalt der Zweimillionenstadt in Südkalifornien*. Leipzig: Bibliographisches Institut.

Waldie, D.J. (1996). *Holy land: A suburban memoir.* New York, NY: St. Martin's Press.
Walks, A. (2012). Suburbanism as a way of life, slight return. *Urban Studies, 50*(8), 1471-88.
Warner, S.B. (1972). *The urban wilderness: A history of the American city.* New York, NY: Harper and Row.
Wu, F., & Phelps, N. (2008). From suburbia to post-suburbia in China? Aspects of the transformation of the Beijing and Shanghai global city regions. *Built Environment, 34*(4), 464–81.

15 Reaching Suburbia: Towards a Socially Just Transit System for Ottawa

CAROLINE ANDREW AND ANGELA FRANOVIC

Introduction

This chapter addresses the suburbanization of Ottawa, Canada, the role of the federal government, and their links with public transit and vulnerable populations. More specifically, the chapter will explain the role of Ottawa as a government city and its impact on urban planning. From there we explain how the city landscape has changed from a simple government town situated in what is now the downtown core and how it flourished beyond the downtown core and beyond its Greenbelt area, into the suburbs of present-day Ottawa. This result is in part due to government planning, in part due to the location and development of other economic sectors, and in part due to the gentrification of Ottawa's core.

From there we aim to explain the impact of this development on Ottawa's current public transit infrastructure. Public transit in Ottawa is widely considered as one of the best bus rapid-transit systems in Canada (mostly for its seamless service to and from the downtown core, and serving its government workers), but it still faces challenges in servicing suburban/rural areas and vulnerable populations that are increasingly residing in those areas.

We emphasize the importance of envisioning policy planning as a bottom-up process: collaborative action between community groups and municipal elected members and staff members that can fundamentally affect the servicing of the needs of marginalized groups. In order to further give a voice to such populations, we introduce the voice of two focus groups with people that reside in suburban Ottawa; one with immigrant women and one with elderly Chinese immigrants. The results give insight into the barriers that these groups face and how public transit succeeds and fails in servicing them.

Ottawa: The Suburbanization of a Government Town

Ottawa is a medium-sized Canadian city with some particular characteristics that make it a larger metropolitan area. It is first of all the capital of Canada and situated on the banks of the Ottawa River. On the other side of the river is the municipality of Gatineau, which is a part of the province of Quebec. The total population of the Ottawa–Gatineau region is 1,282,500 (City of Ottawa, Population, 2015a). Ottawa is the capital of Canada and the seat of government according to the Canadian founding document of 1867. Ottawa was chosen as the capital primarily to avoid having to choose between the two economic centres of Montreal and Toronto although its location situated on the border between Ontario and Quebec added to its attractiveness for the choice of the capital. However, Ottawa and Gatineau together form the National Capital Region (NCR). This was proposed in 1968 by the federal prime minister of the time, Pierre Elliott Trudeau, and agreed to by all the provincial premiers as a way of underlining an enhanced role for the French-speaking Quebec side of the region and making more difficult the establishment of an independent Quebec.

Despite discussion over the last 50 years about the status of the capital and the idea of creating a capital district, Ottawa remains a municipality within the province of Ontario. There is a federal government agency, the National Capital Commission (NCC), which dates back to 1899 as the Ottawa Improvement Commission. The NCC has played a very important planning role in Ottawa in the period up to the mid-1960s. At this point both Ontario and Quebec governments created regional government structures that took on the planning functions and left the NCC to move into a role of creating and managing festivals and events (Chiasson & Andrew, 2009). More directly pertinent for this chapter is the fact that the suburbanization of Ottawa has also taken place by the rapid growth in Gatineau in the past 30 years.

The federal–municipal relationship has not been an easy one, and the NCC did not accept municipal representatives on its board, arguing that it was a national body with a mission to represent Canada to Canadians and to the world. This changed in 2016 as the newly elected federal government decided that the mayors of Ottawa and Gatineau would be included on the board of the NCC, but that until the law could be officially changed, they would sit as observers, without the right to vote.

Ottawa was not created as a government town when it was chosen as the Capital of Upper Canada in 1857; it was a small lumber settlement described by the governor general of the time in terms of "its wild position, and relative inferiority to the other cities named."

It is only with the Second World War that the federal government really expanded its workforce and, more importantly, this larger workforce continued to expand over the postwar period. The way in which the federal government located its employment has an obvious important impact on the spatial development of the region.

> Decisions by the government on the location of its offices have had a distinctive effect on where people work. As will be discussed, these decisions have at times led to suburbanization of work but at other times have anchored jobs in the downtown core. (Andrew, Ray, & Chiasson, 2011, p. 203)

The Gréber Plan and Ottawa's Greenbelt

Suburbanization has certainly taken place, in part related to federal government planning documents and their implementation, in part related to federal politics, and in part related to the space requirements of industries and their employees interested in locating in Ottawa. The publication in 1950 of the *Plan for the National Capital*, better known as the Gréber Plan, underlined the need to decentralize some of the federal government buildings both to decrease traffic congestion but also because of a vision of suburban environmentally attractive areas of employment close to employees' places of residence.

The four drawings below are from the Gréber Plan, illustrating a) the existing rail network; b) the proposed rail network; c) the existing road network; and d) proposed road network (map 15.1). The drawings illustrate the extent of urban development in the immediate postwar period and also the attempt by Gréber to plan for a fairly dense development. The Greenbelt was first recommended by Gréber in his plan in 1950 as a way of controlling urban sprawl and also as an area to build government offices in campus-like pastoral settings. The government of Prime Minister Louis St-Laurent accepted the plan as part of the postwar continuation of the federal government employment expansion and a growth of a Canadian government desire to expand its international image through an expanded vision of its capital.

Expropriations for the Greenbelt started in 1956. The putting into place of the Greenbelt did not stop urban sprawl; it simply jumped across the Greenbelt to build low-density suburban development. It resulted in large suburban areas, situated just beyond the Greenbelt (map 15.2).

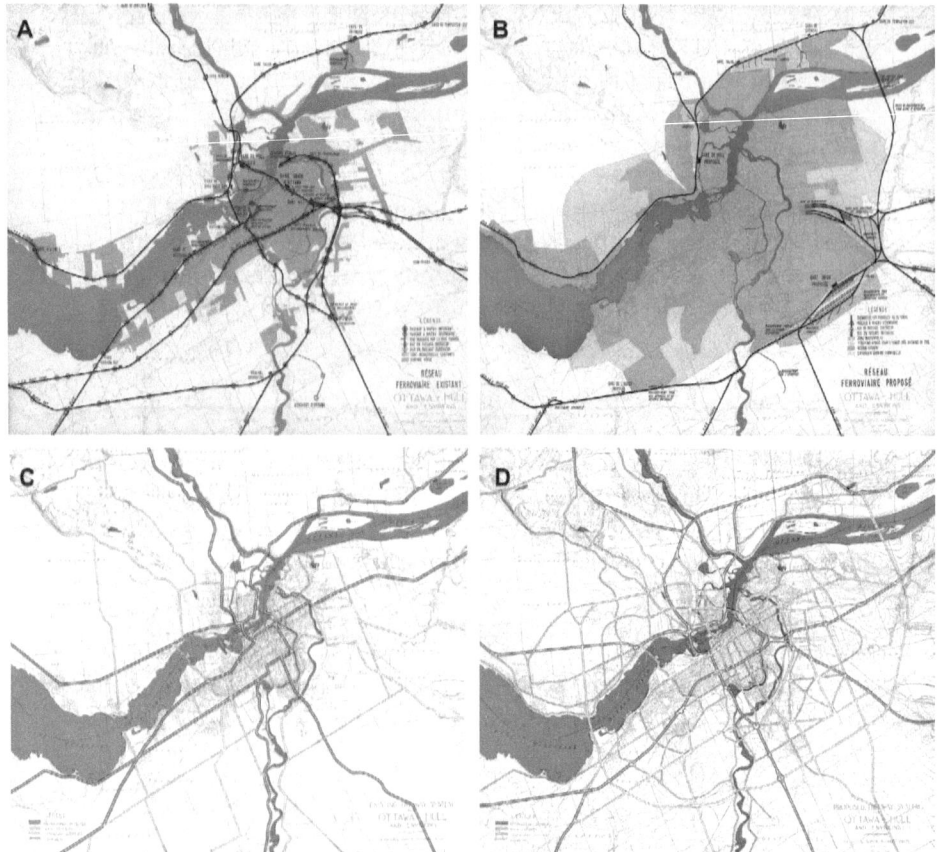

Map 15.1 Jacques Gréber, General Report on the Plan for the National Capital, 1950. Source: Jacques Gréber. *The 1950 Plan for the National Capital Region: General Report*. Ottawa: National Capital Planning Service, 1948. © National Capital Commission / Commission de la capitale nationale

The Suburbanization of Federal Government Buildings

The first examples of office suburbanization were some of the federal government "temporary buildings" built between 1939 and 1945 to accommodate the increase in government workers from 12,000 to almost 36,000. Most of the temporary buildings were built in the core, but a few were built in suburban areas near Dow's Lake (part of Capital ward). A second example of federal office suburbanization was Confederation Heights (River ward) (see map 15.3), where the federal government, in the postwar period and inspired by the planning vision

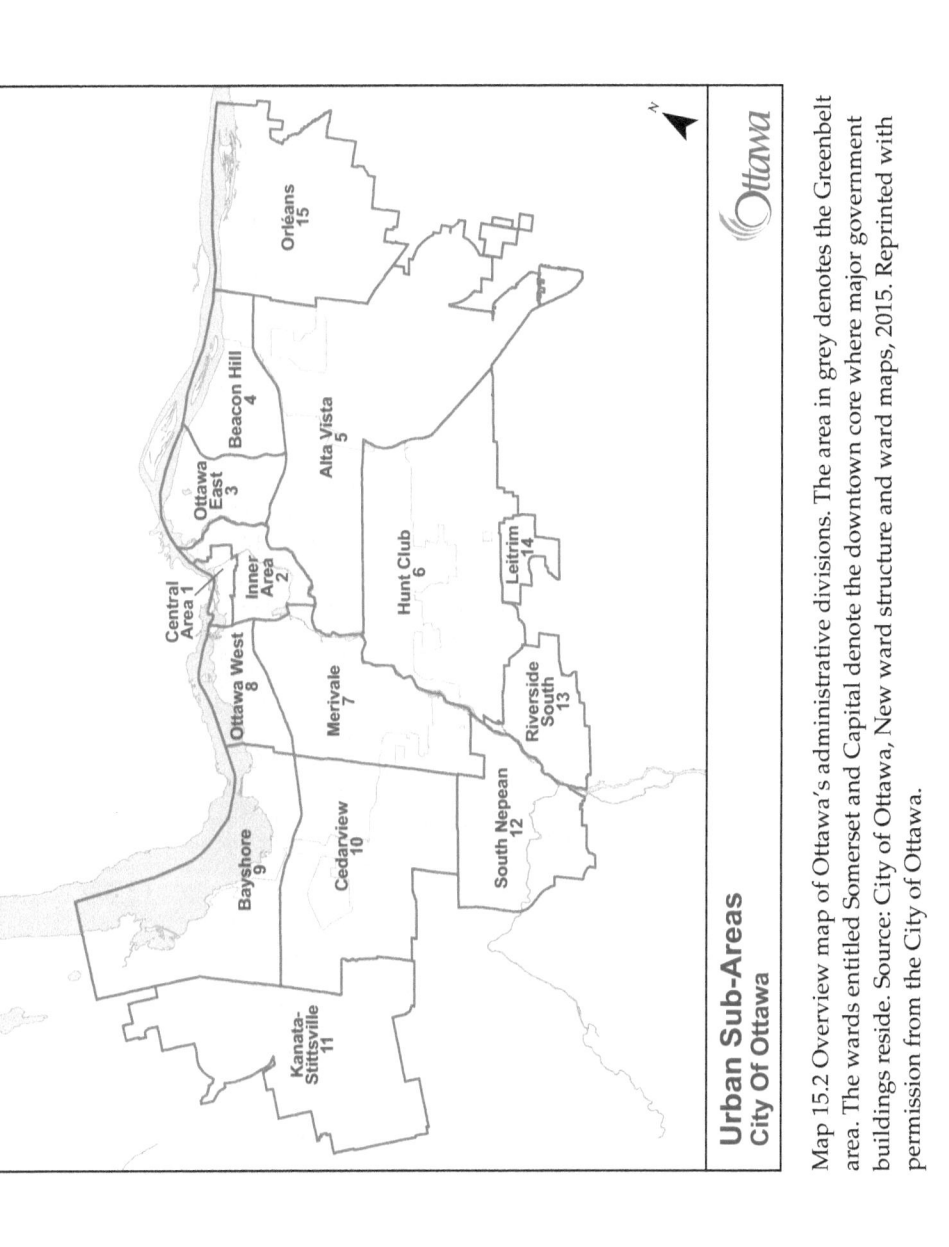

Urban Sub-Areas
City Of Ottawa

Map 15.2 Overview map of Ottawa's administrative divisions. The area in grey denotes the Greenbelt area. The wards entitled Somerset and Capital denote the downtown core where major government buildings reside. Source: City of Ottawa, New ward structure and ward maps, 2015. Reprinted with permission from the City of Ottawa.

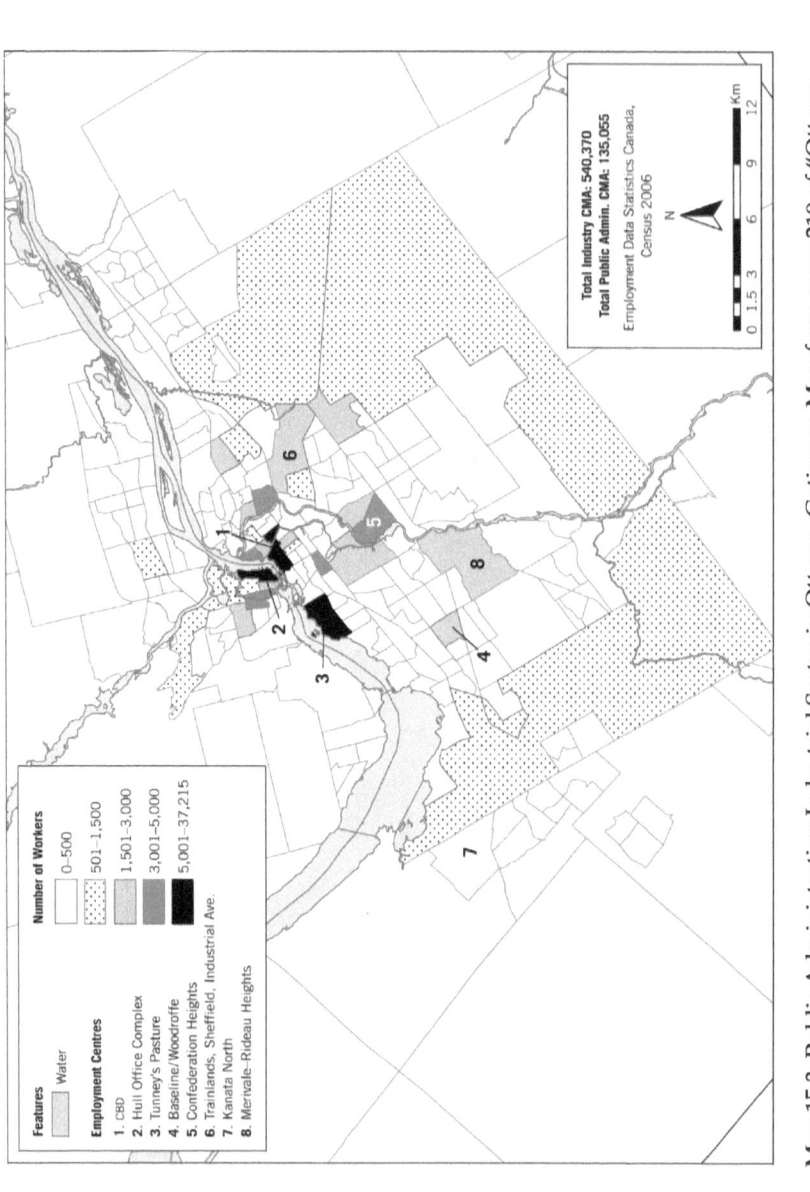

Map 15.3 Public Administration Industrial Sector in Ottawa–Gatineau. Map from page 219 of "Ottawa–Gatineau: Capital Formation," Caroline Andrew, Brian Ray, & Guy Chiasson, in *Canadian Urban Regions*, Larry Bourne, Tom Hutton, Richard Shearmur, and Jim Simmons (Eds.). Reprinted with permission from Oxford University Press.

of the Gréber Plan, located buildings that were at the time federal government buildings (Revenue Canada, the Canadian Broadcasting Corporation, Canada Post, and Public Works) in a suburban location but without much of the environmentally attractive siting that was in the Gréber vision.

This vision was much more front and centre in the planning and building of Tunney's Pasture, inspired by the Gréber Plan, but even more as interpreted by then Prime Minister William Lyon Mackenzie King. King wanted to build a more impressive capital, build pride among Canadians, and build support for the Liberal Party. The increased population of Ottawa had created pressure for space, resulting in a number of municipal amalgamations, including one that encompassed the space of Tunney's Pasture. In 1947 land was expropriated for the building of Tunney's Pasture. Gréber had suggested two buildings that should be located at Tunney's Pasture, two buildings that could be seen as symbolic of the modern state: the National Film Board and Statistics Canada. The government buildings first built in Tunney's Pasture were low rise with lots of parking space and two baseball diamonds, but by the 1960s and 1970s high-rise buildings were added to the site.

The area of the Lebreton Flats, located in the Somerset ward, was also initially expropriated for the construction of federal government buildings, but in fact this land stayed unused for almost 40 years, until the placing of the War Museum.

Gatineau: Suburbanization as a Result of Government Recentralization

Our next case study of the suburbanization of Ottawa is the most directly political and takes us across the river to Quebec. It starts as a recentralization of federal offices in the central core and is directly related to the 1968 federal election, which saw Pierre Elliott Trudeau become prime minister of Canada:

> Trudeau's policy in the capital region was a mixture of territorialized social justice aims and tough intergovernmental politics in that he came to power intending to ensure that the Quebec side of the river would share in the benefits of the federal government presence and, at the same time, that the independence of Quebec would become as difficult as possible. (Andrew, Ray, Chiasson, 2011, p. 228)

Trudeau then had the provincial premiers and prime ministers declare in 1968 that the NCR was composed of Ottawa and Gatineau and proceeded to expropriate land for federal government office buildings. The federal government developed two complexes, Place du Portage and Les Terrasses de la Chaudière, in the central core of what would shortly become the newly amalgamated City of Gatineau (Andrew, Bordeleau, & Guimont, 1981). The Quebec government followed suit with its regional Quebec government office complex. The federal government also suburbanized its office buildings on the Quebec side of the river with the building of the National Archives in suburban Gatineau and taking over the Asticou Centre as a federal government training centre, particularly for French-language training (Canada's Historic Places, 2006).

Continued Suburbanization of Employment to the Western Suburbs

Another example of the creation of suburban employment nodes in Ottawa is Barrhaven, developed in the 1960s on an area immediately outside the Greenbelt. Barrhaven was built on land that had been inhabited by First Nations peoples for a long time and then in the 1960s was developed by the private sector. Residential housing began in the 1960s, and by the 1990s land in Barrhaven was being bought up by companies such as Nortel and JDS Uniphase in the high-tech boom years, when the companies needed space, water, and sewers. At the peak of the boom in 2000, there were 85,000 jobs in the western suburban areas, but with the crash of 2001 and the disappearance of Nortel, high-tech employment has moved back to the core, with small IT companies and / or individuals seeking central locations with easy access to other similar operations.

This led the federal government to decide to buy the former Nortel property in the western suburbs. This was first announced in the fall of 2010: the move to cost $755 million, breaking down to $208 million for the purchase of the Nortel campus, $506 million for refitting, and transition costs of $41 million. At that time officials said the move should save more than $750 million over the next 25 years, for the most part due to leasing fewer buildings. The intention was to consolidate the Department of National Defence's locations from approximately 40 to seven (one in the Ottawa core, two in suburban Ottawa), four in Gatineau (two near the core and two suburban), with the first people to move in late 2015. More recently it became clear that the original timetable for the move would not be met because of structural problems discovered in renovating the buildings. It was announced that these problems will cost approximately $30 million to fix, plus another $6 million to extend leases for Department of National Defence offices around Ottawa. Although some employees have recently moved to the new Department of National Defence building, many have not yet done so due to ongoing construction.

We turn now to the efforts that have been made to create more progressive solutions to public transit in Ottawa, to solve the problems of congestion but also to address problems of marginalization of low-income families living further out from the core. These efforts have mostly been led by community groups exerting pressure from below.

Public Transit in Ottawa

Currently, the City of Ottawa has two main public transit delivery services that serve a population of 870,250 (City of Ottawa, Population, 2015a). Public transit in Ottawa is exclusively serviced by OC Transpo via bus rapid transit (individuals with disabilities are serviced under the Para Transpo banner) and light-rail transit (marketed as the O-Train); making it a multimodal transportation system. The O-Train has been completed (the Confederation Line, which mainly services central Ottawa), and it is projected that by 2024–25 additional stations will be added to the line to the east and west corridors of the city, at which time it will be a fully operational system (O-Train, 2015). The City of Ottawa is the administrative decision-making body for all municipal affairs and as such has the authority over transit issues (Transit Commission) (City of Ottawa, 2010). In 2002, approximately 57% of the transit operating costs were received via fares (which are expected to continuously increase), and the remainder was financed via municipal property taxes (OC Transpo, 2005). The city does not receive any stable funding from provincial or federal governments, but the O-Train project, for example, did receive money from federal and provincial coffers. It is important to note here that although OC Transpo does service part of Gatineau (and vice versa), Gatineau has its own public transit system with no real symbiotic partnership in place. This fact has often been criticized by citizens and politicians alike, especially with the O-Train project and the lack of communication and willingness to include Gatineau as part of any of the currently proposed routes.

The bus rapid-transit system currently in place was initially favoured over light rail due to its cost-efficiency (Richmond, 2001). The present-day bus system in Ottawa is recognized as one of the best public transit (bus) systems in Canada (Al-Dubikhi & Mees, 2010; Cervero, 1998; Rathwell & Schijns, 2002). This success is often attributed to the transit system's ability to provide express routes and service throughout the city and parts of Gatineau, as well as its efficiency and relatively low cost (Al-Dubikhi & Mees, 2010; Cervero, 1998).

In the latest available annual report (Kanellakos, 2014) for OC Transpo, ridership is first and foremost linked with public service employment in the NCR. The report states that "reductions in employment in the federal government have a strong influence on transit ridership as many workers employed in federal government jobs, which are primarily located in downtown Ottawa and Gatineau are transit customers" (Kanellakos, 2014).

This fact is reflected in the physical routes offered and the frequency at which they are delivered. Map 15.4 shows the various bus routes and transit ways throughout Ottawa. While the number of bus routes is impressive, the map shows 1) that bus routes are designed to go to and from the downtown core (and through it); and 2) that the number of routes decreases further away from the downtown core. This idea is echoed in the O-Train plans, in which the Confederation Line would replace the transit line going through the downtown core; running from the east end (Blair Road) to the west end (Tunney's Pasture).

Although the heavy presence of buses in the downtown core remains, presently OC Transpo attempts to meet a variety of needs throughout the city by assessing why the service is required, who the customers are, as well as the location of the services. The system is demand-based and relies heavily on growth and census data.

As a result, the current schematics of OC Transpo have a reliable base network system that spans across the city and would ideally allow for customers to easily walk to a bus stop (within 10 minutes for urban residents) in order to take a direct bus or, at the very least, a bus that will connect them to their final destination, seven days a week (OC Transpo, 2005). In fact, the planning of OC Transpo depends on this notion of connectivity, which is what allows it to be effective as a transit system.

Nonetheless, this notion of connectivity is more easily attainable during peak hours, and routes still serve the downtown core extremely well with several lines that go to and through the downtown core during peak hours (working hours for public servants), to Gatineau (in order to service Place du Portage and Les Terrasses de la Chaudière), and to Tunney's Pasture.

At this time, the presence of buses in the downtown core is overwhelming and has led to problems of traffic congestion on extremely narrow streets. As a result, the goals of the system will be to have buses take customers to the LRT lines (which will service the peak hours and downtown core more easily) and to service off-peak bus lines that will

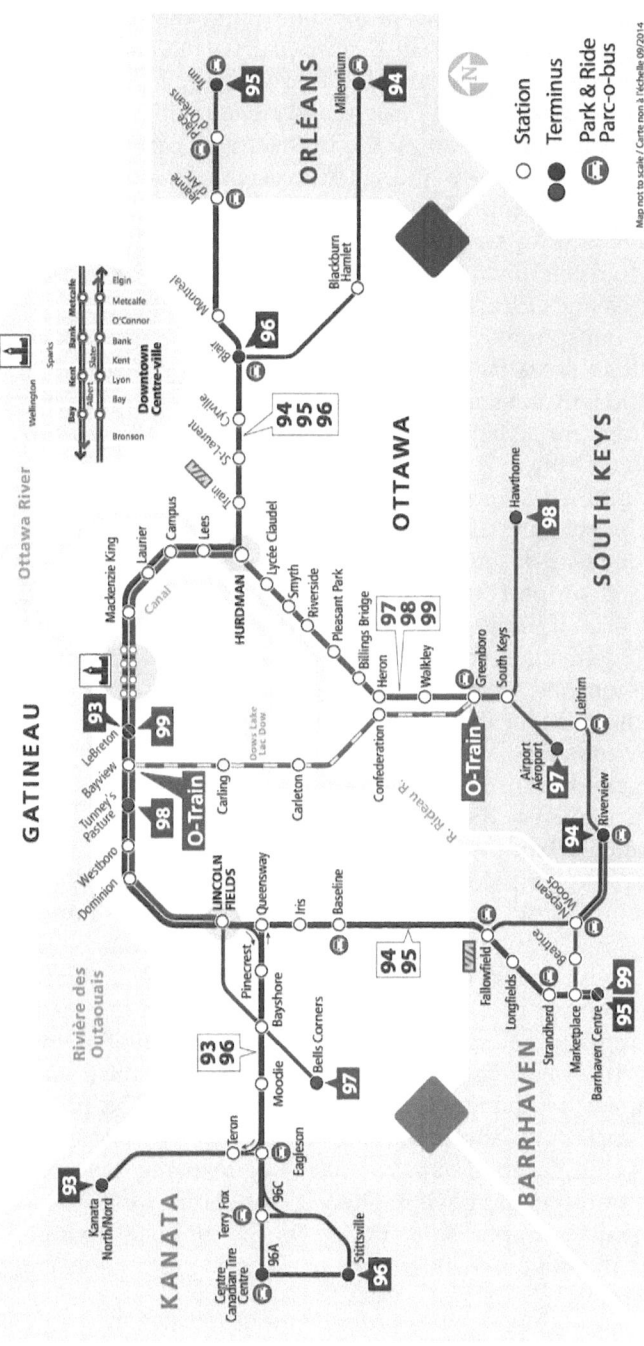

Map 15.4 OC Transpo bus stations in Ottawa, 2015. Reprinted with permission from OC Transpo Transportation Services Department of the City of Ottawa.

remain mostly unchanged except for the slight extension towards the O-Train Confederation Line (i.e., the urban areas of Ottawa).

The fact remains, however, that local services, especially suburban or rural areas in Ottawa and off-peak hours, are left with more infrequent service and longer trips, and all the innovative changes to the transit system are, once again, responding to the needs of the downtown core. According to the *Transit Service and Fare Policy Manual* for OC Transpo, rural areas of Ottawa are served during peak hours, which will allow customers to reach downtown directly, and also by park-and-ride lots located in Kanata, Orleans, Nepean, and South Ottawa.

The 2013 Transportation Master Plan appeared to acknowledge problems suburban areas face as it references the importance of "understanding different transportation needs of urban and rural areas" and "maximizing transportation options for residents of all ages and abilities." (City of Ottawa, 2013). However, no changes have been implemented by mid-2016. According to discussions with several employees from OC Transpo and City of Ottawa officials, connectivity in suburban areas (Greenbelt and beyond) suffers because there are limited resources and support for areas, with not as much growth and variable hours, even if public transit is vital in these areas. Though there is not enough demand in these areas compared to the demand from the downtown core and the most urban parts of Ottawa, OC Transpo does recognize the need for these services and the needs of several vulnerable groups and areas.

OC Transpo officials are aware of the pressure from users and recognize that the users they hear from tend to be "privileged" users (civil servants, urban dwellers), and that they rarely hear from more marginalized groups. However, planners are currently looking at the Vanier area of Ottawa – where there are twice as many francophone residents, plus a hospital that serves the majority of Franco-Ontarians in Ottawa – with a view to whether these needs justify better services. One great example of a needs-based approach came as a result of a council resolution requiring OC Transpo, in its reevaluation of cuts to services in 2011, to use the Equity and Inclusion Lens (a tool that examines the impact that changes in services have on vulnerable populations as well as such populations' specific needs). We will discuss the role and impact community groups can have in enhancing the progressive role public transit could have in Ottawa, but before we do so, we will examine the demographic shifts in Ottawa and the concerns and issues that plague many vulnerable groups in accessing affordable and flexible public transit.

Illustrating the Demographic Shifts: Voices from the Population

According to a study of 15 census metropolitan areas (CMAs) by Dejan Pavlic, the downtown core of Ottawa has had an increase in prosperity (median household income, average value of dwelling, and average gross rent), exclusivity, and rent between 1986 and 2006, while the inner suburbs declined as a result of a decentralizing and gentrification trend found in many American cities and increasingly in Canadian CMAs (Pavlic, 2011). Finally, he found that in the case of Ottawa, prosperity did not increase in the outer suburbs.

Indeed, map 15.5 shows lower-income homes have become more sprawled out in the Ottawa–Gatineau region and are not as centrally located as they once were.

The demographic makeup of where people live and work in Ottawa has changed significantly, with more and more people living past the Greenbelt and downtown core of Ottawa and with much of Ottawa's population now living in the suburbs. Barrhaven, for example, was, as previously noted, an underdeveloped area that now has more than 50,000 citizens living there; Kanata is another example, with a population of more than 60,000 (Ottawa Neighbourhood Study, 2011; Statistics Canada, 2012). Not only are citizens in general starting to reside further away from the centre of Ottawa and taking up residence in what were considered the outskirts of Ottawa, vulnerable populations are mirroring the same phenomenon. For example, citizens between the ages of 60 and 69 years are residing in the East Industrial region (18.8%), Nepean (15.6%), Manotick (14.4%), and so on, all areas that are either in the northwest region of Ottawa or in the east of the city and that would not be directly serviced by the new O-Train Confederation line (Ottawa Neighbourhood Study, 2011; Statistics Canada, 2012). Recent immigrants (2006–11) are scattered throughout the city, with the highest percentage (16.9%) residing in Ottawa's west end, mostly in the Bayshore region, an inner suburban area (Ottawa Neighbourhood Study, 2011; Statistics Canada, 2012). Immigrants that arrived in Ottawa prior to 2006 live throughout the city, but with the highest concentrations in the suburban areas of Kanata, Barrhaven, Greenbelt, Hunt Club, and Heron Gate. All these results indicate that potentially vulnerable populations are increasingly suburbanized, with several highly concentrated nodes in the far west end and the east end of Ottawa. Map 15.6 shows that disadvantaged groups are spread throughout the city.

Map 15.5 Average individual income in Ottawa–Gatineau, 2012. Reprinted with permission from the Neighbourhood Change Study at the University of Toronto.

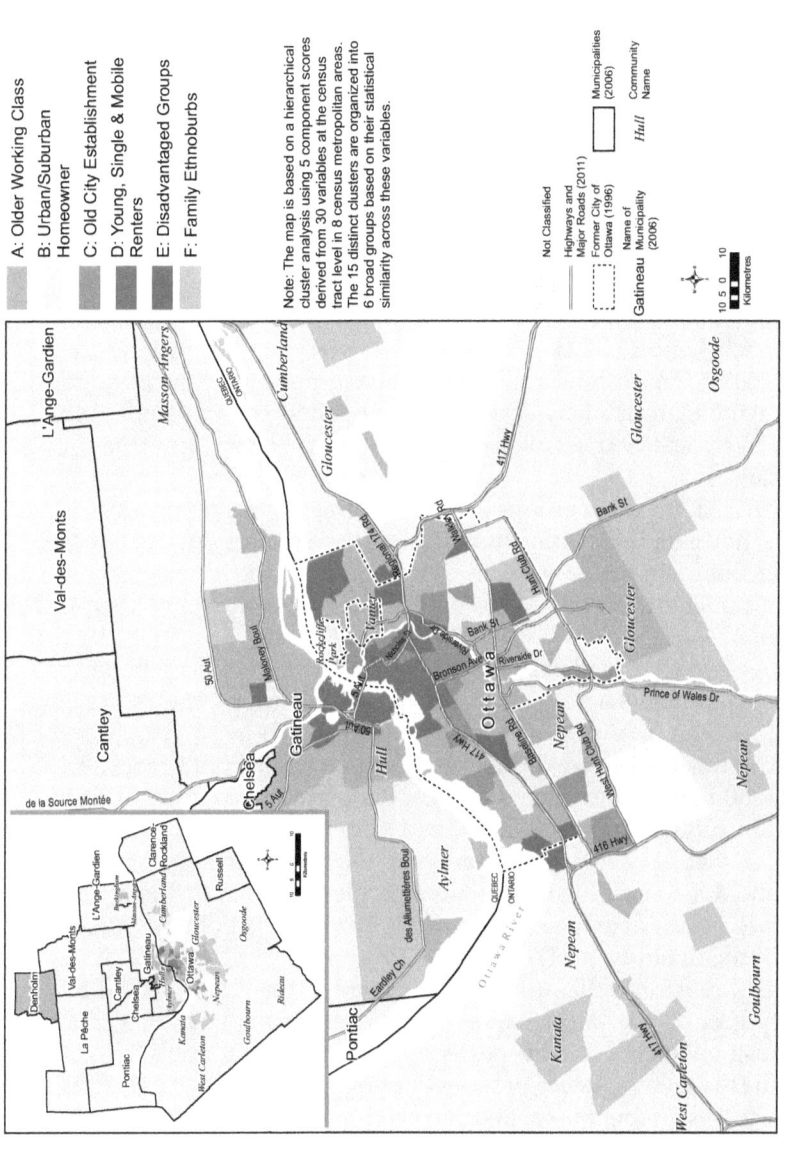

Map 15.6 Typology of neighbourhoods in Ottawa–Gatineau, 2006. Reprinted with permission from the Neighbourhood Change Study at the University of Toronto.

In Ottawa's present geographic constellation, a number of vulnerable groups face realities that do not match the model of a public servant who works a 9-to-5 job. Immigrants, for example, accounted for only 7% of the public administration sector compared to the 18% of Ottawa's regular workforce (Social Planning Council for Ottawa, 2004, p. 14). Instead, many immigrants work in various industries and in manufacturing (Social Planning Council for Ottawa, 2004, p. 14). Consequently, their employment is often not in the downtown core and may not include traditional work hours.

We know that immigrants, women, and the elderly have trouble finding employment, accessing training, finding housing, and accessing public services, including transportation, and that both women and immigrants are more likely to commute using public transit (City for All Women Initiative, 2010a, 2010c; Fraser, 2012; Garrett & Taylor, 1999; Heisz & Schellenberg, 2004; Neufeld, Harrison, Stewart, Hughes, & Spitzer, 2002). The experiences of vulnerable groups are compounded in more remote areas where services are often limited but essential in order to overcome barriers (Fraser, 2012). This is particularly true for public transit.

Public transit for these groups located in rural and suburban areas is available; however the frequency, flexibility, and connectivity are not as advantageous as in the downtown core. Moreover, the increasing cost of taking public transit has made public transit less attractive than it once was.

To gain a better understanding of how public transit directly impacts vulnerable populations, we conducted two focus groups; the first was made up of immigrant Middle Eastern women (30 to 50 years old); the second was composed of immigrants from China, both male and female, who had immigrated as seniors. Both groups represented residents from suburban Ottawa areas, and both regions of Ottawa were chosen for their growing immigrant populations, its mixed makeup of household incomes, the recent cuts in routes (2011) in the area, and an increase in disadvantaged individuals 50 years and older (Ottawa Neighbourhood Study, 2011).

Similarly, regarding the problems identified in the literature and by advocacy groups, the participants' overall complaints were that they were transit reliant but that the current system was not suitable to their needs, due to the infrequency of buses, lack of availability in rural areas, the high costs of using buses, and the need to take two to three buses to arrive to their final destinations. Their needs ranged from picking up their children, going to school, going to work (outside the downtown core and during off-peak hours), going shopping, and attending activities within the city to become better acquainted with their new city and surrounding communities.

We know that to properly ensure an efficient and equitable transit system, all forms of barriers should be removed. Karel Martens argues for the importance for individuals to be able to move in order to more easily access spaces that could lead them to better jobs, services and social activities, and therefore being seen as more productive members of society (Martens, 2016).

Epilogue: The Politics of Planning

In Ottawa, the federal government has been the chief actor in the organization of public transportation. Even if the municipal government is directly involved in servicing the local public transportation, the dominant logic has been to move federal government employees to and from their places of work. There has not been any explicit consideration of the income levels of the different residential neighbourhoods of federal employees. As we have seen, Vanier would not seem to be better served by public transportation despite the fact that, up to very recent indications of Vanier becoming more gentrified, it has definitively been a residential area for less wealthy federal government employees.

The federal government has suburbanized Ottawa by placing buildings in suburban locations and has maintained a strong urban core both by having visions of placing or keeping all the really important things in the core, and by just wanting to find inexpensive office space, wherever it happened to be. We describe this as planning plus politics or, perhaps more accurately, as the inconsistencies of the politics of planning. We suggest that the framework needs to be changed, and public policy for public transit needs to start from a demographic perspective, by understanding the main demographic trends and thinking how public transit could then support and encourage positive trends and mitigate negative trends. To do this, there must be collaborative action between the City of Ottawa and community-based equity-seeking groups advocating for the needs of specific demographic groups. The City of Ottawa defines its major demographic trends as increasing diversity and aging (City of Ottawa, 2006). Using these two demographic trends, relevant community-based and equity-seeking groups are the Ottawa Local Immigration Partnership (OLIP) and the City for All Women Initiative (CAWI) on issues of immigrant integration and belonging and on immigrant women's integration and belonging, and Age Friendly Ottawa on issues of successful aging.

CAWI has worked with the City of Ottawa to co-construct the Equity and Inclusion Lens, which profiles eleven groups seen to be in danger of marginalization and highlights ways to be more inclusive throughout the process of decision making and implementation in Ottawa (City for All Women Initiative, 2010d). Good public transit has been one of the constant themes in CAWI's public campaigns and programs. One of their current programs is called "Making votes count where we live" done jointly with the Coalition of Community Resource and Health Centres and good and inexpensive public transit was one of the issues where objectives were articulated, education was done, and information provided. CAWI has had over the years, some significant successes in terms of policy in relation to public transit: City Council voted to use the Equity and Inclusion Lens to conduct a study on the impact of the September 2011 reductions in bus services; City Council mandated OC Transpo to work with communities to ensure that the new Presto System for bus and O-Train would be accessible for people living on low incomes (2012). Finally, CAWI along with dozens of other advocacy/community groups and individuals successfully petitioned for a low-income pass, which was finally accepted by the City Council and should be implemented by 2017.

OLIP is another community-based group that is sensitive to social justice claims, particularly in relation to recent immigrants. It brings together a very wide coalition of school boards and immigrant serving agencies, and one of its great successes had been to encourage the City of Ottawa to be more active in immigrant integration, in terms of more coordination of city services and more coordinated planning of new initiatives, better communication and interaction with the community and more support for the economic development of the immigrant population, including more access to City employment and more support of immigrant entrepreneurs.

And finally, Age Friendly Ottawa, a local version of the Age Friendly Cities project of the World Health Organization. The eight major criteria for an Age Friendly city are open spaces and buildings; transportation; housing; social participation; respect and social inclusion; civic participation and employment; communication and information; and community and health services (World Health Organization, 2007).

The first three items under transportation are 1) that public transportation costs be consistent, clearly displayed, and affordable; 2) that public transportation is reliable and frequent, including at night and on weekends and holidays; and 3) that all city areas and services are accessible by public transport, with good connections and well-marked routes and vehicles (World Health Organization, 2007).

The active participation of seniors is a crucial part of Age Friendly Ottawa, as is working with the City of Ottawa. Transportation is a central element of this, and there is a very active Ottawa Seniors Transportation Committee with a number of City representatives (OC Transpo, Ottawa Public Health, Para Transpo Service Delivery, Safety Roads Program, Strategic Community Initiatives) and a somewhat larger list of community members, mostly service providers. The committee has a long list of accomplishments, including successfully advocating for Ride-Free Wednesday for seniors over age 65 and a reduced seniors' bus fare.

This policy discussion is really about how to use and combine community-based pressure with that of political leadership from both elected and staff members to bring about social change. The three groups mentioned above already work with the City and have developed tools and practices that have improved the daily lives of populations potentially in danger of marginalization. By focusing on the demographic needs of citizens or by thinking of public policy as a bottom-up process, we can start to move away from considering services, such as public transit, as a mere commodity and see the power it can have as a policy tool in addressing social justice and effectively integrating urban and suburban spaces. Public transit can help increase mobility, remove social barriers, and increase access and social cohesion.

REFERENCES

Al-Dubikhi, S., & Mees, P. (2010). Bus rapid transit in Ottawa, 1978 to 2008. *Town Planning Review, 81*(4), 407–24.

Andrew, C., Bordeleau, S., & Guimont, A. (1981). *L'Urbanisation: une affaire.* Ottawa: Éditions de l'Université d'Ottawa.

Andrew, C., Ray, B., & Chiasson, G. (2011). Ottawa-Gatineau: Capital formation. In L. Bourne, T. Hutton, & R. Shearmur (Eds.), *Canadian urban regions: Trajectories of growth and change* (pp. 202–35). Oxford University Press.

Canada's Historic Places. (2006). Federal heritage building review office. Retrieved from http://www.historicplaces.ca/media/11935/2004-061(e)centreasticou-blocks100-1100.pdf

Cervero, R. (1998). *The transit metropolis: A global inquiry*, Washington, DC: Island Press.

Chiasson, G., and Andrew, C. (2009). Modern tourist development and the complexities of cross-border identities within a planned capital region. In R. Maitland & B.W. Ritchie (Eds.), *City Tourism: National capital perspectives* (pp. 253–63).

City for All Women Initiative. (2010a). Diversity snapshot – Immigrants – Equity and inclusion lens. Retrieved from http://www.cawi-ivtf.org

City for All Women Initiative. (2010b). Diversity snapshot – People with disabilities – Equity and inclusion lens. Retrieved from http://www.cawi-ivtf.org

City for All Women Initiative. (2010c). Diversity snapshot – Women – Equity and inclusion lens. Retrieved from http://www.cawi-ivtf.org

City for All Women Initiative. (2010d). Equity and inclusion lens – A user's guide. Retrieved from http://www.cawi-ivtf.org

City of Ottawa. (2006). A portrait of Ottawa older adults: Demographic and socio-economic characteristics.

City of Ottawa. (2010). Your city government. Retrieved from http://ottawa.ca/en/city-hall/your-city-government

City of Ottawa. (2013). Transportation master plan, 2013. Retrieved from http://ottawa.ca/en/city-hall/planning-and-development/official-and-master-plans/transportation-master-plan

City of Ottawa. (2015a). Population. Available at: http://ottawa.ca/en/long-range-financial-plans/economy-and-demographics/population

City of Ottawa. (2015b). Transit services – 2015. Retrieved from http://ottawa.ca/en/transit-services-2015-budget-briefing-note

Fraser, J. (2012). Rural transportation initiatives: Preventing crime and promoting safety. *Crime Prevention Ottawa*, (August). Retrieved from https://www.crimepreventionottawa.ca/wp-content/uploads/2019/02/Rural-Transportation-Initiatives-Preventing-Crime-and-Promoting-Safety.pdf

Garrett, M., & Taylor, M. (1999). Reconsidering social equity in public transit. *Berkeley Planning Journal,13*(1), 6–27.

Heisz, A., & Schellenberg, G. (2004). Public transit use among immigrants. *Canadian Journal of Urban Research,13*(1), 170–91.

Kanellakos, S. (2014, June 18). OC Transpo annual performance report 2013. Retrieved from https://www.octranspo.com/images/files/reports/2013_annual_report.PDF

Martens, K. (2019). Why accessibility measurement is not merely an option, but an absolute necessity. In C. Silvia, N. Pinto, & L. Bertolini (Eds.), *Designing accessibility instruments: Lessons on their usability for Integrated Land Use and Transport Planning Practices*. Routledge.

Neufeld, A., Harrison, M.J., Stewart, M.J., Hughes, K.D., &Spitzer, D. (2002). Immigrant women: Making connections to community resources for support in family caregiving. *Qualitative Health Research, 12*(6), 751–68.

OC Transpo – Report. (2005). Transit service and fare policy manual.

Ottawa Neighbourhood Study. 2011. Neighbourhood Maps – 2011. Retrieved from neighbourhoodstudy.ca/neighbourhood-maps-2011

O-Train. (2015). Confederation line. Retrieved from http://www.ligneconfed erationline.ca/the-plan/when/

Pavlic, D. (2011). *Fading inner suburbs? A historio-spatial analysis of prosperity indicators in the urban zones of the 15 largest census metropolitan areas.* University of Waterloo. Waterloo, Ontario, Canada.

Rathwell, S., & Schijns, S. (2002). Ottawa and Brisbane: Comparing a mature busway system with its state-of-the-art progeny. *Journal of Public Transportation, 5*(2), 163–82.

Richmond, J. (2001). A whole-system approach to evaluating urban transit investments. *Transport Reviews Journal, 21*(2), 141–79.

Social Planning Council of Ottawa. (2004, December). Immigrants in Ottawa: Socio-cultural composition and socio-economic conditions. Retrieved from www.spcottawa.on.ca/sites/all/files/pdf/2004/Publications/Immigrants -in-Ottawa.pdf

Statistics Canada. (2012). Ottawa, Ontario (Code 3506008) and Ontario (Code 35) (table). Census Profile. 2011 Census. Statistics Canada Catalogue no. 98–316-XWE. Ottawa. Released 24 October 2012. Retrieved from http: //www12.statcan.gc.ca/census-recensement/2011/dp-pd/prof/index .cfm?Lang=E

World Health Organization. (2007). Global age-friendly cities: A guide. Retrieved from www.who.int/ageing/publications/Global_age_friendly _cities_Guide_English.pdf

16 Contested Spaces: Suburban Development in Halifax and Other Midsized Canadian Cities

JILL L. GRANT

Introduction

Planners love to hate the suburbs (Nicolaides, 2006). Stereotyped and criticized as sterile wastelands of big-box stores, oversized houses, manicured lawns, and demographic homogeneity, the suburbs have transitioned in recent years from being seen as the middle class dream to perceived nightmare (Duany, Zyberk, & Speck, 2000; Sewell, 2009). In the 1970s, planners in Toronto and Vancouver began promoting an alternative vision for new development areas: urban intensification. Through the years, that approach increasingly came to dominate Canadian planning policy, supported by the ascendance of New Urbanism, sustainable development, and smart-growth theories (Grant, 2006). Consequently, building policies and practices in many urban-fringe areas began to change. Rather than creating suburban wastelands, planners today aspire to fashioning beloved new urban neighbourhoods (Grant, 2003).
This chapter considers some of the ways in which new planning ideas about suburban development have (and have not) influenced practice in small and midsized cities. While in the early postwar period planners promoted low-density suburban development with segregated land uses and separated transportation networks for cars and people, contemporary planning principles advocate complete communities: that is, compact form, mixed use, mixed housing types, and walkability (Grant, 2002, 2006, 2009; Grant & Scott, 2012). Analysis of Canadian development practices in smaller and midsized cities show that new suburbs in some regions take a relatively urban form: mixed, compact, connected, and walkable. In other areas, however, conventional patterns – segregated, low density, disconnected, and car oriented – persist. What factors help to account for particular suburban development

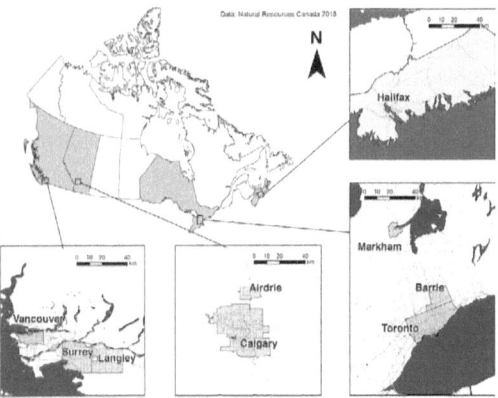

Map 16.1 Overview and reference map of Halifax and other midsized Canadian cities. Map commissioned by author.

trends in various smaller cities? The analysis will pay special attention to Halifax (Nova Scotia), and compares trends found there with patterns seen in smaller cities around three of Canada's largest and most dynamic large cities: Vancouver, Calgary, and Toronto. Developments in Barrie and Markham (Ontario), Airdrie (Alberta), and Surrey and Langley Township (British Columbia) prove especially useful contrasts to the Halifax experience (map 16.1). The Canadian planning policy context for suburban development is remarkably uniform. All the communities profiled have planning policies that promote mix and higher densities. Implementation varies markedly, though (Filion, 2003, 2009; Grant, 2009). After briefly reviewing the approaches planners have been using to influence suburban development in Canada, this chapter describes contemporary trends in Halifax and the comparator cities. The final section discusses some of the key factors that help account for differences in suburban development trajectories of smaller and midsized cities in Canada.

Planning the Suburbs

For much of the first century of town planning in Canada, planners saw suburbs as quiet, leafy places that offered family-friendly living away from the noise and dangers of the city (Harris, 2004; Sewell, 2009). They permitted wide lots, winding streets, and increasingly large single-detached houses. In the last few decades of the 20th century, however, the philosophy of planning shifted, and planners committed to transforming the suburbs into more urbanized places (Grant, 2006).

The transition arguably began with the failures of urban renewal, and decisions in cities such as Toronto, Vancouver, and Halifax to cancel or abandon downtown freeway projects due to public protests (Grant, 1994; Punter, 2003; Sewell, 1993). The 1970s brought a livable city strategy to Vancouver (Hutton, 2011; Punter, 2003) and infill projects such as the St. Lawrence neighbourhood to Toronto (Young, 1995). The Province of British Columbia established the Agricultural Land Reserve in 1973 to protect resources and prevent urban sprawl (Hanna, 1997): with land supply constrained, development practices in Vancouver began to shift to accommodate higher densities. By the 1980s, ideas associated with neo-traditional town planning, transit-oriented development, healthy communities, and sustainable development began to interest Canadian planners, and offered alternative models for developing new areas as compact, walkable, diverse, and vibrant (Grant, 2006; Isin & Tomalty, 1993). In the 1990s and early 2000s New Urbanism, smart growth, and sustainability increasingly dominated planning wisdom (City of Calgary, 1995; Grant, 2006). For instance, the Province of Ontario funded New Urbanism demonstration projects (Grant, 2006) and passed growth-management legislation reflecting smart-growth principles (Ali, 2008; Macdonald & Keil, 2012). The new formula for suburban development was clear: increase densities, mix uses, mix housing types, and generally urbanize the suburbs.

Innovative suburban development projects such as Cornell, in Markham, Ontario, and East Clayton in Surrey, British Columbia won design awards and have been well documented (e.g., Condon & Johnstone, 2003; Gordon & Tamminga, 2002). Less is known about suburban development trends in other small and midsized cities, and few systematic comparisons are found in the literature. This chapter compares the suburban-development experiences of small and midsized cities in four provinces, based on data collected over a 10-year period. Methods used to support this synthesis included in-person interviews with more than two hundred planners, developers, council members, and residents between 2002 and 2012. Field surveys of suburban developments in each of the communities, analyses of planning reports, and examination of census statistics supported the research. The overview presented here draws on a wide range of resources to illustrate the trends found in small and midsized cities, and to theorize about key factors that may explain differences encountered in the kinds of suburbs currently being built.

Understanding Smaller Canadian Cities

The cities profiled here differ in important ways. Settled by Europeans in 1749, Halifax is the oldest of the cities discussed. Halifax is the regional

Figure 16.1 In Halifax, suburbs designated in the regional plan as sites for urban nodes, building nominally higher density housing forms (such as row houses) may not increase overall densities if setback and road standards do not change. Photo by Leah Perrin.

centre and economic hub of Atlantic Canada (Brender & Lefebvre, 2006). As provincial capital of Nova Scotia, it has a significant white-collar workforce in government, health care, and educational sectors. With its large, ice-free port, it hosts Canada's east-coast military base and shipping facilities. The province created Halifax Regional Municipality in 1996 by amalgamating two cities (Halifax and Dartmouth), a town (Bedford), and Halifax County. Pre-amalgamation development patterns were fragmented, leaving a legacy of suburban and exurban developments throughout a large territory. Relative ease of commuting into the city centre makes suburban, exurban, and rural living attractive and affordable for many residents. Consequently, despite policies in regional plans to encourage urban infill, higher densities, and greater mix in land uses and housing, conventional suburban development patterns have proven resistant to change, and the city's suburbs sprawl many kilometres from the city centre (figure 16.1). Perrin & Grant (2014) note that developers cluster particular housing forms, such as townhouses, in "pods," rather than mixing them within blocks. Brewer & Grant (2015) suggest that even in master-planned communities designated as urban nodes in regional planning policies, low-density and segregated land-use patterns continue to appear in Halifax suburbs.

Figure 16.2 The commercial centre of Cornell in Markham was designed for mixed use, but has faced challenges in retaining viable retail uses. Photo by Katherine Perrott.

Located just north of Toronto, and with transit access to Toronto's city centre, Markham has transformed since the end of the Second World War from a quiet village surrounded by farmland to a dynamic and diverse city seen as a hub of New Urbanism and smart growth in Canada. Branded as Canada's high-tech capital, Markham has several large employers of highly skilled workers and hosts Canadian corporate headquarters for firms such as IBM, Toshiba, and Hyundai. Since the 1990s, planning policies have mandated New Urbanism and smart-growth principles that influence the form of suburban growth (Grant, 2006): consequently, new developments show medium densities (Gordon & Vipond, 2005), a fine-grained mix of housing (Grant & Bohdanow, 2008), and significant efforts to mix land uses (Grant & Perrott, 2011) (figure 16.2). Markham is constructing a town centre that will strengthen its urban functions (Filion, 2009). Planners and municipal councillors interviewed in Markham argued that it is not a suburb of Toronto; nonetheless, its proximity to the metropolis entails a level of integration. Markham has experienced challenges in ensuring the success of retail spaces in new suburbs (Grant & Perrott, 2011) and in providing affordable housing, but has become the most ethnically diverse municipality in Canada (Grant & Perrott, 2009).

Figure 16.3 Large homes on large lots on wide streets remain a common form in Barrie suburbs. Photo by Gillad Rosen.

Barrie, Ontario, 90 kilometres north of Toronto, transformed from a relatively quiet, lakefront community in the 1970s to one of the fastest growing cities in the country in the early 2000s. Its population doubled between 1991 and 2011 (Rosen & Brewer, 2013). In 2010 Barrie annexed land to accommodate further growth. Major employers in Barrie include several banking and data centres. About a third of the workforce commutes elsewhere, often to jobs in the northern fringe of Toronto. GO-Train service has facilitated travel to Toronto. Barrie's (2011) plan called for "complete communities" with a mix of land uses and facilities (Barrie, 2011). With housing prices well below Toronto values (CMHC, 2014), however, Barrie has developed a reputation as a suburban destination for young families and for retirees interested in lakefront living (Watson and Associates, 2010). Infill areas near the city centre feature high-rise apartments and condominiums, and townhouses have been built in some suburban areas. At the same time, however, areas of relatively conventional subdivisions of large homes on large lots remain common in Barrie, especially in southern suburbs (figure 16.3).

Some 25 kilometres north of Calgary (Alberta), Airdrie began as a train stop in the late 1800s (Central Alberta Museums, 2015). In the 1970s, construction of a six-lane highway to Calgary – now the heart of the petroleum industry in Canada – drew the small community into the suburban orbit of an economic powerhouse (albeit challenged by the 2014 downturn in oil prices). Recent decades witnessed significant

Figure 16.4 Conventional garage-front houses line the streets of many suburbs in Airdrie. Photo by Troy Gonzalez.

population growth and land annexation for Airdrie (Gonzalez, 2010; Gonzalez & Grant, 2013). Airdrie's planning policy advocates a "balanced community" (Airdrie, 2009, section 2.1); however, Airdrie's small-town character appeals primarily to younger families raising children. Development patterns in Airdrie remain quite conventional, with many detached houses on large lots (figure 16.4). An Airdrie developer interviewed in 2010 explained, "We deliver what people want, not what the theoretical planners downtown want. We deliver what the customers want ... So that is our definition of sustainable. Not some theory about units per acre or densities or those types of things." The suburbs of Airdrie reflect continuing demand for the "Canadian dream" of a house with a yard (Grant & Scott, 2012).

Langley Township (British Columbia), lies 55 kilometres east of Vancouver. The municipal district was primarily a farming community until relatively recently. Hudson Bay Company traders were the first Europeans to settle the area in the 1820s. As the Trans-Canada Highway improved traffic connections with Vancouver in the 1960s, however, suburban communities began to develop through Langley Township. Several gated communities (figure 16.5) appeared during the 1980s and 1990s, often accommodating retirees and creating a fragmented suburban landscape (Grant & Curran, 2007; Grant, Greene, & Maxwell, 2004). In a region of rapid growth, but constrained by provincially protected

Figure 16.5 Some higher density suburban enclaves in Langley Township are gated to limit access. Photo by Dan Scott.

Agricultural Land Reserve lands (Hanna, 1997), Langley Township has adopted a strategy of intensification in designated growth areas around former villages and hamlets (Langley, 2010). Low density, large acreage suburban lots subdivided in the 1960s have been consolidated as policy promotes "complete communities," with a mix of housing types and land uses, and higher densities (Langley, 2010, p. 4). High housing costs contribute to the perception that Langley Township is a relatively exclusive community, even though its suburban development trends have moved away from conventional patterns towards more multifamily units.

Surrey (British Columbia) was incorporated in the 1870s, and during the 1950s began to grow as a bedroom community of Vancouver. In recent years Surrey grew rapidly as light-rail service linked it to the city, some 35 kilometres away. Surrey has long had a working class reputation (Dowling, 1996) and has tried to address concerns over high crime rates (Mason, 2014). It currently has many jobs in clean energy and other forms of manufacturing. Over recent decades it has developed a large suburban immigrant community, with many residents coming from India (Surrey, 2015). Like several of the other cities profiled here, Surrey's plan promotes making it a "complete community" (Surrey, 2010, p. 47). Progressive examples of New Urbanism–style development include East Clayton, a community of mixed housing types, medium densities, and mixed uses (figure 16.6). As in Langley Township, provincial and regional policies protecting agricultural lands, along with strong market demand, are driving intensification in new suburban developments.

Figure 16.6 New neighbourhoods in Surrey typically feature medium-density dwellings. Photo by Dan Scott.

Comparing the Numbers

Halifax is one of the larger midsized cities in the set profiled here. The cities' populations range from fewer than 43,000 in Airdrie to almost half a million people in Surrey. As table 16.1 illustrates, all grew between 2006 and 2011, although at different rates. Airdrie, Alberta (just outside Calgary) grew 10 times more quickly than did Halifax, Nova Scotia (on Canada's Atlantic coast). Halifax had the slowest growth in the set. Municipalities[1] – such as Langley Township and Surrey (British Columbia), Airdrie (Alberta), and Markham (Ontario) – in the commuter sheds of Canada's large cities have been growing at a rapid pace. Barrie (Ontario), which was the fastest growing city in Canada between 2001 and 2006 when it grew 19.2% (Statistics Canada, 2006; Watson and Associates, 2010), slowed considerably in recent years.

The median age of residents in all the communities was below the Canadian average: Surrey, Airdrie, and Barrie have relatively young populations by comparison with Halifax, Langley, Vancouver, and Markham. The suburbs of the communities have younger average ages and more households with children than central neighbourhoods. Table 16.2 indicates that Halifax, Airdrie, and Barrie are similar in that they had relatively low numbers of immigrants, visible minorities, and non-English speakers, especially compared with the large cities – Toronto, Vancouver, and Calgary – near some of them. Toronto, Vancouver, and

Contested Spaces in Halifax and Other Midsized Canadian Cities 337

Table 16.1 Population of areas discussed (2006 and 2011 census)

Cities studied and nearby Census Metro Area (CMA)	Population, 2006	Population, 2011	Change, 2006–2011 (%)
Langley Township, BC	93,726	104,177	11.2
Surrey, BC	394,976	468,251	18.6
Vancouver CMA, BC	2,116,581	2,313,328	9.3
Airdrie, AB	28,927	42,564	47.1
Calgary CMA, AB	1,079,310	1,214,839	12.6
Barrie, ON	128,430	135,711	5.7
Markham, ON	261,573	301,709	15.3
Toronto CMA, ON	5,113,149	5,583,064	9.2
Halifax CMA, NS	372,679	390,328	4.7
Canada	31,612,897	33,476,688	5.9

Source: Statistics Canada (2007, 2012, 2015a, 2015b).

Table 16.2 Population characteristics of areas discussed (2006 and 2011 census)

Cities studied and nearby Census Metro Area (CMA)	Residents born abroad, 2006 (%)	Visible minorities, 2006 (%)	English as language at home, 2011 (%)	Median age, 2011 (years)
Langley Township, BC	17.1	10.0	91.1	40.3
Surrey, BC	38.3	46.1	62.6	37.5
Vancouver CMA, BC	39.6	41.7	68.4	40.2
Airdrie, AB	6.8	3.1	95.6	32.4
Calgary CMA, AB	23.6	22.2	80.3	36.4
Barrie, ON	12.8	6.7	94.0	37.2
Markham, ON	56.5	65.4	52.2	39.6
Toronto CMA, ON	45.7	42.9	67.0	38.6
Halifax CMA, NS	7.4	7.5	94.6	39.9
Canada	19.8	16.2	67.1	40.6

Source: Statistics Canada (2007, 2012, 2015a, 2015b).

Calgary showed high proportions of immigrants, visible minorities, and people speaking languages other than English at home. Surrey (just outside Vancouver) and Markham (near Toronto) appeared to be "ethnoburbs" (Li, 1998), drawing significant numbers of immigrants, speakers of languages other than English, and persons identifying as visible minorities. Grant & Perrott (2009) showed that in Markham the relative proportion of residents of particular ethnicities and of those self-identifying as visible minorities varied markedly from suburb to suburb. By contrast, Airdrie (neighbouring Calgary) and Langley

Table 16.3 Characteristics of households (HH) in areas studied (2006 census)

Cities studied and nearby Census Metro Area (CMA)	Owner-occupied HH (%)	Median HH income	HH with a couple and children (%)	Average HH size (people)
Langley Township, BC	86	$69,805	37	2.8
Surrey, BC	75	$60,168	35	3.0
Vancouver CMA	65	$55,231	29	2.6
Airdrie, AB	89	$78,097	43	2.9
Calgary CMA	74	$68,579	31	2.6
Barrie, ON	76	$64,832	35	2.7
Markham, ON	89	$79,924	46	3.4
Toronto CMA	68	$64,128	34	2.8
Halifax CMA, NS	64	$54,108	26	2.4
Canada	68	$53,634	29	2.5

Source: Statistics Canada (2007).

(near Vancouver) were much lower than the nearby city on markers of diversity. Not all suburbs mirror the composition of the urban region as a whole, as varying degrees of ethnic and racialized clustering is occurring in the suburban fringe of immigrant-receiving communities: marketing agencies recognize the value in documenting the patterns that are developing (Sorensen, 2015). As table 16.3 shows, the communities also differ considerably in household and housing characteristics.

Table 16.4 indicates that Halifax had the lowest density of the communities compared, but the numbers can be misleading. Halifax Regional Municipality covers 5,500 square kilometres and includes vast areas of sparsely settled rural land and wet areas. Haiven (2012) estimates that the Halifax peninsula in 2011 had a population density of approximately 4,500 persons per square kilometre, which would put it above the densities in the other communities. The nature of Halifax's growth, distributed over a vast and fragmented geography, has contributed to relatively low suburban densities (Brewer & Grant, 2015; Millward, 2002). Langley Township, which began urbanizing in recent years but still has a low population density, includes significant tracts of agricultural land. Densities in the other cities shown in table 16.4 were between about 1,300 and 1,750 persons per square kilometre. Single-detached houses ranged from the minority of the housing stock in Surrey, about half the total housing in Halifax, to more than 70% of the total in Airdrie. Older suburbs have a higher proportion of detached housing. CMHC (2011) reported that for the period 2008 to 2010, single-detached housing accounted for only 28% of total housing starts in

Table 16.4 Density and housing characteristics in areas studied

	Density, 2011 (persons/km²)	Single-detached houses, 2011 (%)	Average price of a new single-detached house, July 2014
Langley Township, BC	338.2	58.9	$687,380
Surrey, BC	1,479.9	42.2	$734,900
Airdrie, AB	1,286.0	71.7	$519,621
Calgary, AB (City)	1,329.0	58.7	$635,891
Barrie, ON	1,753.6	62.2	$461,641
Markham, ON	1,419.3	64.1	$795,405
Halifax, NS (CMA)	71.0	51.0	$414,362

Sources: CMHC (2014) and Statistics Canada (2012). Langley and Surrey data from Vancouver Housing Now report. Barrie data from Second Quarter 2014. Markham data from Toronto report.

Langley, but 75% of starts in Airdrie. In other words, cities showed significant differences in development patterns. The suburbs of midsized cities near Vancouver and adjacent to Toronto – with their rapid rates of growth from immigration – have higher densities, fewer detached units, and higher house prices than the suburbs of more remote cities such as Halifax or Barrie, or the Calgary-area community of Airdrie. Average house prices varied in 2014, from a low of $414,000 in Halifax to $735,000 in Surrey (CMHC, 2014). Inflation in the cost of new home prices had become the rule in Canada by 2014 (CMHC, 2014), with critics warning that homes are significantly overvalued (CBC, 2015) and linked to rising levels of indebtedness (Walks, 2013).

Tables 16.1 through 16.4 reveal significant differences in the communities profiled. The final section seeks to identify factors that help explain why conventional suburban patterns remain more resilient in smaller and midsized cities like Halifax, Airdrie, and Barrie than they do in Markham, Surrey, and Langley.

Explaining Variations in Development Patterns

In some of the suburbs of Canada's largest cities, development trends have clearly changed from those that dominated in the 1970s and 1980s. Multifamily units, medium densities, and mixed uses not only are entrenched in policy, but are often being built. The data, however, show mixed trends and sometimes unexpected results. For instance, although growth-management policies and New Urbanism principles are strong in Markham and weak in Halifax, Halifax shows a much lower proportion of single-detached housing starts than found in Markham. What factors affect density patterns and housing and land-use mix? Why do cities such as Halifax, Barrie, and Airdrie prove resilient to some kinds of change? The degree of variability among the six communities examined suggests that size is not the most significant factor in predicting suburban development outcomes. Other factors to consider include policy and implementation contexts, land costs and housing characteristics, geography, and demography.

Although all the communities profiled promote mix and higher densities through their planning policy, their success in achieving targets varies (Filion, 2003, 2009; Grant, 2009). In regions near Toronto and Vancouver, strong provincial policy and regulations help enforce growth management and give local governments stronger tools to increase densities. Markham, Surrey, and Langley Township reflect this reality. The power of planners within municipal governments varies: especially in communities where growth is slow, traffic engineers may still insist on oversized road dimensions, and developers may wield significant influence with city councils. Despite planners' efforts to promote mixed use and neighbourhood commercial uses, big-box retail is proliferating and undermining the success of efforts to mix uses (Brewer & Grant, 2015; Grant & Perrott, 2011). Even though policy consistently promotes urban patterns, the local policy and political context in some of the cities studied can undermine suburban transformation.

Land supply, cost, and market demand are important factors influencing suburban development trends. Growth management and agricultural protection policies limit available land supply around Vancouver and Toronto, encouraging higher density development and inflating land costs. Other factors are also affecting land supply and cost. For instance, a relatively small number of developers monopolize options on land available for development in many markets. With such control, developers can control the supply of developable land released, keeping land values stable or increasing. The ready availability of low-cost credit has increased demand from purchasers in recent years, putting pressure on limited resources. With little affordable housing being built, demand outstrips supply in many areas. With a more abundant land supply, housing prices have increased more slowly in Halifax. Barrie and Airdrie increased their land supply through annexing neighbouring areas, facilitating continued growth with more modest cost increases than otherwise would have occurred. Higher housing costs are encouraging higher density and multifamily forms. At the same time, though, they contribute to the popularity of private and gated condominium or strata communities (Grant & Curran, 2007).

In some cases, factors related to geography and the history of development in a place affect suburban outcomes. Cities far from areas of rapid growth do not experience the spillover effects that transform those surrounding large cities. Markham, Surrey, and Langley Township find their suburban growth influenced by the high prices of housing in two of Canada's largest cities that have become magnets to international migrants: the smaller cities around Toronto and Vancouver are subject to legislation designed to manage growth in the urban regions. Airdrie and Barrie similarly are influenced by their locations within commuting distance of major cities but are at sufficient distance that they attract particular types of households interested in suburban lifestyles. By contrast, Halifax lies well outside the orbit of any large city, in a largely rural region struggling to find achieve growth, unable to attract large numbers of immigrants, and with industries paying relatively low wages by national standards. Until the mid-2000s, housing prices in Halifax were low by comparison with similar sized cities because the province was a major developer of lots for low- and moderate-income buyers. Once the province ended its development program and sold off land holdings, however, suburban housing prices rose rapidly, driving demand for smaller units in multifamily structures and for exurban housing options. Halifax hosts a large transient community (of military personnel and university students), creating high demand for rental units. Relatively low rates of homeownership in Halifax may also influence the choices that developers make in planning new suburban projects. Rental apartments are a significant part of the mix for new suburban areas, but homeowner antipathy for living near renters leads developers to separate housing types into pods rather than mixing them (Perrin & Grant, 2014).

Each of the communities shows the influence of historical development experience. For instance, parts of Langley Township were developed on two-acre lots in the 1960s: today, efforts to consolidate those lots affect where and how new suburban development can occur (Scott, 2010). The ability to blend the best of town and country in a region with a temperate winter climate contributes to suburban demand for condominiums for retirees in Langley. But what accounts, for example, for differences in the ethnic and income profiles of Surrey and Langley Township, both in Vancouver's suburban periphery? Surrey has larger proportions of immigrants, visible minorities, and persons who speak a language other than English at home than does nearby Langley Township. Toronto-area suburbs also show considerable differences in ethnic and racial composition (Agrawal, 2008; Grant & Perrott, 2009; Kumar & Leung, 2005). The demographic profiles of the cities discussed here reveal self-sorting processes at work in urban choices. Larger cities such as Toronto and Vancouver attract immigrants in part because they have a good supply of employment opportunities, but also because they host communities of people speaking common languages, practising similar religions, and with demands for particular goods and services. Markham and Surrey reflect the extent to which housing choices are linked to spatialized cultural practices: people choose to live in communities where they can find others who share key attributes with them. Markham appeals to affluent, large, immigrant households, often speaking Chinese at home and shopping in local Chinese malls. Halifax attracts White, Canadian-born, interprovincial migrants who often choose to rent housing (Grant & Kronstal, 2013). Airdrie and Barrie draw young, White, Canadian-born families buying single-detached houses: central Barrie, by contrast, appeals to retirees looking to live near the lake. Although planning philosophy advocates diverse and mixed communities – complete communities that accommodate a wide range of types of people – those inhabiting the suburbs of midsized cities show continued preference for homogeneity in the residential choices they make. Sikhs choose to live in neighbourhoods with other Sikhs; young families look for areas with other young families; Chinese speakers buy homes near Chinese butchers. Developers understand the dynamics of the local market and, together with those regulating and consuming suburban landscapes, produce locally desirable development forms. In some cases that means they increase densities and provide a mix of housing types and land uses; in other cases, what results is garage-front houses on lots set back from streets without sidewalks.

Planners seeking to increase mix, density, and diversity in new development areas face significant challenges in dealing with cultural expectations and attitudes. Household size continues to drop in many cities, frustrating efforts to increase density. In general people say they want mix and appreciate diversity, yet interviews with residents indicate that people worry about what and who lives next door (Perrin & Grant, 2014). For instance, a Langley Township resident explained her perception that different behaviours complicated neighbourhood life, and kept her from buying in Surrey:

> The demographic of the people here is also predominantly Caucasian, I should add. I think that's important because when you are in Surrey and you've got a predominantly Indo-Canadian community: they have a tendency to have a lot more family members living in one house. And when that occurs of course that results in more cars and everything else.

Suburbs are contested and fluid spaces. Planners want new development to take an urban form, but they encounter varying levels of resistance to their ambitions. From the vantage point of Canada's largest cities, planners may feel they are making considerable progress on key goals of building complete communities. A broader view of development trends in smaller and midsized cities, however, demonstrates wide variation in the kinds of places being built. When their investments are at stake, developers challenge policies that affect the marketability, value, and liquidity of their "products." If they perceive that local markets will not absorb specific types of development, they will not produce them: in such contexts, municipal councils often accept conventional development forms that perpetuate old stereotypes of the suburbs. Residents in some communities resist increased density and housing mix: evidence suggests sorting processes at work that lead residents to choose communities that meet their expectations and allow them to colocate with others like themselves. The continuing development of conventional suburbs suggests that planning policy can only ensure change in contexts where other factors simultaneously contribute to urban transformation.

NOTE

1 Municipalities – including regions, cities, towns, villages, townships, districts, or counties – are incorporated units of governance.

REFERENCES

Agrawal, S.K. (2008). Faith-based ethnic residential communities and neighbourliness in Canada. *Planning, Practice and Research, 23*(1), 41–56.

Ali, A. (2008). Greenbelts to contain urban growth in Ontario, Canada: Promises and prospects. *Planning Practice and Research, 23*(4), 533–48.

Brender, N., & Lefebvre, M. (2006). *Canada's hub cities: A driving force of the national economy.* Conference Board of Canada. Retrieved from http://www.conferenceboard.ca/e-library/abstract.aspx?did=1730

Brewer, K., & Grant, J.L. (2015). Seeking density and mix in the suburbs: Challenges for mid-sized cities. *Planning Theory and Practice, 16*(2), 151–68.

CBC. (2015, April 20). Canadian house prices 35% overvalued, Economist Magazine says. *CBC News* Online. Retrieved from http://www.cbc.ca/news/business/canadian-house-prices-35-overvalued-economist-magazine-says-1.3040698

Central Alberta Museums. (2015). Airdrie settlement. Retrieved from http://www.unlockthepast.ca/Airdrie

City of Airdrie. (2009). *Municipal development plan*. Airdrie, AB: City of Airdrie.
City of Barrie. (2011). *Official community plan*. Barrie: Planning and Development Department.
City of Calgary. (1995). *Sustainable suburbs study*. Calgary: Planning and Building Department.
City of Surrey. (2010). *Official community plan*. By-law No. 12900, Updated to 10 July 2006. City of Surrey, British Columbia.
City of Surrey. (2015). *Community demographic profiles*. Retrieved from http://www.surrey.ca/business-economic-development/1417.aspx
CMHC. (2011). Statistics and data, housing market information. *Housing Now Reports*. Canada Mortgage and Housing Corporation, Ottawa.
CMHC. (2014, July). Housing market information. *Housing Now Reports*. Canada Mortgage and Housing Corporation, Ottawa.
Condon, P., & Johnstone, S. (2003). Collaborative design yields green suburb. *Alternatives Journal, 29*(3), 17.
Dowling, R. (1996). Symbolic constructions of place in suburban Surrey British Columbia. *Canadian Geographer, 40*(1), 70–5.
Duany, A., Plater-Zyberk, E., & Speck, J. (2000). *Suburban nation: The rise of sprawl and the decline of the American dream*. New York, NY: North Point Press.
Filion, P. (2003). Towards smart growth? The difficult implementation of alternatives to urban dispersion. *Canadian Journal of Urban Research, 12*(1, Supplement), 48–70.
Filion, P. (2009). The mixed success of nodes as a smart growth planning policy. *Environment and Planning B: Planning and Design, 36*, 505–21.
Gonzalez, T. (2010). *Airdrie, AB: An overview of development trends*. Retrieved from http://theoryandpractice.planning.dal.ca/suburbs/suburbs_working.html
Gonzalez, T., & Grant, J.L. (2013). Living with density: From "chicken coops" to livability. *Municipal World, 123*(11), 27–30.
Gordon, D.L.A., & Tamminga, K. (2002). Large-scale traditional neighbourhood development and pre-emptive ecosystem planning: The Markham experience 1989–2001. *Journal of Urban Design, 7*(2), 321–40.
Gordon, D.L.A., & Vipond, S. (2005). Gross density and New Urbanism: Comparing conventional and new urbanist suburbs in Markham, Ontario. *Journal of the American Planning Association, 71*(1), 41–54.
Grant, J. (1994). *The drama of democracy: Contention and dispute in community planning*. Toronto: University of Toronto Press.
Grant, J. (2002). Mixed use in theory and practice: Canadian experience with implementing a planning principle. *Journal of the American Planning Association, 68*(1), 71–84.

Grant, J. (2003). Exploring the influence of New Urbanism in community planning practice. *Journal of Architectural and Planning Research, 20*(3), 234–53.

Grant, J. (2006). *Planning the good community: New Urbanism in theory and practice.* London: Routledge.

Grant, J. (2009). Theory and practice in planning the suburbs: Challenges in implementing New Urbanism and smart growth principles. *Planning Theory and Practice, 10*(1), 11–33.

Grant, J.L., & Kronstal, K. (2013). Old boys down home: Immigration and social integration in Halifax. *International Planning Studies, 18*(2), 204–20.

Grant, J.L., & Perrott, K. (2011). Where is the café? The challenge of making retail uses viable in mixed use suburban developments. *Urban Studies, 48*(1), 177–95.

Grant, J.L., & Scott, D.E. (2012). Complete communities versus the Canadian dream: Representations of suburban aspirations. *Canadian Journal of Urban Research: Canadian Planning and Policy, 21*(1, Supplement), 132–57.

Grant, J., & Bohdanow, S. (2008). New Urbanism communities in Canada: A survey. *Journal of Urbanism, 1*(2), 109–27.

Grant, J., & Curran, A. (2007). Privatised suburbia: The planning implications of private roads. *Environment and Planning B: Planning and Design, 34*(4), 740–54.

Grant, J., & Perrott, K. (2009). Producing diversity in a New Urbanism community: Policy and practice. *Town Planning Review, 80*(3), 267–89.

Grant, J., Greene, K., & Maxwell, K. (2004). The planning and policy implications of gated communities. *Canadian Journal of Urban Research, 13* (1, Supplement), 70–88.

Haiven, L. (2012, September 7). Urban density need not be synonymous with height. *Chronicle Herald.*

Hanna, K.S. (1997). Regulation and land-use conservation: A case study of the British Columbia agricultural land reserve. *Journal of Soil and Water Conservation, 52*(3), 166–70.

Harris, R. (2004). *Creeping conformity: How Canada became suburban, 1900–1960.* Toronto: University of Toronto Press.

Hutton, T. (2011). Thinking metropolis: From the "livable region" to the "sustainable metropolis" in Vancouver. *International Planning Studies, 16*(3), 237–55.

Isin, E., & Tomalty, R. (1993). *Resettling cities: Canadian residential intensification initiatives.* Ottawa: CMHC.

Kumar, S., & Leung, B. (2005). Formation of an ethnic enclave: Process and motivations. *Plan Canada, 45,* 43–5.

Li, W. (1998). Anatomy of a new ethnic settlement: The Chinese ethnoburb in Los Angeles. *Urban Studies, 35,* 479–501.

Macdonald, S., & Keil, R. (2012). The Ontario greenbelt: Shifting the scales of the sustainability fix? *The Professional Geographer, 64*(1), 125–45.

Mason, G. (2014, January 3). Surrey's streets have felt unsafe for far too long. *Globe and Mail* online. Retrieved from http://www.theglobeandmail.com/news/british-columbia/surreys-streets-have-felt-unsafe-for-far-too-long/article16193982/

Millward, H. (2002). Peri-urban residential development in the Halifax region 1960–2000: Magnets, constraints, and planning policies. *The Canadian Geographer, 46*(1), 33–47.

Nicolaides, B. (2006). How hell moved from the city to the suburbs. In K.M. Kruse & T.J. Sugrue (Eds.), *The new suburban history* (pp. 80–98). Chicago, IL: University of Chicago Press.

Perrin, L., & Grant, J.L. (2014). Perspectives on mixing housing types in new suburbs. *Town Planning Review, 85*(3), 363–85.

Punter, J. (2003). *The Vancouver achievement: Urban planning and design.* Vancouver: UBC Press.

Rosen, G., & Brewer, K. (2013). *Barrie at a crossroad: Dilemma of a mid-size city.* Retrieved from http://theoryandpractice.planning.dal.ca/suburbs/suburbs_working.html

Scott, D. (2010). *Township of Langley, BC: An overview of development trends.* Retrieved from http://theoryandpractice.planning.dal.ca/suburbs/suburbs_working.html

Sewell, J. (1993). *The shape of the city: Toronto struggles with modern planning.* Toronto: University of Toronto Press.

Sewell, J. (2009). *The shape of the suburbs: Understanding Toronto's sprawl.* Toronto: University of Toronto Press.

Sorensen, C. (2015, March 26). Mapped: 10 years of unprecedented change in Canada's cities. *Maclean's.* Retrieved from http://www.macleans.ca/news/canada/mapped-10-years-of-unprecedented-change-in-canadas-cities/

Statistics Canada. (2006). *2006 Census: Portrait of the Canadian population in 2006. Subprovincial population dynamics.* Retrieved from http://www12.statcan.gc.ca/census-recensement/2006/as-sa/97-550/p13-eng.cfm

Statistics Canada. (2007). *Community profiles, 2006 census.* Statistics Canada Catalogue no. 92–591-XWE. Ottawa: Statistics Canada. Retrieved from http://www12.statcan.gc.ca/census-recensement/2006/dp-pd/prof/92-591/index.cfm?Lang=E

Statistics Canada. (2012). *Focus on geography series, 2011 Census.* Statistics Canada Catalogue no. 98–310-XWE2011004. Ottawa, Ontario. Retrieved from http://www12.statcan.gc.ca/census-recensement/2011/as-sa/fogs-spg/Index-eng.cfm?LANG=Eng

Statistics Canada. (2015a). *Population and dwelling counts, for Canada, provinces and territories, 2011 and 2006 censuses.* Retrieved from http://www12

.statcan.gc.ca/census-recensement/2011/dp-pd/hlt-fst/pd-pl/Table-Tableau.cfm?LANG=Eng&T=101&S=50&O=A
Statistics Canada. (2015b). 2011 *census of Canada: Topic-based tabulations*. Statistics Canada Catalogue no. 98–314-XCB2011049. Retrieved from http://www12.statcan.gc.ca/census-recensement/2011/dp-pd/tbt-tt/Rp-eng.cfm?LANG=E&APATH=3&DETAIL=0&DIM=0&FL=A&FREE=0&GC=0&GID=0&GK=0&GRP=1&PID=103323&PRID=0&PTYPE=101955&S=0&SHOWALL=0&SUB=0&Temporal=2011&THEME=90&VID=0&VNAMEE=&VNAMEF=
Township of Langley. (2010). *Official community plan* (Updated to 3 May 2010). Township of Langley, British Columbia.
Walks, A. (2013). Mapping the urban debtscape: The geography of household debt in Canadian cities. *Urban Geography, 34*(2), 153–87.
Watson and Associates. (2010). *City of Barrie growth management strategy phase 1 summary.* Report prepared for the City of Barrie.
Young, R. (1995). Urban design for intensification: The St. Lawrence test case. *Plan Canada, 35*(4), 35–7.

Epilogue: Suburbs as Transitional Spaces

JAN NIJMAN

The contributions to this volume provide a rich tapestry of North American suburbia in the early 21st century. It is the first collection of its kind as a series of case studies from across the continent, each critically engaging with empirical and theoretical questions about contemporary suburbanization against the backdrop of conventional understandings of the North American suburb. There are plenty of nuggets in each of the individual cases to shed light on the particular suburban dynamics of the metro areas under examination, yet the main value of the collection lies in the insights it provides in more general developments (Nijman, 2015). There is, sometimes, a wrongheaded prejudice against the case study method and it is too often and too easily dismissed as idiosyncratic or empiricist (see Flyvbjerg, 2006, for a compelling deconstruction of such views). Together, the chapters paint the picture of an increasingly differentiated and dynamic sub/urban landscape; one that fundamentally challenges common assertions about suburbia in the aggregate.

The chapters vary obviously in geographical and regional focus and also thematically and in terms of scale (ranging from individual suburbs to metropolitan areas). The book's organization in the four parts allows for easy comparison along these thematic lines, but there are additional important threads running through the various contributions. It is impossible to do justice to all contributions in this epilogue in any kind of detail, but a number of general observations, and their relevance to broader debates, should stand out.

The Vanished Suburban Archetype

Each chapter in this book focuses on suburbia in one particular metropolitan area in North America. Further, the chapters examine that particular case, explicitly or implicitly, against the background of a more general understanding of suburbanization in this part of the world. What may be the first thing to impress the reader is that many strike an exceptionalist tone. That is to say, in most chapters, there is an assumption of a general process of suburbanization or general model of suburbia, but the metropolitan area in question is thought to stand apart. The point is that if all cases deviate in such important ways, there can hardly be an archetype. Ironically, this paradoxical assumption is explicit or implicit in most contributions, and so the conventional notion of the suburb lives on.

That archetype is primarily (and historically) associated with the United States. If most or all contributors to the US chapters find reasons to take exception, the authors on Canada and Mexico, understandably, do so even more. The non-US cases (especially on Mexico City, Toronto, Montreal, and Halifax) provide poignant illustrations that *both* the discourses and material realities of suburbia have crossed international borders but tend to do so in fragmented manner, are mediated in local context, and result in altered, contradictory and sometimes confusing narratives. Both in Canada and Mexico, suburban imaginaries have taken hold *and* are simultaneously rejected, and the metropolitan landscapes are similarly inconsistent. The authors of the non-US chapters situate their cases in national context but also in relation to assumed trends in the United States. It seems that, abroad, the idea of the archetypal US suburb is even more tenacious than it is at home.

Suburbs as Transitional Spaces

One of the interesting things about the imaginary of the classic suburb is that it connoted a settled, stable, situation. Certainly, this was important to the 1950s idea of the suburb where White middle classes families had "arrived." The suburb embodied the achievement of an ideal, the good life; it was harmonious, predictable, and secure, and *change* was not a part of that dreamy constellation (Nijman & Clery, 2015). In reality, however, the suburb as a spatial entity is a momentary piece of an urban puzzle that is always reconfiguring – spatially, economically, socially, and politically.

Perhaps in the United States even more than elsewhere (because of the prevalence of market forces), urbanization and suburbanization are ongoing processes following, in part, the coupled logics of investment and (re)development (Coe, Kelly, & Yeung, 2007). As metropolitan areas grow and expand, "outer" suburbs become "inner" suburbs (and peripheral location as an essential definition of the suburb becomes problematic); as suburbs age (and "newness" as an essential definition of the suburb becomes problematic), along with their populations, housing stock, and infrastructure, they may lose appeal and market value, or become ripe for redevelopment; as central cities repel or attract populations, suburbs are affected; as immigrants seek new homes, suburbs may change colour and cultural fabric; as metropolitan economies move through cycles and restructuring, the economic fortunes and character of suburbs will shift in integral fashion (also seen Hanlon, 2010).

In reality, then, *change* is very much a natural, even essential, part of the suburban constellation, with ceaseless construction of the new and reconstruction of the old. At any given point, the configuration of suburbs is a reflection of the composition of the metropolitan population, of their commonalities and differences along a range of characteristics, and of the political economy of the day. One might say that it is through ceaseless change of suburban constellations that the metropolis at large sustains itself. It is a long way from the (old) idealized notions of the suburb as spaces of stability and tranquility. The life cycle of a suburb maybe be one generation, or two or three, but then change tends to come quickly. While there are exceptions in the sense of long-term stable residential areas (typically upscale, low density, and amenity rich), most suburbs are inherently transitional spaces. This dynamic and even unsettled nature of suburbia comes to the fore in various chapters, from the post–civil rights suburbs in Phoenix to Atlanta's inner-ring suburbs or the outer suburbs of Ottawa. It is especially salient in Wilson's contribution on Chicago, where the makeover of several inner suburbs, undergoing economic restructuring and large-scale immigration, remains very much in the balance. The chapter on Edmonton, applying a longer term historical perspective, exposes the fluid nature of suburbia even more, and shows how that reality tends to remain hidden from prevailing contemporary popular narratives.

The Suburbia Aggregation Problem

The contributions to this volume underscore the growing diversification of suburbs (e.g., the chapters on the Bay Area, Miami, Piedmont region, Los Angeles, Halifax). While individual suburbs remain quite homogenous within, the range of suburbs has widened very considerably over the past decades. This is in some ways a reflection of growing intrametropolitan inequalities and an ongoing sorting process that is in part by choice and in part by lack of choice (Bishop, 2008; Sampson, 2011). Suburbs can differ enormously in terms of demographics; income; ethnicity and race; housing type, ownership, and prices; density; amenities, etc. This means, in turn, that comparisons of central cities and *suburbs in general* have lost meaning and have actually become quite deceptive because they conceal much more than they reveal.

This suburbia aggregation problem is particularly relevant to common observations about the so-called convergence of central cities and suburbs. The problem is that there is often only one (or a few) central cities but there are many very different suburbs. For example, the often-heard suggestion that central cities and suburbs have become more alike because lower income residents and poverty have shifted to the suburbs while more affluent residents have chosen to live in central cities (e.g., Brookings Institution, 2011) – is a rather meaningless generalization because "the" suburbs constitute a bewildering collection of different areas and suburban poverty tends to be spatially concentrated. Lower income residents have shifted to very specific suburbs, and in others their entry has undoubtedly been precluded. Recent inequality and poverty trends across metropolitan areas have been well documented, especially in regards to housing (e.g., Desmond, 2016; Greenberg, 2017; Immergluck, 2018; Ronald, 2018).

The same aggregation problem occurs in discussions about ethnic and racial diversity. For example, a recent Urban Land Institute (ULI) report (2016, p. 10) states that "diverse suburbs are already the norm. Over one-half of the residents of economically challenged and stable middle-income suburbs are minorities ... This finding highlights the widespread diversity of the suburbs." Here, even while the ULI report focuses on selections of suburbs, the aggregation problem is evident: taken together, poor suburbs have diverse populations, but individually they tend to be considerably more homogeneous. Further, these kinds of aggregations tend to perpetuate, in effect, the stereotype of majority-minorities inhabiting challenged suburbs. The cases of Vancouver, Miami, Phoenix, and Atlanta show otherwise, with either recent immigrants or African American populations constituting majorities in various well-to-do suburbs.

This is not the place for any kind of global discussion, but there is considerable evidence for growing inequality and segregation trends worldwide. Certainly, as a number of chapters in this volume indicate, they can be observed across North America outside the United States (e.g., Mexico City, Vancouver). Recent evidence from Europe affirms the same, even if the contrasts are presently not as sharp as in the United States (Musterd, Marcińczakb, van Ham, & Tammaru, 2017).

In the United States, the growing diversity and inequality among suburbs is solidified in the enormous proliferation of sublocal government and governance structures (Addie, 2016; Hamel & Keil, 2015; Nijman & Clery, 2015; Nelson, 2009). Nationally, there are about 90,000 local governments, including municipalities, towns, townships, school districts, water-management districts, and so on. The combined number of municipalities and town(ships) is about 19,500, up from 16,800 in 1952. Most of the municipal incorporations in the past half century entailed secession from existing municipalities or new independence within a county. At the sublocal residential level, the trends in the United States are quite astonishing: between 1970 and 2011, the number of association-governed residential communities rose from 10,000 to 314,200. Today, more than 62 million nationwide reside in association-governed communities. Association-governed communities include homeowners' associations, condominiums, cooperatives, and other planned communities.

In the United States, municipal incorporation and the formation of forms of local governance are typically driven by consideration of fiscal independence and/or spatial exclusion: what Tilly refers to as "opportunity hoarding" (Tilly, 1999). Just from the perspective of these ingrained and embedded governance institutions, it is hard to imagine an end to the continuing forces of fragmentation and spatial differentiation of the US metropolis. Interestingly, in New York and Los Angeles, where this fragmentation and disparity may be more advanced than elsewhere, some countermovements have emerged. The two chapters in this volume dealing with those cities document political organization, which has met with some degree of success, across suburbia of populations that are separated in space and by formal political boundaries.

Transforming Sub/urban Landscapes

Most of the chapters engage, directly or indirectly, with questions about the ways in which suburbs are understood or defined in comparison to central cities (explicitly so, for example, in the chapters on Vancouver, Miami, Toronto, Atlanta, and the Bay Area). As noted in the introduction to this volume, the simple city–suburb dichotomy that supposedly prevailed in the early postwar years had already disappeared towards the end of the 20th century. The contributions to this volume indicate how the dichotomy has been further eroded in recent years. Suburban and urban landscapes, and the distinctions between them, have been transformed in several ways.

First, the residential landscapes in selected suburbs have changed, adding to a wide range of suburban environments in terms of class, race, ethnicity, demographics, and overall cultural fabric. Across the board, metropolitan areas today include inner and outer suburbs with populations that used to be concentrated in central cities, especially lower income and minority populations. This is evident in most of the metro areas discussed in this volume, including Atlanta, Phoenix, Chicago, Ottawa, Edmonton, the Bay Area, Miami, and Los Angeles. And yet, it should be pointed out that this may not be quite as new as it is often made out to be. To some extent, at least, the suburbanization of the working classes and of ethnic minorities in the past was very real, but tends to be crowded out of the prevailing suburbanization narratives. Wiese (1993) wryly observed that "historians have done a better job excluding African Americans from the suburbs than even white suburbanites."

Second, the economic activity patterns in selected suburbs have been changing for several decades now (though here, too, a warning against a-historicism is warranted – see, e.g., Walker & Lewis, 2004). By the mid-1990s, about twice as many people commuted to work within suburbs as commuted between them and cities (Sharpe & Wallock, 1994, p. 2). The classical monocentric city belonged to the past; the polycentric metropolis had arrived, and the notion of edge cities became commonplace (Garreau, 1991). Subsequently, Lang (2003) introduced the concept of "edgeless cities," pointing to what he considered a more widespread phenomenon: free-form clusters of office space of various sizes and configurations that can be found across suburbia. He emphasized that edgeless cities are *not* "edge cities waiting to happen" but constitute a crucial dimension of the 21st-century metropolis:

> Suburbia's economy reached an unprecedented diversity by the 1980s, as specialized service enterprises of every kind were established outside central business districts ... Yet even as they become more urban, suburbs maintain a distinct pattern. A new metropolitan form therefore has emerged in the past several decades: low density, automobile dependent, and dispersed. Not quite the traditional city, suburb, or exurb, but with elements of all three, it is the still-emergent America on the mall, the beltway, the subdivision, the multiplex movie theater, the drive-through fast-food outlet, the low-rise office cube, and the shopping strip. (Lang, 2003, p. 9)

More recently, the suburban economic (and residential) landscape has been further enriched by so-called "*ersatz* urbanism" (The Economist, 2016a): the kind of instant, ready-made, high-rise, and high-density mixed developments that have appeared in various suburbs across North America. Such developments further contribute to the decentralized urbanization of suburbia, purportedly bringing a sense of urbanity to suburban environments. They are sometimes part of suburban transit-oriented development projects, suggesting the connectivity to central cities remains important.

Third, the landscape of central cities has changed profoundly in the wake of what Ehrenhalt (2013) has termed the "great inversion": the opposite of suburbanization in the sense of a notable new influx of residents into city centres; these new "urbanites" are relatively young, tend to be highly educated, have relatively high incomes, and are predominantly White. This trend has been accompanied with an economic resurgence and redevelopment of many inner cities. The idea of inversion rests, of course, on the prior history of suburbanization (and White flight) and urban decay that was especially characteristic of US cities. It does not mean that suburbs have been growing any less quickly (Kolko, 2017), but it does imply central-city revival and renewed growth (see the chapters on the Bay Area, Vancouver, Toronto, and Miami, for instance). While the reversal of trends applies especially to the United States, similar developments have been observed in Canada and other parts of the world.

In the United States, the great inversion and accompanying gentrification of the central city has notable racial/ethnic dimensions and, as such, is again reminiscent of suburbanization trends in the 1960s and 1970s. From New York to Atlanta to San Francisco, the "return" to the city tends to exclude minorities (and lower income earners) and, as such, is causing a racial (and political) makeover of central cities. At the same time, as is pointed out in the chapter on New York, there is concern in certain suburbs about the departure of these cohorts to the urban centres. Those suburbs may at the same time witness an influx of foreign migrants or the very people who are being displaced at the centre.

Thus, central cities have in the past couple of decades been subject to redevelopment, gentrification, and the construction of private residential associations. In a way, suburbanization has come to the central city. Presently, the population of older suburbs can be more established than that of redeveloped downtowns. For some, the suburban dream has been replaced with dreams of a condo in the central city.

Fourth, the "great inversion" has been entwined with a renewal of the urban economy. This is not what some used to call the "new urban economy" about 20 years ago (e.g., Brookings Institution, 2001) – which was associated with suburbanization and decentralization of economic activity (cf. the emergence of edge and edgeless cities mentioned above) – it is, instead, about the reassertion of central cities as spaces of economic activity, another kind of "big sort" but in the corporate realm.

One facet of this new centralized urban economy concerns large company headquarters moving back from suburbs to urban cores (Economist, 2016b). Examples include General Electric in New York, Motorola in Chicago, and Biogen in Boston. A good number of young tech firms started up and are still located in central cities, such as Salesforce in San Francisco or Amazon in Seattle. The difference with suburban-based headquarters and other economic activity in the suburbs is that the city locations tend to have a smaller workforce that is higher skilled ("technologically fluent"), higher paid, and has a preference for urban living.

Some have argued that the logic of agglomeration in this new urban economy has shifted to an emphasis on knowledge transfers and networking in a high density and high circulation environment – the city as "the office," especially for growing numbers of self-employed and freelancers – but also on spatially clustered opportunities for *consumption* (e.g., Glaeser, Kallal, Scheinkman, & Shleifer, 1992; Carlino, 2005; also see Scott, 2017). This consumption economy, too, is intricately embedded in the urban environment of the central city, where urbanity itself is part of the consumption experience. Arguably, this even applies to the archetypal suburban phenomenon of the shopping mall, which is now reinvented in urban form. In the words of Stefan Al (2017), "a new breed of shopping centre is integrating so seamlessly into its urban surroundings that it can be difficult to draw any line between city and mall whatsoever." Suburban malls are said to be closing down in growing numbers (in part because retail has gone online), but the urban "mall" is being reinvented. Examples include Westfield World Trade Centre in Manhattan and downtown Miami's Brickell City Center, each providing mall-like environments that appear woven into the street life of a city. Less convincing versions abound (e.g., Atlanta's Ponce City Market) but the common denominator is clear: these are meant to feel like urban spaces, albeit safe, convenient, and sanitized. It is all carefully manufactured, to be sure, and the ultimate effect is a renewed commodification of urban space. Thus, the new consumption economy plays out differently in central cities and suburbia.

Finally, and following on the point above, the distinction between cities and suburbs has been blurred and changed through the evolving discourses of design, planning, and through narratives of what constitutes a desirable sub/urban environment. This matter is dealt with most explicitly in the chapters on Vancouver and Halifax. Notions of the New Urbanism, which has traditionally been set in suburban or even small-town environs, survive in some smaller metropolitan areas, but so do the traditional suburban ideals of large single-family homes in low-density developments in peripheral locations. These narratives have been emphatically joined by a discourse that celebrates urbanity, one that applies to the resurrection of central cities but also to new urban forms (cf. the *ersatz* urbanism mentioned above) that have been brought to suburbia (Phelps, 2015). The ULI (2016, p. 31), addressing an audience mainly of developers and realtors, refers to the "willingness" of metropolitan populations, whether in central cities or suburbs, "to pay a premium for more urban living," adding that it has "become somewhat of a luxury good that many households will not be able to afford."

The Unavoidable Central City

Various chapters in this book direct attention to the broader dynamics of metropolitan areas at large and to city–suburb relations (e.g., Vancouver, Atlanta, New York, Toronto, the Bay Area, Ottawa). Each case suggests that while the overall landscape has come to be dominated by suburbs, the central cities play a critical role in the overall metropolitan constellation and dynamics. At the same time, the various cases underscore that central cities, just as much as suburbs, are transitional spaces. The emergence of polycentrism, in varying degrees from one metro to the other (also see, e.g., Pereira, Nadalin, Monasterio, & Albuquerque, 2013) implies that centres can emerge in formerly suburban terrain and that central cities can wax or wane in terms of their role in the metropolitan region. Nonetheless, there can be little doubt about the resurgence of central cities in recent years.

These observations are confirmed in recent innovative research in "urban analytics" that conceptualizes and measures movement over time across metropolitan areas. Most sub/urban research focuses on place attributes – as do most of the contributions to this volume – and not on movement or circulation (with the exception, mainly, of commuter flows). The rapidly increasing availability of data on various types of movement (e.g., public transit travel, Uber and similar services, cell phone communications, etc.) can provide new insights in the role of *centrality* in metropolitan constellations. For example, in a longitudinal study of travel patterns across Singapore, Zhong et al. (2017) find growing polycentrism, but at the same time the main central city has gained in importance. The rise of the consumption economy, concentrated in highly urbanized spaces, no doubt contributes to this centrality in movement patterns.

Such observations seem at odds with speculations in a strand of theoretical literature on the "implosion" of urban centres in a process of "planetary urbanization" (Brenner, 2014; Keil, 2017) – unless this 'simply' means that central cities, like suburbs, are transitional spaces. There is no question that relatively neatly bounded cities of the preindustrial past have given way to urbanizing regions; that central cities were fundamentally transformed during industrial times and again during postindustrial times; and that dynamic polycentrism has become a standard feature of metropolitan landscapes. It also seems clear that the relations between central cities and suburbs should not be understood as asymmetric dependencies. But none of that negates centrality as a critical element in the working of, and in our understanding of, metropolitan regions.

Sub/urbanism and the Challenge of Modelling the Metropolis

Since the early days of the Chicago school, urbanists have attempted to construct spatial models of the city. The various contributions to this volume and the discussion above indicate that this challenge has become ever more daunting against the background of the growing size and complexity of metropolitan regions and the transitional qualities of the constituent parts. Not surprisingly, recent research finds that traditional models are hard to fit to present-day metropolitan realities (Beveridge & Halle, 2011). The field of urban studies is famously pluralistic, and some would not hesitate to discard urban modelling altogether to the dust heap of disciplinary history. Others continue the quest. At any rate, this collection of case studies illustrates the difficulty of the challenge.

One part of the problem is the seemingly endless variation in suburban environments; another lies in the rapid transformations and various directions of change of certain suburbs and central cities; and a third pertains to the substantial overall variation among metros across North America, and even within Canada or the United States, as this volume testifies.

Generally, it is agreed that metropolitan spatial structure hinges on the two dimensions of morphology and functionality (see, e.g., Burger & Meijers, 2012), a conceptual perspective akin to the basic distinction in geography between formal and functional regions. The former denotes place attributes such as population characteristics, density, economic activity, and so on; the latter refers to linkages between places, connectivity, movement, and networks.

The modelling challenge, accordingly, is twofold. First, the measurement of morphological attributes (especially of suburbs) must be fine-grained and dynamic (i.e., must account for direction – and, ideally, speed – of change). The chapter on the Piedmont region in this volume provides a good example (also see, e.g., Delmelle, 2016). This type of research is making promising progress and shows that common and very general distinctions between, for instance, "inner" and "outer" suburbs have decreasing validity. But these kinds of studies, typically case studies of one or a few metropolitan areas, also indicate such levels of differentiation and intricacy that are very difficult to reduce to more general models, not least because the patterns also vary quite substantially from one metro area to the next (Delmelle's research focuses on Chicago and Los Angeles).

The second challenge is to incorporate movement (linkages, flows, networks) in the models. Cleary, this is not without conceptual and methodological challenges. What is the underlying theoretical framework and how to avoid the limitations of traditional structural functionalism? How to employ recent insights from social network analysis? How to define and delineate the intrametropolitan spaces among which linkages are investigated? And not least, how to interpret the evolving linkages and relate these empirical observations back to theory?

The pluralistic nature of the field of urban studies is worth embracing for obvious reasons. However, the current disconnect between abstract theoretical assertions on sub/urbanism and rapidly developing methodological approaches in urban analytics is painfully obvious and, arguably, leaves the study of North American (sub)urbanization wanting. In the meantime, the theoretical–empirical case studies in this volume should help to fill the void.

REFERENCES

Addie, J. (2016). Theorizing suburban infrastructure: A framework for critical and comparative analysis. *Transactions of the Institute of British Geographers, 41*(3), 273–85.

Al, S. (2017, March 16). All under one roof: How malls and cities are becoming indistinguishable. *The Guardian*.

Beveridge, A.A., & D. Halle. (2011). The rise and decline of the L.A. and New York schools. In D.R. Judd & D. Simpson (Eds.), *The city, revisited: Urban theory from Chicago, Los Angeles and New York* (pp. 137–69). Minneapolis: University of Minnesota Press.

Bishop, B. (2008). *The big sort: Why the clustering of like-minded America is tearing us apart*. New York: Houghton Mifflin.

Brenner, N. (2014). *Implosions/explosions: Towards a study of planetary urbanization*. Jovis Publishers.

Brookings Institution. (2001). *The new urban economy: Opportunities and challenges*. Washington, DC. https://www.brookings.edu/research/the-new-urban-economy-opportunities-and-challenges/

Brookings Institution. (2011). *The state of metropolitan America: Suburbs and the 2010 census*. Washington, DC.

Burger, M., & Meijers, E. (2012). Form follows function? Linking morphology and functional polycentricity. *Urban Studies 49*(5): 1127–49.

Carlino, G. (2005). The economic role of cities in the 21st century. *Business Review*, Q3, 9–15.

Coe, N., Kelly, P., & Yeung, H. (2007). *Economic geography: A contemporary introduction*. Malden, MA: Blackwell.

Delmelle, E.C. (2016). Mapping the DNA of urban neighborhoods: Clustering longitudinal sequences of neighborhood socioeconomic change. *Annals of the AAG, 106*(1), 36–56.

Desmond, M. (2016). *Evicted: Poverty and profit in the American city*. Crown Publishers.

The Economist. (2016a, April 30). Ersatz urbanism: Instant, ready-made downtowns bulldoze the distinction between city and suburb. *The Economist*.

The Economist. (2016b, September 3). Leaving for the city: Lots of prominent American companies are moving downtown. *The Economist*.

Ehrenhalt, A. (2013). *The great inversion and the future of the American city*. Vintage.

Flyvbjerg, B. (2006). Five misunderstandings about case-study research. *Qualitative Inquiry, 12*(2), 219–45.

Garreau, J. (1991). *Edge city: Life on the new frontier*. New York: Anchor Books.

Glaeser, E., Kallal, H.D., Scheinkman, J.A., & Shleifer, A. (1992). Growth in cities. *Journal of Political Economy, 100*(6), 1126–52.

Greenberg, M. (2017, August 17). Tenants under siege: Inside New York City's housing crisis. *The New York Review of Books*.

Hamel, P., & Keil, R. (Eds.). (2015). *Suburban governance: A global view*. University of Toronto Press.

Hanlon, B. (2010). *Once the American dream: Inner-ring suburbs of the metropolitan United States*. Philadelphia, PA: Temple University Press.

Hanlon, B., Short, J.R., & Vicino, T. J. (2010). *Cities and suburbs: New metropolitan realities in the United States*. Oxford: Routledge.

Immergluck, D. (2018). Renting the dream: The rise of single-family rentership in the sunbelt metropolis. *Housing Policy Debate*. doi:10.1080/10511482.2018.1460385

Keil, R. (2017). Extended urbanization, "disjunct fragments" and global suburbanisms. *Environment and Planning D: Society and Space*. doi:10.1177/0263775817749594

Kneebone, E., & Berube, A. (2013). *Confronting suburban poverty in America*. Washington, DC: Brookings Institution Press.

Kolko, J. (2017, May 22). Seattle climbs but Austin sprawls: The myth of the return to cities. *The New York Times*.

Lang, R.E. (2003). *Edgeless cities*. Washington, DC: Brookings Institution Press.

Musterd, S., Marcińczakb, S., van Ham, M., & Tammaru, T. (2017). Socioeconomic segregation in European capital cities. Increasing separation between poor and rich. *Urban Geography, 38*(7), 1062–83.

Nelson, R.H. (2009). *The rise of sublocal governance* (Working Paper #09–45). Washington, DC: Mercatus Center, George Mason University.

Nijman, J. (2015). The theoretical imperative of comparative urbanism: A commentary on "Cities beyond Compare" by Jamie Peck. *Regional Studies* 49(1): 183–6.

Nijman, J., & Clery, T. (2015). The United States: Suburban imaginaries and metropolitan realities. In P. Hamel & R. Keil (Eds.), *Suburban governance: A global view* (pp. 57–79). University of Toronto Press.

Pereira, R., Nadalin, V., Monasterio, L., & Albuquerque, P. (2013). Urban centrality: A simple index. *Geographical Analysis, 45*(1), 77–89.

Phelps, N. (2015). *Sequel to suburbia: Glimpses of America's post-suburban future*. Cambridge: MIT Press.

Ronald, R. (2018, April 6). *Generation rent. Inaugural lecture*. University of Amsterdam.

Sampson, R.J. (2011). *Great American city: Chicago and the enduring neighborhood effect*. University of Chicago Press.

Scott, A.J. (2017). *The constitution of the city: Economy, society, and urbanization in the capitalist era*. London: Palgrave Macmillan.

Sharpe, W., & Wallock, L. (1994). Bold new city or built-up 'Burb? Redefining contemporary suburbia. *American Quarterly, 46*(1), 1–30.

Tilly, C. (1999). *Durable inequality*. Berkeley: University of California Press.

Urban Land Institute. (2016). *Housing in the evolving American suburb*. Washington, DC.

Walker, R., & Lewis, R. (2004). Beyond the crabgrass frontier: Industry and the spread of North American cities, 1850–1950. In R. Lewis (Ed.), *Manufacturing suburbs: Building work and home on the metropolitan fringe* (pp. 16–31). Philadelphia, PA: Temple University Press.

Wiese, A. (1993). Places of our own: Suburban Black Towns before 1960. *Journal of Urban History, 19*(3), 30–54.

Zhong, C., Schlapfer, M., Muller Arisona, S., Batty, M., Ratti, C., & Schmitt, G. (2017). Revealing centrality in the spatial structure of cities from human activity patterns. *Urban Studies, 54*(2), 437–55.

Contributors

Jan Nijman (Editor) is Distinguished University Professor and founding director of the Urban Studies Institute at Georgia State University and a professor of geography at the University of Amsterdam. His expertise is in urban and regional development and comparative urbanism, with regional interests in North America, South Asia, and West Europe. Recent (edited) books include *Amsterdam's Canal District: Origins, Evolution, and Future Prospects* (University of Toronto Press, 2020).

Caroline Andrew is the former director of the Centre on Governance and a retired professor in the School of Political Studies at the University of Ottawa. She is a nationally recognized authority on urban and feminist studies, as well as on cultural diversity. She was named Distinguished University Professor in 2006.

Derek Brunelle is an urban planner and development manager for Toronto Community Housing Corporation. In addition to his career in the public sector, he pursues research and writing on housing, local economic development, and urban history. He is a graduate of York University's Master of Environmental Studies program.

Tom Clery is an independent scholar and teacher with a Master's degree in geography from the University of Miami. His research concentrates primarily on issues of urban inequality and cultural representation in the United States. He has published in *Environment and Planning A*, the *International Journal of Applied Geospatial Research*, and with the University of Toronto Press.

James DeFilippis is a professor at the Bloustein School of Planning and Public Policy, Rutgers University. His research focuses on the politics and economics of cities and communities, with a special interest in processes of social change and questions of power and justice in cities.

Angela Franovic is a PhD student at the Department of Political Studies at the University of Ottawa. Her main research interests focus on administrative reforms in postcommunist and postconflict regions and on questions of social equity in local governance.

Liette Gilbert is a professor in the Faculty of Environmental Studies, York University. Her expertise is in urban and environmental politics, migration and citizenship, and North American border politics.

Diane Gillespie has a BA in Anthropology, a BA in Urban Planning and a Master's degree in Healthy Promotion, concentrating on healthy communities and the built environment. Her career focuses on the nexus of community development, urban planning, and affordable housing.

Jill L. Grant is professor emeritus in the School of Planning, Dalhousie University. Her work examines the cultural context of community planning, exploring the values that planners, developers, and residents bring to the places they design and inhabit. She is an elected fellow of the Canadian Institute of Planners.

Pierre Hamel is professor of sociology at Université de Montréal and the former editor of the sociology journal *Sociologie et sociétés*. He is also a former chair of Canadian studies at Paris 3—Sorbonne Nouvelle. In 2018 he published an edited book with Louis Guay, *Les aléas du débat public: Action collective, expertise et démocratie* (Québec : Les Presses de l'Université Laval).

Katherine Hankins is professor of geography at Georgia State University and studies the politics of neighbourhood change. As an urban geographer, her work on Atlanta neighbourhoods examines ways in which social inequalities around race and class are expressed in and addressed through urban spaces and the interplay between urban spatial change, neighbourhood activism, and other forms of urban politics.

Richard Harris is an urban historical geographer at McMaster University. His research is on housing, housing policy, and the social geography of cities and suburbs in North America and British colonies in the

Contributors 365

20th century. Past president of the Urban History Association (2017–18), he is a fellow of the Royal Society of Canada and of the Royal Canadian Geographical Society, and he is a former Guggenheim fellow.

Steven R. Holloway is professor of geography and director of Urban and Metropolitan Studies at the University of Georgia. He studies the intersections between race and class in neighbourhood spaces. His work examines evolving patterns of segregation and diversity as related to housing finance market processes, suburbanization (especially of economically precarious populations), the growth of immigrant communities, and the increased importance of mixed-race households.

Roger Keil is professor and former director of the CITY Institute, York University. The author of *Suburban Planet* (Polity, 2018) and coeditor (with K. Murat Güney and Murat Üçoğlu) of *Massive Suburbanization: (Re)Building the Global Periphery* (University of Toronto Press, 2019), he was the principal investigator of the Major Collaborative Research Initiative on Global Suburbanisms, from which this volume emerged. He researches global suburbanization, urban political ecology, cities and infectious disease, and regional governance.

Paul Knox is University Distinguished Professor and founding dean of the Honors College at Virginia Tech. His most recent books include *Metroburbia: The Anatomy of Greater London* (Merrell, 2017), and *London: Architecture, Building, and Social Change* (Merrell, 2015).

Kieran Moran completed a BA in Sociology from the University of Alberta. He works in the field of community engagement, specifically around issues of infill housing and community economic development in inner-city communities.

Christopher Niedt is associate professor of applied social research in the Department of Sociology, Hofstra University. His main interests are the effects of metropolitan growth and decline on race and class inequality. He is the editor of *Social Justice in Diverse Suburbs: History, Politics, and Prospects* (Temple University Press, 2013).

Jamie Peck is professor of geography and the holder of the Canada Research Chair in Urban and Regional Political Economy at the University of British Columbia. A global professorial fellow at the Institute for Culture and Society at Western Sydney University, he is the managing editor of *EPA: Economy and Space*.

Deirdre Pfeiffer is associate professor in the School of Geographical Sciences and Urban Planning at Arizona State University and a member of the American Institute of Certified Planners. Dr. Pfeiffer is a housing planning scholar, with expertise on housing as a cause and effect of growing social inequality and the role of housing planning in meeting the needs of diverse social groups. Her current research explores the interconnections between housing and health, the housing experiences of US millennials, and the effects of single-family rentals on US neighbourhoods.

Claire Poitras is a professor of urban studies at Urbanization Culture Society Research Centre of the National Institute of Scientific Research in Montreal, Canada, of which she is the former director. Between 2010 and 2019 she directed the network Villes Régions Monde. She writes about the built environment and urban/suburban history. Her recent publications have also addressed the changes of working class neighbourhoods in the Montreal area.

Alex Schafran is assistant professor in the School of Geography, University of Leeds. His research focuses on the contemporary restructuring and retrofitting of urban regions, with a particular emphasis on the changing dynamics of race, class, and segregation across space and place. He recently published *The Road to Resegregation: Northern California and the Failure of Politics* (Berkeley: University of California Press, 2018).

Rob Shields is the Henry Marshall Tory Chair, and professor in human geography and in sociology at the University of Alberta. His work spans architecture, planning, and urban and regional geography. He is the founding editor of *Space and Culture* journal and has authored or edited a variety of books on urban issues, including *Spatial Questions, Strip-Appeal, What Is a City?* and *Places on the Margin*.

Elliot Siemiatycki is a special advisor in the Ontario Ministry of Economic Development, Job Creation and Trade, based in Toronto, Canada. He holds a PhD in Urban and Economic Geography from the University of British Columbia.

Richard Walker is professor emeritus of geography at UC Berkeley. His research has covered economic geography, urban and regional development, capitalism and politics, resources and environment, class and race. His most recent book, which expands on themes in the article in this collection, is *Pictures of a Gone City: Tech and the Dark Side of Prosperity in the San Francisco Bay Area* (Oakland: PM Press, 2018). He also directs The Living New Deal Project, documenting and mapping all public works produced in the 1930s.

Fang Wei obtained a PhD in planning at Virginia Tech and is presently associate professor of regional urban planning at Zhejiang University. Her interests are in urbanization and suburbanization, urban economics, and land-use and environment planning.

David Wilson is professor of geography at the University of Illinois, Urbana-Champaign. His interests are in urban political economy in contemporary redevelopment of urban and suburban places in North America and the global west, and in the racial dynamics of urban politics. He (co-)authored many books, including most recently *Chicago's Redevelopment Machine and Blues Clubs* (New York: Palgrave Macmillan, 2018) and *Urban Inequalities across the Globe* (London: Routledge, 2015).

Elvin Wyly is professor of geography at the University of British Columbia. His research focuses on urban inequality, racial discrimination, housing finance, and gentrification.

Index

Page numbers with *f*, *m*, and *t* indicate figures, maps, and tables, respectively.

Aboriginal people. *See* Indigenous peoples; Mill Woods (Edmonton)
Act to Preserve Agricultural Land, 153, 162–3
affordability/accessibility: Atlanta, GA, 237; Ciudad Santa Fé (Mexico City), 54; Ciudad Satélite (Mexico City), 50–1; and esthetic devaluation, 9; as inherent to suburbs, 70–1; Toronto, ON, 36, 37; in 21st century, 5; Vancouver, BC, 144–5. *See also* "drive until you qualify"
affordable housing: and class segregation, 186–7; and land trust in NJ, 189–90; Mill Woods (Edmonton), 258; New Jersey, 175; New York, 181, 183, 188–9
African Americans: in Bay Area, 122–3; in Charlotte, NC, 89, 90; in Durham, NC, 89–90; and foreclosures, 187, 189, 217; and historians, 354; in Los Angeles, CA, 289, 298; in Miami region, 77f, 80, 81, 83; in New York, 174–6, 176f, 178–82, 179–80t, 192;

and Phoenix post–civil rights suburbs, 202, 209–15, 211t, 212t; and Phoenix postwar suburbs, 205–6, 211t, 212t; in Piedmont, NC, 93–102 passim, 93t, 94t, 97f, 99t, 101t. *See also* Atlanta, GA; immigration/immigrants; people of colour
Age Friendly Ottawa, 323, 324
aging residents: in Barrie, ON, 342; and condos in Langley, 341; in Ottawa, ON, 319, 321m, 323–5; in Piedmont, NC, 93–102 passim, 93t, 94t, 97f, 99t, 101t
Agricultural Land Reserve, 330, 334–5
agriculture: in Edmonton, AB, 256; in Miami region, 67, 72; in Montreal region, 153–5, 154f; in Phoenix, AZ, 204, 214
Agritopia, 214
Aiken, Walter, 226–7
air conditioning, 68
Airdrie, AB, 333–4, 334f, 336–8, 337f–9f, 340, 342
airports, 30, 82–3

Al, Stefan, 356
Alameda Corridor Jobs Coalition (ACJC), 299
Alonso, W., 88
amalgamation, 3
amenities: housing booms in Mexico, 48–9, 56; libraries in Montreal, 165–6; and Maryvale (Phoenix), 206; and Montreal developers, 160–1; in Toronto, ON, 34. *See also* infrastructure; services
Amiskwaciy Waskahikan, 252, 260
antisuburban utopia, 131
Antonioni, Michelangelo, 294
Arcade Fire (band), 166
artists, 32–3
Asian residents: in Bay Area, 123; and foreclosures, 187; in Markham, ON, 342; in New York, 178, 179–80t; and Phoenix post–civil rights suburbs, 209–11, 211t, 212t, 213, 214–15; and Phoenix postwar suburbs, 211t, 212t
association-governed communities, 353
Atlanta, GA: first Black suburbs, 225; and foreclosures, 240; mayors, 235, 236; mortgages, 227; overview, 223–4, 241–2; population distribution, 225m, 230m, 231m, 233, 234, 237–9, 238m; population growth, 236, 239; poverty, 239–41, 240f; property values, 235–6, 239; race relations, 224, 225m, 228–9, 232; South DeKalb County, 237, 239; suburban expansion, 229–35, 239, 241–2
"Atlanta Blacks Losing in Home Loans Scramble" (Dedman), 233–4
Atlanta Life Insurance Company, 230
automobiles: in Bay Area, 113; in Edmonton, AB, 247, 249; as luxury items, 6; in Ottawa, ON, 309; and streetcars, 8; in Toronto, ON, 35–6; in Vancouver, BC, 136
Avondale (Phoenix), 213–14, 215f, 217

bailouts, 118–19
Bakhtin, Mikhael, 279, 283
Banham, Reyner, 294
banks, 8, 226, 230. *See also* mortgages
Bargen, P.F., 245
Barrhaven (Ottawa), 314–15, 317m, 319, 333f
Barrie, ON, 333, 336–8, 337f–9f, 340–2
Bartholomew, Harland, 132–3
Bay Area, 111m; booms/busts, 118–20, 122; defining, 110–12, 121; and developers, 114–16; employment/economy, 112–13; greenspace, 116–18, 117m; homeownership, 115f; housing styles, 116; new housing, 120m; overview, 109–10, 123–4; public transit, 113; racial geography, 122–3
Bay Area Rapid Transit (BART), 113
Beasley, Larry, 137, 143
Bélanger, A., 161
benevolence, 277
Benner, C., 298
Berwyn and Cicero (Chicago), 272m, 273t; employment, 273–4, 275; and globalization, 278–80, 283; historic designation, 275–6; immigration rhetoric, 274–5; inclusive-speak, 275–7; marginalization of immigrants, 280–1, 282–3; and Mexican immigration, 271, 273; overview, 271–2, 282
big-box retail, 340
Biscayne Bay, 68, 69m, 71, 72
Black Gold Suburbs, 293

Black mecca. *See* Atlanta, GA
Bloch, Robin, 289
Boddy, Trevor, 142
book overview, 3–4, 9–18, 349–59
boomburbs, 163–4, 214
booms/busts: California, 118–20, 122; Edmonton, AB, 257, 258; Uniondale (Long Island), 190; Vancouver, BC, 143–4. *See also* foreclosures; Great Recession
Brace, Charles Loring, 172
Brewer, K., 331
Bronxification, 174
Brossard, QC, 157–60
building industry, 48, 54–5, 205, 263n3. *See also* developers; house construction
Burgess, E.W., 88
buses, 315–16, 317m
Bus Riders Movement, 296, 298
Bus Riders Union (BRU), 298, 300–1
Byrne, Jane, 275–6

Calgary, AB, 329m, 337, 337f–9f. *See also* Airdrie, AB
Campbell, Bill, 236
Canada Mortgage and Housing Corporation (CMHC), 27
Canclini, García, 53
Capron, G., 49, 50
Carranco, Moreno, 51, 54
Carver, Humphrey, 27m
Cascade Heights (Atlanta), 232, 236
Census Bureau, 121
census designated places (CDP), 83–4
census metropolitan areas (CMAs), 319
census tracts, 247
central business district (CBD), 30, 31, 173. *See also* downtowns
Centro de Derechos Laborales, 191–2
Century City (Los Angeles), 53

Chalco Lake, 57
Chalco Solidaridad (Mexico City), 57–8, 58f, 59
Charlotte, NC, 89
Chicago, IL, 269–70, 273, 281–2. *See also* Berwyn and Cicero (Chicago)
Chicago School, 88, 102, 104, 144
Chinese residents, 342
Cicero (Chicago). *See* Berwyn and Cicero (Chicago)
cities vs. suburbs: and centrality, 357–8; city decay, 7; Ciudad Satélite (Mexico City), 50; first instance of, 7; in Miami region, 74–5, 79t; overview, 353–7; and poverty, 352; prewar, 6; in Toronto, 28–9; visiting, 31
Citizens Trust Bank, 226, 228, 230, 234, 242n1
City for All Women Initiative (CAWI), 323–4, 325
City of Quartz (Davis), 294
city planning: biases, 300; contemporary, 329–30 (*see also* Halifax, NS); design awards, 330; as elite project in New York, 171–4; government offices and transit in Ottawa, 323; history in Vancouver, BC, 132–5; inclusion/exclusion of immigrants [*see* Berwyn and Cicero (Chicago)]; master-planned communities, 213–14, 331
Ciudad Santa Fé (Mexico City), 51–4
Ciudad Satélite (Mexico City), 49–51, 51f
Civil Rights Acts, 202, 233
Clark, S.D., 28, 28–9
cluster analysis, 91–2, 93, 93t, 94t, 95f, 96f, 97f
coastal resorts. *See* waterfront communities
college graduates, 181

Collier Heights (Atlanta), 229–30, 231m, 236
colonial displacement, 132, 251–5
colour blindness, 275
community benefits agreements (CBAs), 300, 301
Community Benefits Network, 299
community groups, 323–5. *See also* social justice/movements
Community Reinvestment Act, 202
commutes: in Bay Area, 112–13, 120–1; in Edmonton, AB, 247, 249; in Halifax, NS, 331; in Los Angeles, CA, 298; to Mexico City, 56; in Miami region, 74; in Montreal, QC, 153, 154f, 155; in Vancouver, BC, 143; within suburbs, 354
Concord, CA, 123
condos: as commodities, 143–4, 145–6; in Langley Township, BC, 341; as legal entity, 137–8; in Miami region, 76; in NY/NJ, 186; in Toronto, ON, 31; in Vancouver, BC, 130–1, 138–9, 142–3, 145
Connecticut, 177, 179t, 185t, 188, 192
connectivity, 316, 318, 322, 328
construction lobby, 6
consumption, 356–7. *See also* retail activities
Coral Gables, FL, 67, 70, 71, 75
corporate headquarters, 52, 54, 356
cottage country, 38
court cases, 174–5
Crawford, T., 90
creative class, 32–3
crime, 34, 35, 208, 335
Cubans, 80. *See also* Hispanic residents; Latino residents
cultural genocide, 255

Dadeland (Miami), 83, 84
Davis, Mike, 294

day labourers, 182–3, 273–4
de Alba, M., 49, 50
Death and Life of Great American Cities, The (Jacobs), 9
DeBlasio, Bill, 192–3
debt, 339
deindustrialization, 7, 173, 176–7, 184, 281
densification initiatives, 49, 76, 155. *See also* Toronto, ON; Vancouver, BC
density of population: of Airdrie, AB, 339f; of Atlanta, GA, 75, 233, 234m; of Barrie, ON, 339f; of Calgary, AB, 339f; and commuting, 153; and developers, 116; of Edmonton, AB, 256, 257; *ersatz* urbanism, 354–5; of Halifax, NS, 331, 331f, 338, 339f; and household sizes, 342; and housing costs, 340; of Langley Township, BC, 335, 335f, 339f; of Laval, QC, 157; of Markham, ON, 339f; of Miami region, 75–6, 79t; of Montreal, QC, 153, 155–6; of Phoenix, AZ, 75, 203–4; and public transit, 247; and suburban stereotype, 23; of Surrey, BC, 336f, 338, 339f; of Toronto region, 340; of Vancouver region, 136, 340
Department of National Defence, 314–15
design awards, 330
Detroit, MI, 188
Deux-Montagnes (Montreal), 154f
developers: and agricultural land in Montreal, 163; in Airdrie, AB, 334; in Atlanta, GA, 225, 226–7; in Bay Area, 114–16; controlling land, 340; economies of scale in Mexico, 48, 55; Edmonton, AB, 258; enlarging opportunities, 6;

illegal development in Mexico City, 47, 49; in Los Angeles, CA, 292; marketing material, 160; and mortgages in Mexico, 48; and zoning and regulatory frameworks, 8. *See also* building industry; house construction; land development patterns, 339–43
Devil Dogs, 214
discrimination: in Atlanta, GA, 228, 233–4; in Edmonton, AB, 250; in Levittown (Long Island), 175, 176; and Phoenix post–civil rights suburbs, 202, 213; in Phoenix postwar suburbs, 205–6; in Yonkers, NY, 174–5. *See also* racism
diversity: in Berwyn (Chicago), 274–7; and ethics, 262; in Mill Woods (Edmonton), 259; in Montreal suburbs, 156–7, 159, 165–6; in New York, 181–2; as normal, 352; and Phoenix post–civil rights suburbs, 202–3, 214; in Piedmont, NC, 97–8; and reality, 342; in Toronto, ON, 24, 28, 33. *See also* people of colour
Dolan, Xavier, 166
Domaines de la Rive-Sud, 160
Dominick, Larry, 279–80
Donald, Dwayne, 260
Don Mills (Toronto), 26
#donthaveamillion, 145
Doral (Miami), 78, 79t, 82–3
downtowns: Chicago suburbs and globalization, 277–9; gentrification in Los Angeles, 290; gentrification overview, 355; Miami, FL, 68, 76, 85–6; Ottawa, ON, 316, 317m, 318, 319; post-industrialized, 270–1; re-corporatized Toronto, 30; residentialization/de-downtownification, 142–3, 355; revitalizations in NY/NJ suburbs, 186; Vancouver, BC, 130, 137, 138–43. *See also* central business district (CBD)
"drive until you qualify," 9, 145, 155–6. *See also* affordability/accessibility
Durham, NC, 89–90

East Bay, 110, 123
economic development, 270, 274, 277–9, 281, 355–6. *See also* gentrification
edge cities: Bay Area, 112; Coral Cables, FL, 75; Dadeland (Miami), 83; Kendall (Miami), 85; Piedmont, NC, 91; Toronto suburbs, 30
edgeless cities, 75, 85, 91, 354
Edmonton, AB: density of, 256–7; growth pattern, 257; history of, 252–3, 254–5, 262n2; Indigenous peoples residing in, 249–50, 251; Indigenous title assertion, 251–4, 253m; investment in, 257; oil discovery, 256; overview, 249; population growth, 251. *See also* Mill Woods (Edmonton)
Ehrenhalt, A., 355
Eidelman, G., 258
elections, 34
The Electors' Action Movement (TEAM), 135–6
elites, 4–5, 6, 47, 228. *See also* Atlanta, GA; Ciudad Satélite (Mexico City)
elitism, 9
eminent domain, 188–9
employment: barriers, 299; in Bay Area, 112; in Chicago suburbs, 270–1, 273, 273t; in Edmonton, AB, 249, 250; and escaping city centre, 5; in Los Angeles, CA, 299–300; in New York, 184–6,

185t; office work in suburbs, 8; in Ottawa, ON (*see* Ottawa, ON); in Phoenix, AZ, 204, 218; in Piedmont, NC, 90; and public transit, 316; in Toronto suburbs, 29–30; in Vancouver, BC, 139, 143
England, 4
entrepreneurs, 224, 226–7, 228, 279
environment, 132, 291
environmentalism, 117, 298
environmental issues, 208
Equity and Inclusion Lens, 318, 324
ERASE Racism, 175
Erickson, Arthur, 137–8
ersatz urbanism, 354–5
Essex Community Land Trust (ECLT), 189
Essex County, NJ, 189–90
ethnoburbia, 259, 261, 290, 337
Evans, B. (Chicago planner), 278–9, 279
Everglades National Park, 67, 68, 69m
evictions, 60. *See also* foreclosures
exoticization, 275, 282
Expo '86, 137
exurbia, 36, 331. *See also* penurbia

Fair Housing Act, 202, 213
False Creek North (Vancouver), 137
False Creek South (Vancouver), 136
Federal District (Mexico City), 47
Federal Housing Administration (FHA), 202, 227–8
festivals, 274–5, 308
Filion, P., 155
financial crisis, 47–8. *See also* Great Recession
FIRE (Finance, Insurance, Real Estate) sector, 177
Fishman, Robert, 3, 5, 8, 172, 293
Florida, Richard, 32

Florida City (Miami), 67, 79t, 80, 81, 81f, 85
Flowers, Theodore "Tiger," 226
Forcier, André, 166
Ford, Rob, 34, 37
Fordist/post-Fordist regulation, 293
foreclosures: in Atlanta, GA, 240; in Bay Area, 122, 123; in Los Angeles, CA, 295, 295f; in New York region, 187–8; in Phoenix, AZ, 215, 216m; and social justice, 188–90
foreign ownerships, 145–6
Forsyth, A., 247, 249
Fort Lauderdale, FL, 67
Fortin, Andrée, 156
francophones, 161–3, 162t, 164m, 166–7, 318
Franklin, Shirley, 236
freeways. *See* highways
French flight, 161–3, 166–7

garbage dumps, 53
gated communities: in Barrie, ON, and Airdrie, BC, 340–1; in Langley Township, BC, 334, 335f; in Montreal, QC, 160; in New York, 193
Gatineau, QC, 308, 313–14, 315–16, 317m. *See also* Ottawa, ON
Geddes, Patrick, 172
gender, 7
General Electric, 8
General Motors, 8, 298
General Telephone & Electronics Corporation (GTE), 177
gentrification: in Atlanta, GA, 232; in Bay Area, 122; in Chicago suburbs, 279–80; and immigration, 270; in Los Angeles, CA, 290; in New York City, 181; overview, 355; in Piedmont, NC, 98; in Toronto, ON, 30–1; in Vancouver, BC, 139

geographic fetishism, 84–5
Geography of Nowhere, The
  (Kunstler), 9
ghettoization, 270, 273
Gilbert (Phoenix), 213–14, 217
globalization, 177, 277–8, 283
Gordon, Jennifer, 191–2, 247
government: and immigrant
  policies, 270; locations of, 309,
  310–15, 312m, 323
Grant, J.L., 331, 337
great inversion. *See* inversions
Great Recession: and Atlanta, GA,
  239; and Bay Area, 122, 123; and
  California, 122; and Los Angeles,
  CA, 295; and New York region,
  187–8; and Phoenix, AZ, 215–18,
  216m
Great Reversion, 186
Greater Uniondale Area Action
  Coalition (GUAAC), 190
Gréber Plan, 309, 310m
Greenbelt (Ottawa), 309, 311m
greenspace, 116–18, 117m
GTA (Great Toronto Area). *See*
  Toronto, ON

Haiven, L., 338
Halifax, NS, 329m, 331f, 337f–9f;
  density of, 338; housing
  prices, 340–1; overview, 331–2;
  population growth, 336
Hanchett, T.W., 89, 102–3
Hank González, Carlos, 53
Hardwick, Walter, 135–6, 137
Hartsfield, William B., 229
Hartshorne, T., 236
Hayden, D., 4, 5
Hayeur, Isabelle, 166
hedge city, 146
hegemony, 142–3, 144. *See also*
  homogeneity

Henry, A. (Chicago planner), 273,
  275, 277–8
Hern, M., 143
Hialeah (Miami), 78–80, 79t
Hiebert, D., 137
high-rise apartments, 26, 34
highways: Ciudad Santa Fé (Mexico
  City), 52; Ciudad Satélite (Mexico
  City), 50–1, 51f; Halifax, NS, 330;
  Toronto, ON, 330; Vancouver,
  BC, 131, 135, 136, 330. *See also*
  parkways
Hispanic residents: in Miami region,
  77f, 80, 81, 83, 84, 85; in New York,
  179–80t; in Piedmont, NC, 90. *See
  also* immigration/immigrants;
  Latino residents; people of colour
historically Black colleges and
  universities (HBCUs), 227, 228
Hohokam people, 204
home modification, 56
Home Mortgage Disclosure Act, 202
Home Owners' Loan Corporation
  (HOLC), 227–8
homeownership: in Bay Area,
  114, 115f; and Black banks, 226,
  230, 234–5; and condos, 31; and
  landowners, 229; in Montreal, QC,
  149; in Piedmont, NC, 93, 93t, 94t,
  97; and regularization in Mexico,
  59–60; and rental properties in
  Atlanta, 235. *See also* affordability/
  accessibility; foreclosures
Homestead, FL, 67, 72
homogeneity, 56, 160. *See also*
  hegemony
HoSang, D., 192
house construction, 119f, 120m, 258,
  263n3. *See also* building industry
household appliances, 8
housing bubbles, 76, 110, 118–19,
  143–4. *See also* booms/busts

Hoyt, H., 88, 96
Hudson County, NJ, 177
Hume, Christopher, 37
Hunter Hills (Atlanta), 228, 231m

ideals, 23, 28–9, 30–7, 49
Ihlanfeldt, K.R., 236
"illegal" land occupation. *See* informal settlements
immigration/immigrants: in Canada, 259; in Chicago suburbs [*see* Berwyn and Cicero (Chicago); Chicago, IL]; and citizenship, 270; in City of Laval, 157; and festivals, 274–5; inclusion/exclusion [*see* Berwyn and Cicero (Chicago); Chicago, IL]; in Los Angeles, CA, 289–90; and low-wage sectors, 270–1, 273–4, 275; in Markham, ON, 342; in Miami region, 65, 68, 74; in Montreal, QC, 161; in New York, 181–2, 190; in Ottawa, ON, 319, 322, 324; in Piedmont, NC, 90–100 passim, 93t, 94t, 97f, 99t, 101t; and population growth, 339; and social justice, 183, 190, 191f; stereotypes, 276; suburbanization of, 182–3; in Surrey, BC, 336; and Toronto attraction, 342; Toronto settlement patterns, 27m, 33–4, 33m; Toronto suburban ideal, 36; Toronto's first generations, 24; using transit, 322–3, 324; in Vancouver, BC, 342; in various midsize cities, 336–8. *See also* people of colour
inclusion/exclusion. *See* Berwyn and Cicero (Chicago); Chicago, IL
income disparity: in Bay Area, 121–2; Indigenous peoples, 250; in Mexico City, 54; in Miami region, 71; in New York, 192; in North Carolina, 89; in Ottawa, ON, 320m; and suburban living, 29; in Vancouver, BC, 143–5
income parity, 212
incorporation, 80, 82, 241, 290, 353
Indian Act, 255, 262n1
Indigenous peoples: in Edmonton region [*see* Edmonton, AB; Mill Woods (Edmonton); Papaschase Cree people/reservation]; historical presence, 254–5, 260–1; in Los Angeles region, 290–1; in Mexico City, 57; in Ottawa, ON, 314; in Phoenix, AZ, 204; suburbs as settlement, 261; terms, 262n1; title assertion, 251–4; in Vancouver, BC, 132
industrial areas, 5
industrial conglomerates, 8
Industrial Revolution, 4
Infonavit, 48
informal settlements, 57–61
infrastructure, 48–9, 203. *See also* amenities; services
International Congress of Modern Architecture, 294
inversions, 289, 290, 355
involutions, 289
Iracheta Cenecorta, A., 60
Ixtapaluca (Mexico City), 54, 55f

Jackson, K., 4
Jackson, Maynard, 235, 236
Jacobs, Jane, 9, 24
Janzen, M., 247
Jersey City, NJ, 177
Jewish residents, 36
Jim Crow, 228. *See also* segregation/stratification
Jones, P. (Chicago planner), 278, 280
judgments, 23, 28, 30–1

Kanata (Ottawa), 317m, 319
Kendall (Miami), 70–1, 79t, 83–4, 85
Kendall West (Miami), 78, 79t, 83–4
Kennedy, Liam, 277
King, Martin Luther, Sr., 230
King, Peter, 183
King, William Lyon Mackenzie, 313
Kingston-Galloway (Scarborough), 35
Kirkwood (Atlanta), 233
Kruse, K., 233
Kunstler, James Howard, 9

labour movement, 192, 296
Lake, R., 235
Lakewood Plan, 290
L.A. Live, 300, 301
land: agricultural into residential, 5; in Bay Area, 115–16, 119; and Black homeownership, 229–30; developing in Mexico City, 47; Expo '86 in Vancouver, 137; and Indigenous peoples (*see* Indigenous peoples); and informal settlements, 57–60; price increase in Canada, 258; public/private in Canada, 259, 314; redeveloping in Toronto, 26; supply of, 340–1. *See also* Mill Woods (Edmonton); privatization
land trusts, 189–90
Lands, L., 224
Lang, R.E., 354
Langley Township, BC, 334–5, 336, 337f–9f, 338, 340–1
language, 161–2, 162t
Latino residents: in Bay Area, 123; in Chicago suburbs [*see* Berwyn and Cicero (Chicago)]; and foreclosures, 187, 189, 217; in Los Angeles, CA, 289, 298; in Maryvale (Phoenix), 208; in New York, 178–81, 179–80t, 181, 182–3, 191–2, 191f; and Phoenix post–civil rights suburbs, 209–15, 211t, 212t; and Phoenix postwar suburbs, 205–6, 211t, 212t. *See also* Hispanic residents; immigration/immigrants; people of colour
Laval, QC, 157, 158f
Leduc-Nisku Business Park, 249
Legoretta, Ricardo, 53
LeJeune, Jean-Francois, 68
Levittown (Long Island), 7–8, 175, 176f
Ley, David, 136, 137, 144–5
libraries, 165–6
Li Ka-Shing, 137
Living First policy, 137, 139, 142, 143
loans, 48. *See also* mortgages
Long, John F., 206, 213
Long Island, NY, 173f, 180t; and discrimination, 175; foreclosures, 187; and immigration, 183, 191f; and manufacturing, 172–3; redevelopment, 186; and regional planning, 174; and young people, 178, 181
Los Angeles, CA, 292m, 297f; and Great Recession, 295; history and geography, 290–2, 293–4; as horizontal, 293–4; manufacturing jobs, 296; 1992 rebellion, 299; overview, 289–90, 301; as postmodern, 102; postsuburban politics, 295–6; privatization, 290, 293, 297; segregation, 289; social justice, 296, 297–302; vs. Vancouver, BC, 130; water, 291, 298; working class, 296
Los Angeles Alliance for a New Economy (LAANE), 300
Los Angeles School, 102, 290
Los Angeles: The Architecture of Four Ecologies (Banham), 294

Lovero, Robert, 279–80
Lower Mainland Regional Planning Board of British Columbia, 133–4
low-income residents: in Chicago suburbs, 273–4; in Mexico City, 45, 46, 48, 53, 57, 60–1 (*see also* informal settlements); in Montreal, QC, 164; in Ottawa, ON, 320m, 32–1m; in Piedmont, NC, 93–103 passim, 93t, 94t, 97f, 99t, 101t; and renting in Phoenix, 217; suburbs vs. central cities, 352; transit in Los Angeles, 298
Lucero, Marcelo, 182–3

Mann, Eric, 298
Maricopa County (Phoenix), 215, 218
Markham, ON: awards, 330; Cornell, 332f; as ethnoburb, 337; growth management, 339–42; and New Urbanism, 36; overview, 332; population growth, 336; populations/characteristics of, 337f–9f
Marois, G., 161
Martens, Karel, 322–3
Maryvale (Phoenix), 206–7, 208, 208f, 217
mass-produced homes, 116, 142
master-planned communities, 213–14, 331
mayors, 235, 236, 279–80, 308
Merrick, George, 70
Métis peoples. *See* Indigenous peoples; Mill Woods (Edmonton)
Mexican immigrants. *See* Berwyn and Cicero (Chicago)
Mexico, 48, 51, 55, 60
Mexico City, Mexico: Chalco Solidaridad, 57–8, 59; Ciudad Santa Fé, 51–4; Ciudad Satélite, 49–51; informal settlements, 57–61; Ixtapaluca, 54–5; overview, 45–9; peripheralization, 45; urban planning, 61
Miami, FL, 69m; agriculture, 67, 72; cities vs. suburbs, 74–5, 79t; city hall, 70; Dadeland, 83, 84; demographics, 74, 74t, 79t; Doral, 78, 79t, 82–3; downtown, 68, 76, 85–6; Florida City, 67, 79t, 80, 81, 81f, 85; Hialeah, 79, 80; history of, 65, 66–8, 75; immigration, 65, 68, 74; Kendall, 70–1, 79t, 83–4, 85; Kendall West, 78, 79t, 83–4; overview, 84–6; Pinecrest, 79t, 81, 82f, 85; population growth, 65, 67f, 68; suburb as term, 70–1; urban development boundary (UDB), 72; waterfront communities, 71–2
Micallef, Shawn, 34
Michelson, William, 28, 29
middle class: in Mexico City, 45, 47 [*see also* Ciudad Satélite (Mexico City)]; in Phoenix, AZ, 205; in Piedmont, NC, 93–102 passim, 93t, 94t, 97f, 99t, 101t. *See also* Atlanta, GA
middle-income, 212t, 219n1–2
migrants: in Mexico City, 47, 60; in New York, 182–3; in Phoenix, AZ, 205; in Piedmont, NC, 90
military, 262n2, 293
Mill Woods (Edmonton), 246f, 248m; diversity in, 259; employment, 249; Indigenous peoples' historical presence, 260–1; Indigenous peoples residing in, 251, 259–60; overview, 245–7; planning of, 257–8; suburbs as settlement, 261; transportation, 247, 249, 261–2. *See also* Edmonton, AB
Mirabel (Montreal), 162–3

missions, 290–1
modelling, 358–9
modernization, 48, 51, 52–3
Montreal, QC: agriculture, 153–5, 154f; art in, 166; boomburbs, 163–4; City of Brossard, 157–60; City of Laval, 157, 158f; commutes, 153, 154f; densification initiatives, 155–6; employment, 139; francophones leaving centre, 161–3, 166–7; French as only/main language, 162t, 164m; geographic and social characteristics, 150–7, 151m, 152t; history of, 149; and homeownership, 149; Mirabel, 162–3; overview, 149–50, 166–7, 167n2; renters, 164–5; sprawl, 153, 155
mortgages: as industry, 48; and people of colour, 202, 205, 227, 233–4; and White residents, 227–8
Moses, Robert. *See also* parkways
Mossiman, Dean, 279
Mount Laurel Township, NJ, 175
Muller, P.O., 6
Mumford, Lewis, 172
municipality incorporation, 67, 67f
Murdie, R.A., 88, 102
Mutual Federal Savings and Loan Association, 230, 234–5
mythic journeys, 7

NAACP, 174–5
National Capital Commission (NCC), 308
National Workers' Housing Fund, 48
neoliberalism, 280
Neutra, Richard, 294
New Deal, 227, 228
New Jersey: deindustrialization, 176–7; discrimination, 175; downtown revitalizations, 186–7; foreclosures, 188; manufacturing jobs, 185t; public transit, 172; and young people, 178, 179t
"New Suburban History" (Jackson and Surgrue), 227
New Urbanism, 36, 330, 332, 335, 339–40, 356–7
New York: and DeBlassio, 192–3; downtown revitalizations, 186–7; early history of, 171–2; and globalization, 177; immigration, 181–3; and manufacturing, 172–3, 177, 184–6, 184f, 185t; natives of, 182–3; overview, 170–1, 193; rebuilding suburbs, 183–4; and regional centres, 176–7; regional government, 174; regional planning, 171–4, 177; segregation, 174–5; size of, 170; and social justice, 188–93; and younger residents, 178–81
Newark, NJ, 176, 188
NIMBY (Not In My Back Yard), 26, 175
Nortel, 314
North American Free Trade Agreement (NAFTA), 47–8, 52
North Bay, 110
nostalgia, 278–9
nuclear family, 5

Oakland, CA, 110, 112, 114–16
obesity, 36
Occupy movement, 122
Ochs, Mary, 299
OC Transpo, 315–18, 317m, 319, 322–5
office sectors: in Ciudad Santa Fé (Mexico City), 52–3; and edgeless cities, 354; in Edmonton, AB, 249; Markham, ON, 332; in Miami region, 75; regional centres around

New York, 177; suburbanization of work, 8. *See also* central business district (CBD)
oil industry, 256, 258, 263n2, 293, 333
Oliver, Frank, 255
opportunity hoarding, 353
Orysiuk, B.R., 245
othering, 275–6, 280–1
O-Train, 315, 316, 319
Ottawa, ON, 310m, 311m; Barrhaven, 314–15, 317m, 319; demographics, 319–22, 321m; Greenbelt, 309, 311m; immigrants in, 319, 322, 324; incomes, 320m; Kanata, 319; location of government, 309, 310, 312m, 313, 314–15, 323; overview, 307–8; public transit, 309, 315–18, 317m, 322–5
Ottawa Local Immigration Partnership (OLIP), 323, 324, 325
Ottawa Seniors Transportation Committee, 325
outmigration, 7, 32, 229

Pani, Mario, 50
Papaschase Cree people/ reservation, 252–6, 260, 263m
Parker, Dorothy, 130
parkways, 173. *See also* highways
Pastor, M., 298
Paterson, NJ, 176, 188
Peck, J., 289
pentimento, 254, 262
penurbia, 166. *See also* exurbia
people of colour: and foreclosures, 187–8, 189, 215, 240; and gentrification, 355; in Mill Woods (Edmonton), 250–1, 259; and mortgage loans, 202, 205; and Phoenix post–civil rights suburbs, 210–15, 211t, 212t, 219; and Phoenix postwar suburbs, 207, 211t, 212t; and renting in Phoenix, 217; in Toronto, ON, 35; transit in Los Angeles, 298; in various midsize cities, 337–8, 337f, 341–2; and youth drain, 178–81, 179–80t. *See also* immigration/immigrants; segregation/stratification; *various ethnicities*
peripheralization, 45
peripheral urbanization, 294
Perrin, L., 331
Perrott, K., 337
Perry, Clarence, 256
Perry, Heman, 224, 226
Peyton Wall, 232
Philips, Art, 135
Phoenix, AZ: Avondale, 213–14, 215f, 217; and foreclosures, 215, 216m; Gilbert, 213–14, 217; overview, 203–4, 218–19; population growth, 204–5, 204f, 209; post–civil rights suburbs, 202, 209–14, 211t, 212t; postwar suburbs, 201–3, 205–6, 207–8, 211t, 212t, 217; property values, 215, 218; unemployment, 215, 218
Piedmont, NC: data and methods of study, 91–2, 92f; demographics, 93t; diversity in, 97–8; employment structures in, 90; family structures in, 90; low-income residents, 103; neighbourhood succession, 99t; overview, 89, 102–3; socioeconomic change, 88–91; spatial patterns of neighbourhood distribution, 94–5, 94t–7f, 97f; spatial patterns of neighbourhood growth, 100–1, 101t; spatial patterns of neighbourhood

succession, 98–9; upper-income residents, 103
Pinecrest (Miami), 79t, 81, 82f, 85
Plane, A., 275–6
Plan for the National Capital. See Gréber Plan
Plaza Satélite, 50
pollution, 208, 291–2, 294, 298
polycentrism, 357–8
population density. *See* density of population
population growth: Atlanta, GA, 236; Edmonton, AB, 251; Halifax, NS, 336; and immigration, 339; in Mexico City, 46–7, 57; Miami region, 65, 67f, 68, 73, 73m; Montreal, QC, 151–2, 152t, 162–3; Phoenix, AZ, 204f; Piedmont, NC, 90; in Toronto, ON, 32m; Vancouver, BC, 132, 138, 138t, 139t
Portillo, José Lopéz, 53
post–civil rights suburbs, 202, 209–14
postsuburbanization, 29, 295–6, 297f
postwar suburbs, 24–9, 201–3, 205–8, 211t, 212t, 217
poverty: in Atlanta, GA, 239–41, 240f; in Chalco Solidaridad (Mexico City), 60; in Chicago suburbs, 281–2; deconcentration, 186–7; in Edmonton, AB, 250; in Los Angeles, CA, 289; in Miami region, 77f, 85; in Montreal, QC, 164; and neighbourhood dynamics, 219n2; in Newark, NJ, 176; perpetuating, 273; and Phoenix post–civil rights suburbs, 209, 211, 211t, 212t; and Phoenix postwar suburbs, 211t, 212t; in Piedmont, NC, 93–102 passim, 93t, 94t, 97f, 99t, 101t; and public transit, 34; suburbs vs. central cities, 352; and Toronto immigrants, 34–5
Pratt, G., 137
privatization: land in Canada, 259; in Los Angeles, CA, 290, 293, 297; in Mexico, 47–8; Vancouver Expo '86 land, 137
property values: in Atlanta, GA, 235–6, 239; and homogeneity, 160; in NJ, 188; in Phoenix, AZ, 205, 215, 218
public housing, 26
public transit: to Barrie, ON, 333; in Bay Area, 113; and density, 247; in Edmonton, AB, 247, 249, 261–2; and employment, 316; in Los Angeles, CA, 296, 298, 300; in Montreal, QC, 149, 155, 159; in New York, 172, 186; in Ottawa, ON, 309, 315–18, 317m, 319, 322–5; in Surrey, BC, 335; in Toronto, ON, 26–7, 34; Transit-Oriented Development (TOD), 178, 181; as unsuitable, 322–3

Quartier DIX30, 159
Quebec, 308, 313–14

race: Bay Area American Babylon, 122–3; colour blindness, 275; and exurbia, 36; and foreclosures, 240 (*see also* people of colour); "negro expansion" in Atlanta, 228–9; 1950s "sitcom" suburbs, 7; and youth drain, 178, 179–80t. *See also* people of colour; segregation/ stratification; *various ethnicities*
race riots, 224
racial transitions, 235, 239
racism: and anti-immigration, 183; in Atlanta, GA, 229, 233; in Bay Area, 123; in Edmonton, AB, 250;

and HOLC maps, 227–8; Latinos in Chicago suburbs, 275, 280–1, 282–3; and Phoenix post–civil rights suburbs, 213, 214; and Phoenix postwar suburbs, 202, 205–6; trivializing, 275. *See also* discrimination; segregation/stratification; White flight
Ragsdale, Lincoln, 206
Ranciere, 281
Rathie, Bill, 133–5
real estate. *See* building industry; developers; land
real estate surcharges, 145–6
realities, 23, 30–7, 54
redevelopment. *See* gentrification
redlining, 227–8
Reed, Kasim, 236
regional government, 174, 308, 353
Regional Plan Association (RPA), 172, 174, 186
regionalism, 299–301
regularization processes, 59–60
Relph, Ted, 28
renters: in Bay Area, 121–2; in Halifax, NS, 342; in Montreal, QC, 164–5; in Ottawa, ON, 321m; in Phoenix, AZ, 217–18; in Piedmont, NC, 93–100 passim, 93t, 94t, 95–6, 97f, 99t, 101t
Research Triangle (Piedmont), 90–1
reservations, 245, 252–6, 253m
retail activities, 150, 159, 340, 356
Rhodes, Ray, 277
Right to the City Alliance, 300
Rogers, Marie Lopez, 214
Roy, Ananya, 61
Russell, Herman, 235

Sainte-Marthe-sur-le-Lac (Montreal), 165f
Saint-Joseph-du-Lac (Montreal), 154f
San Buenaventura (Ixtapaluca), 55–6, 55f
San Fernando Valley (Los Angeles), 291
San Francisco, CA, 110, 111–16, 118–19, 123
San Francisco Bay Area. *See* Bay Area
San Jose, CA, 112–13
Sassen, S., 270–1
schools, 232–3
Scobey, D., 171
segregation/stratification: in Atlanta, GA, 224, 227, 233, 237–9; in Bay Area, 122–3; in Chicago suburbs, 273, 280, 281, 283–4; in Florida City, FL, 80; as growing trend, 352–3; in Los Angeles, CA, 289; in Montreal, QC, 162; in NY/NJ, 174–5, 184, 186–7; in Phoenix postwar suburbs, 201–2, 205, 207–8, 217; in Piedmont, NC, 89, 90; rationalization, 270. *See also* discrimination; race; racism
Séguin, Anne-Marie, 160
self-built housing, 48, 58, 60
services: in Atlanta, GA, 241; in Doral (Miami), 82; as government responsibility, 8; in Los Angeles, CA, 290; in Mexico, 59–60; in Mexico City, 48–9, 56; in Pinecrest (Miami), 81. *See also* amenities; infrastructure; privatization
settler society, 251
shopping malls, 159, 356
Siemiatycki, Elliot, 289
Sikh residents, 342
Silicon Valley, 110, 112, 119, 123
single-family houses: in Atlanta, GA, 228; in Bay Area, 114; in Phoenix, AZ, 205; in Vancouver, BC, 133, 145; in various midsize cities, 339f

single mothers, 164
single-person households, 167n1
680 Corridor (Outer East Bay), 112
skinheads, 214
smaller/midsize cities, 328–9. See also Airdrie, AB; Barrie, ON; Halifax, NS; Langley Township, BC; Markham, ON; Surrey, BC
smart growth, 330, 332
Smith, A. (Chicago planner), 276
Smolka, M.O., 60
social housing: in Mexico, 48–9, 55; in Vancouver, BC, 137; in Yonkers, NY, 174–5
social justice/movements: civil rights in Atlanta, 232–3; environmentalism in Los Angeles, 298; immigrant rights in Chicago, 284; immigrant rights in Long Island, 183; Los Angeles, 296, 297–302; New York, 188–93; regionalism, 299–301
social mobility, 5
Soja, E., 300
South Asian residents, 36, 342
South DeKalb County (Atlanta), 237, 239
South Phoenix, 217
sprawl: in Bay Area, 113, 118; in Edmonton, AB, 256–7; in Halifax, NS, 331; in Los Angeles, CA, 293, 297; in Montreal, QC, 153, 155; in Ottawa, ON, 309, 319; in Phoenix, AZ, 207; Vancouver, BC against, 131, 133–6, 134m, 330
squatting, 60
St. Laurent, Louis, 309
Stamford, CT, 177, 192
Stamford Organizing Project (SOP), 192
Standard Life Insurance, 224
standardization, 26, 142–3

stereotypes, 251, 276. See also suburbs, stereotypes
Stoll, M.A., 90
Strategic Actions for a Just Economy (SAJE), 300
stratification. See segregation/stratification
streetcars, 5, 8, 113–14
subaltern urbanism/suburbanism, 61
subprime lending, 187–8. See also foreclosures
subsidies, 6, 189
suburbanization: of cities, 182; dichotomies, 65, 84, 85–6 (see also cities vs. suburbs); of downtowns (see Vancouver, BC); of government (see Ottawa, ON); history of, 4–9; of immigration, 182–3; rate of increase, 6; stages and outcomes, 84; US/Canada vs. Mexico, 60; of work, 8 (see also office sectors)
suburbocentrism, 130–1, 136–46
suburbophobia, 132–6, 134m
suburbs: aggregation problem, 351–3; as changing, 269, 289, 328, 350–1, 358; as diverse, 9 (see also diversity); end of, 3, 289; as exclusive, 6; loss of, 8; modelling, 358–9; as settlement, 261; term use, 4, 7, 29, 70–1, 247, 249; vanishing archetypes, 349, 350
suburbs, stereotypes: as banal, 8–9; countering, 34–5; and creative class, 33; ideals, 8, 23, 28–37, 49, 163, 349; judgments, 23, 30–1; 1950s "sitcom," 3, 7, 78, 205, 247, 351; people of colour and poverty, 352; realities, 23, 30–7; and renting in Phoenix, 217; stereotypes as vanishing, 349, 350;

in Toronto, ON, 27–8, 34–5, 37; as wastelands, 328
Surrey, BC, 330, 335, 336, 336f, 337f–9f, 340–2
Sustainable Communities Consortium, 186

taxes: in Quebec, 155; for services, 8, 81, 82; and sprawl, 133–5
technology, 90, 113, 204, 293, 314. *See also* Silicon Valley
Temporary Coordinating Committee on Housing (TCCH), 227, 228
Thompson, Robert, 229–30
Tilly, C., 353
title assertion, 251–4
Toronto, ON, 329m, 337f–9f; city vs. suburbs, 28–9; and condos, 31; as diverse, 24, 26, 33; early postwar decades, 24–9; employment in, 29–30, 139; gentrification, 30–1; growth management, 340–1; high-rise apartments, 26; as immigrant city, 24; inner neighbourhoods, 24; Metro, 24–6, 25m; as modern metropolis, 29–36; politics, 34; population, 24, 32m; population growth, 339; poverty in, 34–5; public transit, 26–7; residential geographies, 30; settlement patterns, 27m, 33m; social/housing characteristics, 31t; suburb stereotypes, 27–8, 34–5, 37; as three zones, 35–6, 37; as unusual, 37–8
Torontonians, The (Young, Phyllis), 28
townhouses, 331
toxic pollution, 208
tract housing: in Mexico City, 48; in Phoenix, AZ, 205, 206–7; reinventing, 203; as sanitized, 205; and stereotypes, 27–8

Transit-Oriented Development (TOD), 178, 181
transnationalism, 129, 131
transportation. *See also* automobiles; public transit
Trudeau, Pierre Elliott, 307–8, 313
Truth and Reconciliation Commission, 255, 262
Tunney's Pasture, 312m, 313, 316, 317m
Tweed, William Magear (Boss), 172

unidad habitacional model, 50
Uniondale Community Land Trust (U-CLT), 189–90
unions. *See* labour movement
United States: history of suburbanization, 4–9; influencing other countries, 50, 53; mythic journeys in, 7; suburbs rate of increase, 6
University of Alberta, 262n2
upper-income residents, 91–103 passim, 93t, 94t, 97f, 99t, 101t. *See also* elites
urban development boundary (UDB), 72, 76
urbane, 4
urbanization, 354–5
Urban Land Institute (ULI), 352
Urban League, 229
urban renewal. *See* gentrification
urban studies, 358–9
utopia, 5, 171

vacancies, 49
Valle de Chalco. *See* Chalco Solidaridad (Mexico City)
Vancouver, BC, 130, 329m, 337f–9f; affordability, 144–6; colonial displacement, 132; commutes, 143; downtown, 138–43; early

history of, 132–3; employment, 139–41, 140t, 141f, 143; Expo '86, 137; False Creek North, 137; False Creek South, 136; geographies of population change, 138–9, 138t, 139t; growth management, 340–1; Living First policy, 137, 139, 142, 143; overview, 129–32, 144–5; population growth, 339; against sprawl, 131, 133–6, 134m, 330; wealth in, 143
Vancouverism, 129, 130f, 137, 141–2
Victorian houses, 114, 116

waterfront communities, 67, 71–2.
Weitz, J., 90
Westchester County, NY, 174
Westmount (Montreal), 164–5
West Philadelphia, 5
Whitaker, Matthew, 205
White flight: in Atlanta, GA, 232–3; in Miami region, 85; in Montreal, QC, 161; in New York, 173; in Phoenix post–civil rights suburbs, 202; in Phoenix postwar suburbs, 207; in Toronto, ON, 36
White residents: in Airdrie, AB, and Barrie, ON, 342; in Atlanta, GA, 224, 229, 232–3, 235, 239; in Bay Area, 123; and foreclosures, 217; in Los Angeles, CA, 289; in Miami region, 74t, 76, 79t; in Mill Woods (Edmonton), 259; and mortgage loans, 227–8; in New York City, 180t, 181, 183; Phoenix homeownership, 205; in Phoenix post–civil rights suburbs, 202, 211–13, 211t, 212t; in Phoenix postwar suburbs, 211t, 212t; in Piedmont, NC, 89; and suburban stereotypes, 7, 11; in Toronto exurbia, 36. *See also* race; racism; segregation/stratification
White supremacy, 214
"Why Pay Rent" campaigns, 5
Wiese, A., 226, 228, 354
Wirth, L., 142
Wood, R., 174
work. *See* employment
working class: in Bay Area, 122, 123; and city centres, 4; identity and suburbs, 6; in Los Angeles, CA, 296, 298; in New York, 172, 174
Workplace Project, 191–2
Wyly, Evlin, 289

Xia, Evelyn, 145
Xico (Chalco Solidaridad), 58f

Yan, Andy, 146
Yonkers, NY, 174–5
Young, Andrew, 236
Young, Phyllis, 28
younger residents: in Edmonton, AB, 250, 260; and NJ, 178, 179–80t; in New York, 178–81, 179–80t, 181, 183

Zabriskie Point (film), 294
zombie homes, 187–8, 190

GLOBAL SUBURBANISMS

Series Editor: Roger Keil, York University

Published to date:

Suburban Governance: A Global View / Edited by Pierre Hamel and Roger Keil (2015)

What's in a Name? Talking about Urban Peripheries / Edited by Richard Harris and Charlotte Vorms (2017)

Old Europe, New Suburbanization? Governance, Land, and Infrastructure in European Suburbanization / Edited by Nicholas A. Phelps (2017)

The Suburban Land Question: A Global Survey / Edited by Richard Harris and Ute Lehrer (2018)

Critical Perspectives on Suburban Infrastructures: Contemporary International Cases / Edited by Pierre Filion and Nina M. Pulver (2019)

The Life of North American Suburbs: Imagined Utopias and Transitional Spaces / Edited by Jan Nijman (2020)

www.ingramcontent.com/pod-product-compliance
Lightning Source LLC
Chambersburg PA
CBHW022210090526
44584CB00012BA/378